After Sputnik

50 Years of the Space Age

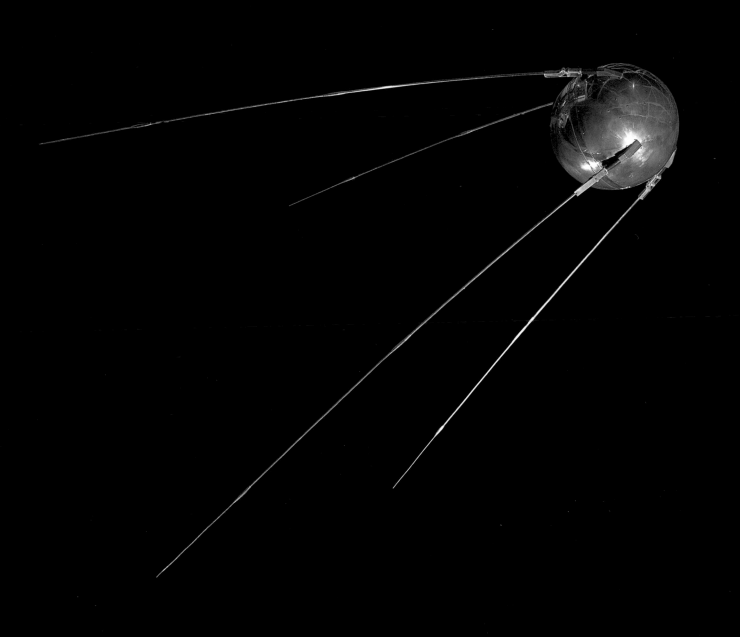

After Sputnik

50 Years of
the Space Age

Edited by Martin Collins
Smithsonian National Air and Space Museum

✳ Smithsonian Books

◖ Collins
An Imprint of HarperCollinsPublishers

PAGE 1: Sputnik Model

PAGES 2–3: Astronaut Neil Armstrong's Apollo 11 Spacesuit

After Sputnik: 50 Years of the Space Age

HarperCollins books may be purchased for educational, business, or sales promotional use. For information please write: Special Markets Department, HarperCollins Publishers, 10 East 53rd Street, New York, NY 10022.

Produced by Charles O. Hyman, Visual Communications, Inc., Washington, D.C. Designed by Kevin Osborn, Research & Design, Ltd., Arlington, Virginia

Published 2007 in the United States of America by Smithsonian Books in association with HarperCollins Publishers.

Library of Congress Cataloging-in-Publication data has been applied for.

ISBN-10: 0-06-089781-3

ISBN-13: 978-0-06-089781-9

Printed in China.

07 08 09 10 TP 10 9 8 7 6 5 4 3 2 1

Contents

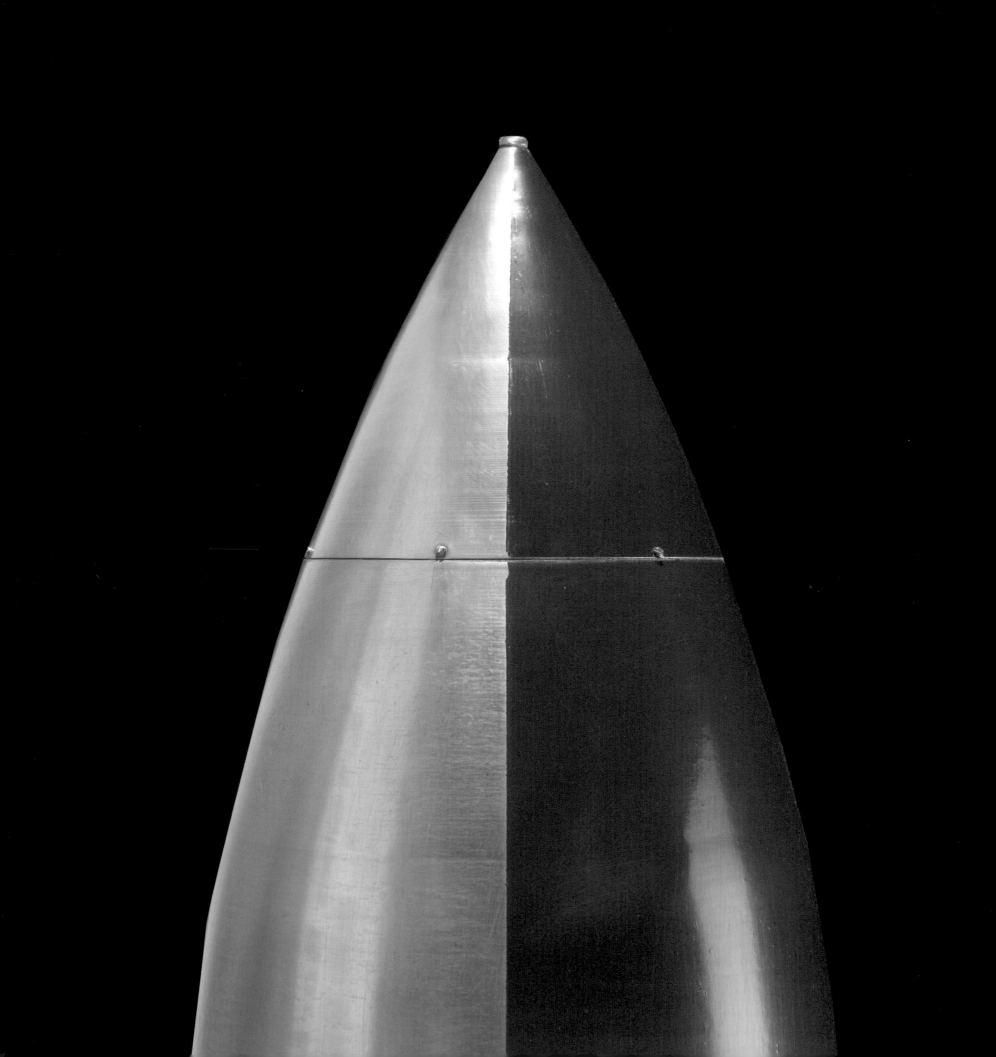

Acknowledgments

THE BEAUTY of life in the Museum is the possibility, indeed, the normal everyday expectation, of creative collaboration with people of exceptional talent. It is central to the Museum's identity and mission. This book benefited tremendously from and depended fundamentally on this institutional virtue. From beginning to end, *After Sputnik*, as an endeavor of many hands, drew on this culture of the "whole is more than the sum of its parts."

My colleagues in the Museum's Division of Space History provided the intellectual energy for the project. They are a national asset, essential players in preserving the heritage of the space age. Their years of thoughtful engagement with the byways and deeper meanings of spaceflight provide the backbone of this book. Each contributed essays (see the credits at the back); my thanks to: Paul Ceruzzi, James David, David DeVorkin, Roger Launius, Cathleen Lewis, Valerie Neal, Allan Needell, Michael Neufeld, Jennifer Skomer, Margaret Weitekamp, Frank Winter, and Amanda Young. As an essay contributor, I also thank Tom Crouch, from the Museum's Aeronautics Division, who graciously joined in this effort. From these ranks, I must give special recognition to Jennifer Skomer. On a day-to-day basis, she was my intellectual sidekick and a remarkably energetic force in keeping the many pieces of the project organized and on track—and blessedly tolerant of my shortcomings. This book would not have been possible without her. In the first months of the project, Margaret Weitekamp (expecting her first child) shaped the book's presentation of the intersection of the space age and culture. May her now born and vibrant son Xavier live long and prosper after listening to the murmurings of space historians in his most formative moments.

As is evident throughout this publication, rich, beautiful images of the Museum's artifacts carry the story. These are the product of our very snap-savvy in-house photographers: Mark Avino, Eric Long, Dane Penland, and Carolyn Russo. Their special eye and creativity add immeasurably to the presentation of the artifacts, bringing them alive as tangible things invented, used, or lived in. Eric Long, especially, gave much of himself in this project, shooting the great bulk of the photography. His gifts are obvious and his collegiality and insights improved us all.

The Museum's Collections Division provided unwavering assistance in preparing artifacts for photography, often on short deadline and at inconvenience to their other responsibilities. In our wide-ranging research, the Museum's Archives and the Smithsonian Institution Libraries were generous with their time and support.

Last but not least were our colleagues external to the Museum. Copyeditor Debby Zindell turned the often creaky prose of academics into something better. Chuck Hyman, producer of the book, and Kevin Osborn, designer, added crucial and elegant expertise born of long years of experience. They gave the book its "book-ness," a thing of immediacy and art. They were boon companions throughout and I learned much from them. Caroline Newman, our editor from Smithsonian Books, believed in us—amazingly, thoroughly—and offered the perfect balance of nudging and razor-sharp insight, guiding us down the most fruitful paths. She made the long road of work seem a romp and an undertaking with heft.

And, always a nod to my beautiful sons, Gus and Flynn— ever my teachers.

OPPOSITE: Detail of nose cone of Robert Goddard's "1935 Series-A" rocket

Preface

WITH THE launch of Sputnik in October 1957, spaceflight became a fact and a rich canvas for human action and imagination. We are now 50 years into humanity's historic effort to reach beyond Earth and explore, comprehend, and use the vast domain of space. In recounting this effort, spaceflight often is regarded as a story of firsts and triumphs leavened with the occasional failure and missed opportunity. Having lived through five decades of the space age, we might now look to add more texture to this story and ponder the details of well- and little-known acts as well as the larger meanings and effects of our venture into space.

Seen from this vantage, spaceflight, especially in the United States, clearly emerges as a central thread in national life, intertwining with politics, business, foreign affairs, the worlds of science and technology, popular culture, and our very sense of who we are and our place on the Earth and in the universe. In those 50 years, spaceflight in all it guises—from Earth-orbiting satellites of various kinds, from probes to the planets, to human exploration— has traversed from the exceptional to the everyday. Spaceflight and its achievements literally surround us—in how we think, consider the future, and in how we live minute-by-minute in each day.

This book will explore this transformation through the distinctive eye of a national museum and its storehouse of artifacts. The Smithsonian National Air and Space Museum holds in trust nearly 14,000 artifacts relating to rocketry and spaceflight. These objects range from nineteenth-century military rockets, to treasures of the Apollo journey to the Moon, to SpaceShipOne, winner in 2004 of the Ansari X Prize for demonstrating the possibility of privately financed human spaceflight. Through a selection of more than 140 artifacts from the Museum collection this book seeks to convey the broad sweep of the space age. The book presents each artifact as an individual story, seen in detail but connected to larger outlines of American history and spaceflight's role in shaping our world.

Telling the history of spaceflight through artifacts is not a mere convenience. The "real stuff" of history offers a distinctive way to engage the past. In 1881, George Goode Brown, a leader in shaping the Smithsonian, offered that "objects … [are] permanent landmarks of the progress of the world in thought, in culture, or in industrial achievement." Their value is the same today. For scholars and visitors alike the "real stuff" of history pushes us to think and explore how and why an object was created, used, and participated in changing our lives. All have a story, wide-ranging and revealing of a particular historical moment. Each arose from a complex saga of individual creativity and inspiration, politics and ideologies, deep cultural beliefs, and more. Each bears the stamp of specific challenges, successes, and missteps. Each is a slice of the world. In its sturdy, undeniable presence, each artifact asks us, in unique fashion, to reflect on choices seen and made. Individually and in total, the artifacts in this book invite the reader to reflect on the space age as a lived, flesh-and-blood undertaking and a venture that has remapped the human experience.

OPPOSITE: Missiles displayed at the Smithsonian's Arts and Industries building, 1960s

Inventing Spaceflight: The Years before Sputnik

SPUTNIK marked the beginning of the space age—but spaceflight, as a mix of imaginings and technological investigations, took shape in the prior decades. By the 1920s and 1930s, spaceflight had emerged as a cultural phenomenon, with scientific papers, technical treatises, the new genre of science fiction, inventor-heroes, newspapers, "pulp" magazines, amateur "interplanetary" societies, films, world's fairs, and occasional government support giving substance to the idea— or, its close, essential cousin, rocketry. From this activity rose the first examples of space-related technology, the fruits of hard-won invention: liquid-fueled rockets, improved solid-fuel rockets, test stands and test sites, and apparatus such as fueling cans and protective gear for humans. By the latter part of this period, the Smithsonian Institution's National Museum already had collected one of the era's prize symbols of innovation: American rocket pioneer Robert Goddard's "1935 Series-A Goddard Rocket."

This new technology was developed in experimental laboratories and remote test sites—some run by professionals, others by amateurs—familiar to the early twentieth-century public. American newspapers and magazines had been chronicling the achievements of inventors such as Thomas Edison and Alexander Graham Bell and the contributions of the industrial age's research corporations, such as General Electric, since the nineteenth century. Early in the twentieth century, a new genre of "boys' magazines" appeared that encouraged the young to turn their basements, garages, or bedrooms into laboratories and join the world of invention. Prior to the late 1930s, however, specialized facilities devoted to rocketry were few and poorly funded, compared to the many corporate and government laboratories dedicated to other areas of research. The tangible results were

mostly modest: Robert Goddard struggled over more than two decades (from about 1915 through the 1930s) to turn the theory of the rocket into a reliable technical reality. The images of the laboratory and the inventor-at-work did, however, carry strong cultural associations. As the foundation of invention and progress, these dynamos of change were constantly remaking the everyday world—through the introduction of electricity, the airplane, the car, radio, telephone, and, now quite possibly, the rocket and spaceflight.

Through a vigorous interplay of scientific experimentation, public enthusiasm, and rapidly expanding media, large segments of the population came to know about spaceflight as idea and possibility. Newspapers, for example, were quick to alert readers to any development in technology or theory. By the mid-1920s, the "big three" spaceflight theoreticians and experimentalists— Russian Konstantin Tsiolkovsky, German Hermann Oberth, and Robert Goddard—became widely known. They were the hero-inventors who might actually create a way for humans to literally break free of Earth: developing a rocket, sufficiently powerful and reliable, to lift itself, humans, and supplies onto a journey into space.

In the United States, Europe, and the Soviet Union spaceflight became part of "Building the World of Tomorrow"—the theme of the 1939 New York World's Fair—but in a broader sense reflected a deeply rooted belief in progress, common then and still today. This core belief translated into struggles over political ideologies— progress, as a fact and as an aspiration, raised the question of which political systems best integrated science and technology with notions of a "good" society. Spaceflight, as possibility and as one emblem of progress, became a lens through which to ponder the broad scope of the human condition—as individuals, communities,

OPPOSITE: On July 24, 1950, this V-2 with a Bumper rocket as a second stage lifts off, the first launch at historic Cape Canaveral.

societies, on Earth or traveling to the stars, now or in the future. The conflicting ideologies of capitalism and communism each claimed a spaceflight pioneer as its own and aligned this futuristic undertaking with its respective political-cultural outlook.

FROM today's perspective, one might say future possibility exceeded accomplishment in the early twentieth century. But those ideas of spaceflight colored the response to Sputnik and the exploration that followed. Consider Hugo Gernsback, publisher of the first science fiction magazine, *Amazing Stories*, which debuted in April 1926. In this publication, Gernsback proclaimed the creation of a new form of imaginative expression: "Scientifiction … a charming romance intermingled with scientific fact and prophetic vision." The monthly "pulp" featured men (and women in supporting roles), machines, and exotic landscapes, with the action almost always in the future. The magazine covered a broad sweep of science and technology. Space-themed adventures were prominent (the lead story in the first issue was Jules Verne's "Off on a Comet—or Hector Servadac," which inspired an eye-catching cover featuring the planet Saturn) but were seen as just one topic among many that chronicled the seemingly endless and omnipresent "amazing" changes associated with innovation and new knowledge.

The merging of science and fiction seemed perfectly natural and timely to Gernsback. Verne, H.G. Wells, and Edgar Allan Poe already had built a following for the genre in their science-infused fiction. For Gernsback and his readers, "scientifiction" captured a new relationship among culture, science, and imagination. "It must be remembered," Gernsback noted, "that we live in an entirely new world. Two hundred years ago stories of this kind were not possible. Science, through its various branches of mechanics, electricity, astronomy, etc., enters so intimately into all our lives today, and we are so much immersed in this science, that we have become rather prone to take new inventions and discoveries for granted. Our entire mode of living has changed with the present progress." *Amazing Stories*, thus, was not mere entertainment, a commercial frill to the serious business of scientific and technological advance, but a partner in this vast cultural change.

The idea of progress permeated life in this period and helps us to understand the emergence of spaceflight concepts in the United States, Europe, and the Soviet Union. Spaceflight embodied the

"modern" worldview that humans, through the exercise of reason, are the makers of the world here and now and will be in future, which contrasted with the prior, "pre-modern" worldview that favored custom, tradition, and dogma. Spaceflight in the late nineteenth and early twentieth centuries, whether expressed culturally in the literature, pulp magazines, comics, and films of the day, or technologically in the first attempts at theorizing and experimentation, were two sides of the same coin, each an attempt to formulate the future through reason and imagination. Thus, for Gernsback and many of his contemporaries, spaceflight was part of this sweeping current of change, in which the advance of science and technology remade humanity's "entire mode of living."

But to move from broad ideas to real, working technologies required applying to spaceflight a crucial invention of the industrial age: the research facility, well-funded, well-organized, and dedicated to solving complex problems—in this case those involved in reaching space, exploring, and returning. The challenge of spaceflight was beyond the means of the individual inventor, amateur group, or corporation. Only government might undertake such an effort—and for practical reasons of state. In the late 1930s, the Nazis first took this approach, but they had little interest in spaceflight, focusing instead on perfecting the rocket as a formidable weapon.

The work in Germany, undertaken from the late 1930s through World War II and led by charismatic engineer Wernher von Braun, though intended to produce "vengeance weapons," had a profound effect on the course of spaceflight. With the production of hundreds of V-2s, the Peenemünde team made the rocket a practical device and created a model for a research organization designed to overcome complex problems. Most important, perhaps, the Nazi's success helped make warfare the rationale for further development of the rocket. After World War II, the rocket and the atomic bomb became prominent symbols of a new age—one in which terrifying acts of war could occur over transcontinental distances with vast destruction. With these new technologies, a war could conceivably begin—and end—in a matter of minutes or hours, not years. No person, any place on the globe, was immune to this threat. Political and military leaders in the United States and other nations believed that constant ongoing military preparedness, sustained by relentless innovation in science and technology, stood as the only and best defense. The prewar notion of progress as a broad dynamic between innovation and society, creativity and human development, took on a new, intensified connotation as a product of military need.

In the years that followed, rocket and spaceflight technology took shape within this military-driven context but still remained infused with prewar notions of adventure and a human destiny in the stars. No person embodied this intertwining of perspectives more completely than von Braun, whose career arced from German amateur rocket society participant to Nazi engineer to U.S. Army star to NASA's leading rocket man—from military pragmatist to visionary spaceflight romantic.

As THE U.S. and Soviet Union entered into the Cold War in the late 1940s, the idea of "preparedness" was central in both nations, leading to a decades-long arms race, one thread of which led to Sputnik and the 1960s Space Race. From the perspective of the early twenty-first century, the intensity and breadth of the post–World War II commitment to preparedness may be hard to fathom. Take one symbolic marker. In November 1945, months after the end of the war, Henry "Hap" Arnold, Commanding General of the Army Air Forces, collaborated with *Life* magazine (the most widely read national magazine of the period) on a feature entitled "The 36-Hour War." Through a series of graphic artist's illustrations, the magazine took the reader through a sequence of events possible in the "next great conflict." It begins "with the atomic bombardment of key U.S. cities" with a visual revealing "a shower of white-hot rockets [falling] on Washington, D.C.," complete with a mushroom cloud ready to envelop the U.S. Capitol. The article encapsulated a broadly shared view that the threat of war would remain a constant of life—and that science and technology were both cause and salvation.

In the U.S., such thinking led political and military leaders to establish thousands of research and development projects—in government laboratories, universities, and industry—at first, haltingly, in the late 1940s and then with increased urgency and funding after the Korean War began in 1950. Much of this effort was "classified," held as a government secret. The public's knowledge of the details of this extensive research and development effort represented only the tip of the proverbial iceberg.

A fraction of the overall work—which was still sizable in terms of funding, personnel, and numbers of institutions—was devoted to rocketry and spaceflight. By comparison with the prewar era, progress was extraordinary. In the 1920s and 1930s, spaceflight ideas were primarily expressed through fiction and speculative essays by rocket enthusiasts. In the 1940s through the mid-1950s, the dominant genre was the technical report—thousands upon thousands emerged from all the government-sponsored work, investigating the broad range of problems raised by rocket and spaceflight. In the prewar years the examples of technology were few and rudimentary; in the postwar years the examples multiplied (primarily focused on the rocket) and showed increasing technical sophistication.

Before Sputnik, the idea of spaceflight, with its military implications, moved from possibility to a prospect-in-waiting, supported by a vast network of technical expertise. In 1955, the United States secretly committed to developing a spy satellite. Later that year, the government publicly announced a program to launch a scientific satellite, called Vanguard, as part of the International Geophysical Year scheduled to begin in 1957. These two announcements highlighted the fact that knowledge about developments in spaceflight varied widely. Government, university, and industry personnel privy to secret efforts and plans realized the depth of technical knowledge at the ready. The general public, forming their view of spaceflight from newspapers, magazines, and films (whose accounts ranged from the serious to the speculative) had a less certain view of near and long-term prospects. Even so, by the mid-1950s the subject of spaceflight permeated American culture. A 1952–1954 *Collier's* magazine series, with von Braun as primary consultant, detailed the technologies and scenarios of human space exploration, emphasizing their plausibility. From 1955 to 1957, Walt Disney aired a series of three shows on television ("Man in Space," "Man and the Moon," and "Mars and Beyond") that followed in the footsteps of the *Collier's* series and also involved the energetic von Braun. In 1957, just months before Sputnik, *Life* magazine ran a cover story on the Vanguard satellite project, offering an inside look that included plenty of photos of U.S. satellites nearly ready to go. When the Soviets launched Sputnik in October 1957, the surprise to the U.S. public was not that it happened but that the Soviet Union did it first.

Robert H. Goddard's Liquid Oxygen Flask and Carrier
1923–1924

"I recognized a long time ago [the need], and I began, as soon as I could, to carry out researches with liquid propellants, which had several times the heat energy of powder."

—ROBERT H. GODDARD, 1928

THE American rocket pioneer Robert H. Goddard was probably the first man in the world to experiment with liquid rocket propellants. His initial experiments, begun in 1915, used solid propellants, which were simple to work with and inexpensive. But, in 1921, he switched to liquids. Theoretically, they possessed far greater energy content than solids and promised superior performance.

Goddard realized that the combustion of liquid propellants in an engine could be controlled by valves opening or closing, thereby regulating the flow of fuel. In contrast, once a solid-fuel rocket ignited, it could not be throttled down or shut off—it burned until the fuel was exhausted. (Today, there are devices that cut off the burning when desired.)

In both solid and liquid propellants, an "oxidizer" is needed—a substance that contains oxygen for the fuel to burn. In solid propellants, such as gunpowder, the fuel and oxidizer are mixed together. In liquid fuel systems, the oxidizer and fuel are separate substances, each contained in its own tank. They must be brought together and then ignited for combustion. Together, the fuel and oxidizer are called the "propellant."

Goddard chose liquid oxygen for his oxidizer and ordinary gasoline for fuel. But this choice of oxidizer posed a serious problem. As a liquid, oxygen is extremely cold, -183°C (-297°F). But Goddard benefited from a prior invention. In 1872 the Scottish chemist James Dewar invented a special container that maintains the temperature of either very hot or very cold liquids. This container, now called a Dewar flask, is usually a double-layer glass bottle, the narrow space between the walls a near vacuum to prevent the transmission of heat. In addition, the inner and outer walls have a silver or other reflective coating to reflect thermal energy.

The ordinary Thermos® bottle works on the same principle.

The device shown here is one of the original small Dewar flasks probably used by Goddard during his experiments in 1923 and 1924. Later, as his rockets grew in size and became more sophisticated, he used much larger liquid oxygen containers.

A Dewar flask is extremely fragile—this one is made of hand-blown glass. For this reason, and the fact that liquid oxygen makes the outer wall of the flask intensely cold, the manufacturer provided thick felt padding (that seen here is original) to protect and handle the flask.

Goddard needed to transport the fuel for his experiments and flights so built his own carrier made out of wood. The flask fits easily between the slats, the lips resting on the slats and preventing the flask from falling out. Goddard made the carrier with handles at each end, enabling two people to carry the delicate flask. The Dewar flask and homemade carrier seem very crude by our standards, but they worked.

Today, our increased understanding of cyrogenics (the behavior of very cold substances) has enhanced our ability to handle and work with extremely cold liquids. The J-2 rocket engines of the second and third stages of the Saturn V and the Shuttle main engine used liquid hydrogen, which, at -259°C (-434°F) is far colder than liquid oxygen. These advances highlight the daunting challenges that confronted Goddard as he experimented with rocket technology. The Dewar flask was one small part of years-long work to fulfill his dream of reaching space.

Mrs. Robert Goddard, who witnessed many of her husband's rocket test activities, donated the flask and carrier to the Museum in 1959.

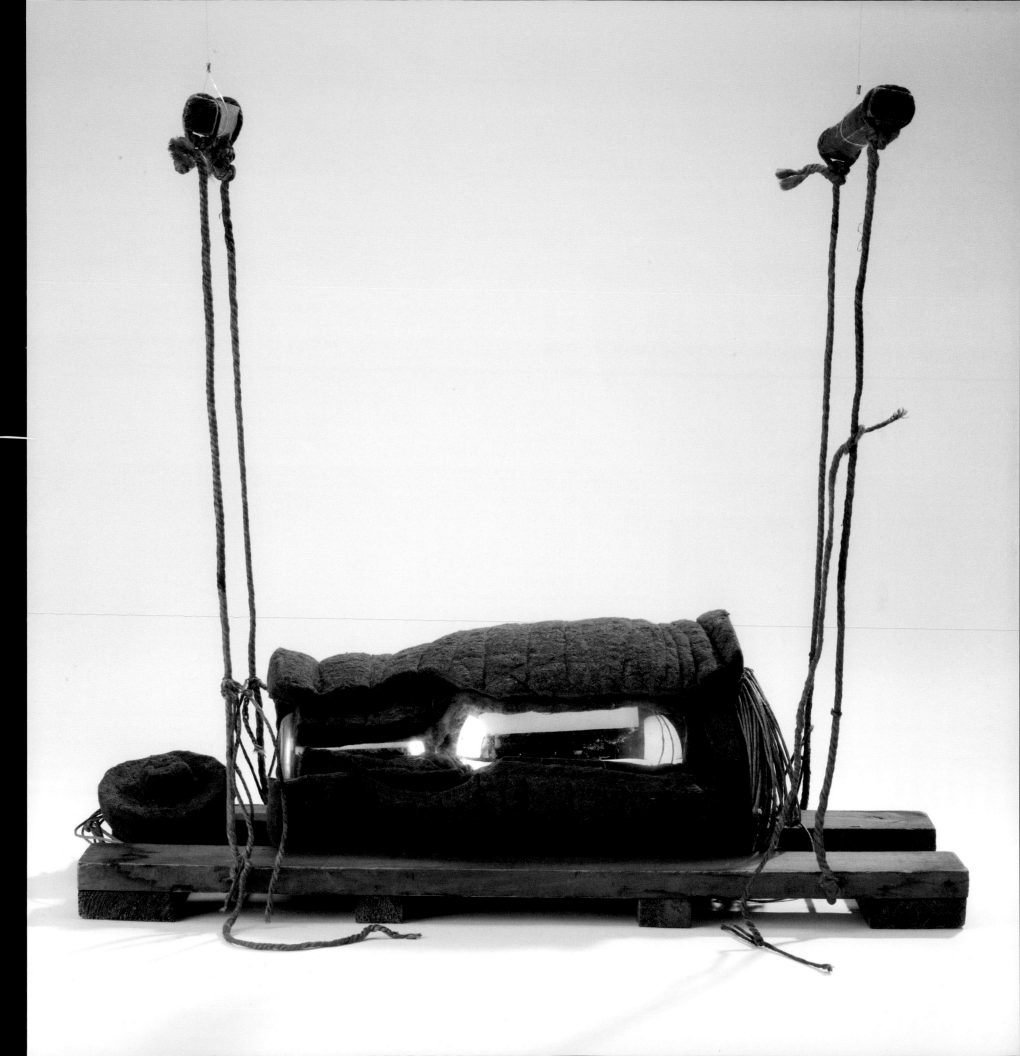

"One-Stick" Repulsor Rocket Nozzle
1931

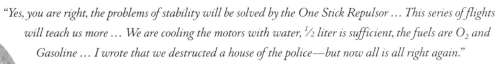

"Yes, you are right, the problems of stability will be solved by the One Stick Repulsor … This series of flights will teach us more … We are cooling the motors with water, ½ liter is sufficient, the fuels are O₂ and Gasoline … I wrote that we destructed a house of the police—but now all is all right again."

—WILLY LEY, LETTER TO G. EDWARD PENDRAY, 1931

IN 1923 Hermann Oberth, an obscure German-Romanian high school teacher, published a small theoretical work in German, *The Rocket into Interplanetary Space,* that gradually inspired a small band of true believers in the feasibility of spaceflight. On July 5, 1927, a group of them met in a smoky tavern in Breslau, Germany (now Wroclaw, Poland), to found the Verein für Raumschiffahrt (Society for Space Travel or VfR). Oberth and his followers sparked a fad for rocketry and spaceflight greater than any other country except Bolshevik Russia, one that reached its apogee in 1928–1929 with numerous stunts involving gunpowder-rocket-powered vehicles. In the fall of 1929, famous director Fritz Lang premiered his realistic space travel movie, *Woman in the Moon*, with Oberth as its scientific advisor.

The rocket stunts contributed little to the development of liquid-fuel rocketry, however, which was the only technology Oberth believed would provide sufficient energy to propel humans into space. In the aftermath of the film, he and a few assistants in the VfR carried out some small liquid-fuel rocket experiments in Berlin during the summer of 1930. Oberth went back to Romania, but his associates formed an amateur group that came to be called the Raketenflugplatz (Rocketport) Berlin. One of its members was the engineering student Wernher von Braun.

In 1931, the Raketenflugplatz launched its first small liquid-fuel rockets, propelled by liquid oxygen and gasoline. Calling them repulsors, after a spaceship in a classic German science-fiction novel, these rockets were "two-stick" versions: the nozzle was at the head of the rocket and two long, skinny

tanks ran down the side. In August 1931, the group followed that with the "one-stick" Repulsor IV, in which the two propellant tanks were in one long body attached to the side of the "nose-drive" rocket, a design that imitated the classic black-powder rocket stabilized by a stick. Rockets of this type were fired as high as about 1.5 kilometers (about a mile) and came down by parachute, when they worked correctly.

One of the group's members was the young engineer Herbert Schaefer, who kept this rejected nozzle as a souvenir. Machined out of aluminum, it would have been surrounded by a water jacket to absorb the heat of combustion when the engine was burning. The fuel may have been alcohol instead of gasoline, as the Raketenflugplatz group shifted over to this combination recommended by Oberth. As the Great Depression deepened in 1932, however, the finances of the ill-funded group deteriorated, along with the political situation in Germany. In 1933, after Hitler came to power, the VfR and the Raketenflugplatz slowly fell apart owing to money troubles, political disagreements, and pressure from the army and the Nazi regime to shut down experimentation. Meanwhile, at the end of 1932, Wernher von Braun left to work on secret army experiments, leading eventually to the V-2 ballistic missile of World War II.

When Schaefer emigrated from Germany to the United States in 1935 to escape the Nazis, he took this little nozzle with him and then donated it to the Museum 1978. Virtually everything else salvaged from the Raketenflugplatz was lost in World War II, with the result that this object may be the only surviving rocket artifact from that historic effort. Conceived as part of an almost utopian effort to find a way to travel into space, instead it led first to the development of the rocket as a weapon.

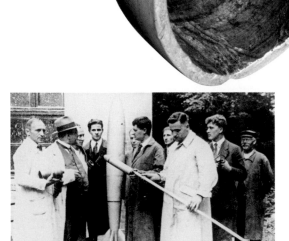

ABOVE: German rocket pioneers, including Rudolph Nebel (far left), Hermann Oberth (center, in profile), and a young Wernher von Braun (second from right), test a rocket engine in Berlin, July 1930.

Buck Rogers Trading Card Number 444
1936

"The Mongolians, with overwhelming fleets of great airships, and a science that far outstripped that of crippled America, swept in … annihilating American aircraft, armies and cities with their terrific disintegrator rays."

—PHILIP FRANCIS NOLAN, *Armageddon—2419 AD*, 1928

MOST adult Americans struggling to survive the Great Depression of the 1930s must surely have regarded spaceflight as errant nonsense. Their offspring often took a very different view, thanks largely to the growing power of new media. There were pulp magazines, like *Amazing Stories*, the first issue of which hit the newsstands in April 1926. Earlier pulps, inexpensive publications printed on the cheapest paper and bound in garish and usually suggestive covers, had offered lurid tales of crime, Wild West adventure, aerial derring-do, and romance. Publisher Hugo Gernsback sent the genre spinning in an entirely new direction, filling *Amazing Stories* with tales of "scientifiction," reflecting his own boundless enthusiasm for a bright utopian future shaped by technology.

The August 1928 issue of *Amazing Stories* carried a tale typical of the fare Gernsback offered his young readers—*Armageddon—2419 AD*, by Philip Francis Nolan. The story involved Anthony Rogers, a young engineer who was trapped in a Pennsylvania mine in the year 1975 and rendered unconscious by "radioactive gas." He awakes in the year 2419 to find America in ruins and under attack by the evil Mongolian "air lords" of the Han Empire. Our hero hooks up with a female heroine, Wilma Deering, acquires a brilliant sidekick named Dr. Huer, and enlists in the ultimately successful battle against the forces of Asian tyranny.

Intrigued, John C. Dille, operator of a newspaper syndicate, thought that the story might serve as the basis for a comic strip. He enlisted Nolan to write the strip but suggested a new, snappier name for the hero—Buck Rogers, a moniker inspired by a western movie star. *Buck Rogers* premiered in newspapers on January 7, 1929, and was an immediate success. A color strip for the Sunday papers followed on March 30, 1930. The *Buck Rogers* radio program commenced in 1932 and aired four times a week for the next 15 years.

The Buck Rogers phenomenon was a marketer's dream. The radio series offered a variety of listener extras, including space ranger badges. Stores offered Buck Rogers disintegrator ray guns, model rocket ships, games, Big Little Books, pop-up books, and trading cards, like this example from the Museum collection, donated by an individual collector, Michael O'Harro, in 1997. The rocket belt, shown here, was an essential part of the Buck Rogers persona. The original franchise inspired some highly successful copycats, notably Alex Raymond's *Flash Gordon* strip, which first appeared in 1934.

The first film, *Buck Rogers in the 25th Century: An Interplanetary Battle with the Tiger Men of Mars*, starred Dille's son and premiered at the Century of Progress Fair in Chicago in 1933. A 12-part movie cliff-hanger starring Buster Crabbe hit the theaters in 1939. Buck Rogers made his television debut in a live series that aired in 1950–1951. NBC launched updated versions of Buck and Wilma as primetime fare during the 1979–1980 seasons. Video games and graphic novels kept the original interplanetary hero alive and well into the 1990s.

Like all products of popular culture, Buck Rogers was a reflection of his times. His struggle to turn back the Mongol hordes, and Flash Gordon's battles with the merciless Emperor Ming, seemed natural enough to a generation that regarded the "Yellow Peril" as the principal threat to Western civilization. Buck, though, remains very much a part of the American lexicon. Wild-eyed techno-logical dreams are still dismissed as "Buck Rogers stuff," and the phrase "No bucks, no Buck Rogers," a line from the film *The Right Stuff* (1983), continues to serve as a shorthand statement of one of the great truths of spaceflight.

Liquid-Oxygen Can and Protective Helmet, American Rocket Society
1930s

"A good rule for rocket experimenters to follow is this: always assume that it will explode."

—*Astronautics* (JOURNAL OF THE AMERICAN ROCKET SOCIETY), OCTOBER 1937

I N THE 1930s, members of the fledgling American Rocket Society (ARS) used these tools to launch experimental rockets. Although working independently, they joined rocket pioneer Robert Goddard as the first dedicated U.S. investigators of this new technology. And, like Goddard, they saw the rocket as means to an end: the eventual human exploration of space. These objects—a simple steel watering can and protective helmet—testify to the improvised nature of their effort and their level of enthusiasm, and reflect America's idealism and love of amateur tinkering.

Each tool served a purpose. The can funneled liquid oxygen ("lox") into rocket motors. Accompanying tongs allowed experimenters to handle the can, super-cooled by the lox. The World War I vintage helmet offered protection during tests of the rockets. The crude rocket motors frequently exploded—an unnerving possibility for the amateur experimenters, who as they lit the engines crouched behind makeshift barriers about 15 meters (50 feet) away. The helmets were painted white to help observers locate the testers as they conducted their work in fields around New York.

Eleven men and one woman founded the amateur association, originally organized as the American Interplanetary Society, on April 4, 1930, in New York City. Most of the founding members were science fiction writers or fans, young idealists caught up in the spaceflight movement of the period, and with few exceptions, had no technical background. Dedicated to the idea and cause of spaceflight, they promoted their interest through talks, articles in their mimeographed *Bulletin of the American Interplanetary Society*, and meetings. In April 1934, in an effort to make the group more "professional" and to attract engineers, the members renamed themselves the American Rocket Society.

By the summer of 1932, the group began crude experiments with liquid-propellant rockets. In the midst of the Depression, though, funding was scarce. They relied on makeshift equipment,

including this lox can, tongs, and helmet, and had no permanent testing site, using instead available vacant fields.

Nonetheless, between November 1932 and September 1934, the group attempted four launches, two of them successes. ARS Rocket No. 2, their first flight, flew in May 1933 at Great Kills, Staten Island, New York, and went up to 76 meters (250 feet). The second, launched in September climbed to 116 meters (382 feet) and was hailed as "spectacular." Yet, the society's Experimental Committee realized that rocket motors fired on static test stands provided better technical data than the occasional launch. Thus, by 1935, the ARS ceased flying rockets and built their first static rocket test stand.

The society's improvised amateur experimentation in the 1930s eventually did contribute to the development of American rocketry. Soon after the bombing of Pearl Harbor in 1941, ARS member James Wyld (with three other members) formed Reaction Motors, Inc. (RMI). Wyld's ARS rocket design of a regeneratively cooled engine aided the development of Jet-Assisted-Take-Off (JATO) rockets—boosters designed to assist aircraft during liftoff. Of more significance, after the war, RMI adapted Wyld's design to produce the 6000C-4 rocket engine that powered the Bell X-1 in 1947, the first plane to fly faster than sound.

The ARS itself flourished and underwent phenomenal growth after the war years, emphasizing professional standards attractive to engineers and scientists. Spaceflight ideas and discussions joined those devoted to rocketry in the society's journal, in its national meetings, and its "space symposiums." These undertakings drew increasing public interest and made the society a focal point in the emerging world of space exploration.

G. Edward Pendray, one of the founders of the American Interplanetary Society, donated these artifacts to the Museum in 1967.

ABOVE: ARS members conduct a static test of a liquid rocket engine at their Test Stand No. 2 in Midvale, New Jersey, 1941.

OPPOSITE: G. Edward Pendray and David Lasser, leaders in the ARS, pour liquid oxygen into a rocket tank prior to a test in Stockton, New Jersey, 1932.

Impacts and Influences

Early Science Fiction and the First Rocket Researchers

Jules Verne's *From the Earth to the Moon*, 1865

ABOVE: An illustration by Henri de Montaut in *From the Earth to the Moon* depicting the "project vehicle."

"My interest in space travel was first aroused by the famous writer of fantasies, Jules Verne. Curiosity was followed by serious thought."

—Konstantin Tsiolkovsky, Russian teacher and theoretician

I n 1865, the year that the American Civil War ended, Frenchman Jules Verne wrote *De la Terre à la Lune* (From the Earth to the Moon), an extraordinary novel describing three men on a voyage to the Moon. Like many of Jules Verne's writings, this book was science fiction in the truest sense: a story that relied on the latest science for its backbone, believability, and drama. Verne drew upon physics, chemistry, engineering, metallurgy, and astronomy to build a story that not only captivated readers, it ignited the imaginations of some of the world's rocketry pioneers.

Layers of technical details fueled the book's success, some fantastic and others that actually hint at aspects of modern space travel. In the novel's 28 chapters, members of the Baltimore Gun Club forged a giant cannon to shoot a bullet-like capsule carrying three adventurers into space. All aspects of the vehicle and its journey were thoroughly explained, down to the telescopes tracking the vehicle. Verne determined that the forces that would be experienced within the fictional "projectile vehicle" would be compensated for with an elaborate shock absorber below the vehicle's floor that would recoil downward, expelling water through external valves as "a kind of spring"; interior walls featuring a "thick padding of leather" offered extra protection. Anticipating the need to replenish the vehicle's breathable air, he suggested that different mixes of chemicals would release oxygen into the depleted air, as well as absorb the noxious "carbonic acid" exhaled by the travelers, and he even calculated the amounts required to sustain three men (and two dogs) for 24 hours.

The intrepid space vehicle naturally needed to be tested before launch. One Gun Club member volunteered to live inside the capsule for a week, sealed off from all Earthly provisions. This fictional decision actually paralleled later "plugs out" tests done by NASA to ensure that spacecraft functioned without physical connections to Earthly power sources and communications—as well as the isolation testing done during early astronaut examinations. "Shouting a boisterous hurrah," the fictional club's volunteer emerged, having "grown fat" in the hospitable—and presumably space-worthy—environment of Verne's capsule.

The novel's realistic science (complete with equations and calculations throughout the text) was reinforced by detailed illustrations. Henri de Montaut's images of the Moon's phases gave readers the latest information about Earth's planetary satellite, while an engraving of the capsule shooting skyward in the cannon's fiery blast captured the launch's excitement.

Jules Verne's space adventure was received with great enthusiasm, some of its readers among the world's most respected scientists and thinkers. Russian teacher and spaceflight theoretician Konstantin Tsiolkovsky credited Verne's novel with inspiring some of his own serious thinking about multi-stage rockets and reactive motion. Years later and miles away, Hermann Oberth, the Transylvania-born rocketry pioneer, recalled that, as a child, he read Verne's space books "at least five or six times and, finally, knew [them] by heart." Robert Goddard, the American inventor of liquid-fueled rockets, scribbled comments and corrections in the margins of his copy. Goddard's own investigations later corrected Verne's unworkable cannon-as-launch-vehicle premise, substituting the rocket as the best way to reach outer space. In many ways, objects such as Goddard's 1935 A-Series rocket owe a debt to the scientific realism of Jules Verne's nineteenth-century novel.

Goddard 1935 Series-A Rocket
1935

"The rocket is now in that most interesting period of discovery where the shorelines are unplotted and the future limited only by imagination."

—CHARLES LINDBERGH, JULY 1937

ROBERT H. Goddard first became smitten with the idea of spaceflight in 1899, at the age of 17. He was an avid reader of a serialized version of H.G. Wells's story, *War of the Worlds*, in the *Boston Post* newspaper, and was profoundly inspired by it. Soon after, according to his autobiography, as he trimmed a cherry tree in his backyard, the possibility of spaceflight came to him in a vivid daydream, in which he envisioned himself ascending to Mars. After the daydream he vowed to devote his life to solving the "spaceflight problem"—a vow he commemorated each year on October 19, the date of his life-changing dream.

One of the basic problems of spaceflight was propulsion—finding a means to escape Earth's gravity and also operate in space. For years Goddard studied different theoretical possibilities before settling on the rocket in 1909. It took the young visionary more years to work out the calculations and to find that liquid-fuel rockets would be far superior in performance and controllability than solid-fuel types.

In 1915, Goddard began experimenting with solid-fuel rockets, which were simple and inexpensive, and then, in 1921, he switched to liquid-fuel rockets. He was able to carry out these experiments through grants from the Smithsonian Institution and, later, from the Daniel and Florence Guggenheim Foundation for the Promotion of Aeronautics. Yet, he kept his dreams largely to himself, and his stated goal was to develop an upper atmospheric research rocket, which he saw as the necessary first step toward spaceflight.

Goddard made the world's first flight of a liquid-fuel rocket on March 16, 1926. It was, by today's standards, a primitive affair. Achieving high-altitude rocket flight was difficult, but he continued his experiments.

Thus, in the A-series, undertaken between September 1934 and October 1935, Goddard built and flew simple rockets using pressurized nitrogen (instead of pumps) to force the propellants into the combustion chamber, and gyroscopic control to maintain steady, vertical flight. Parachutes lowered the spent rockets back to the ground. This series of rockets produced his most successful flights in 30 years of experimentation. The fourth flight of an A-series rocket, on May 1, 1935, reached an altitude of 2,286 meters (7,500 feet), Goddard's highest flight to date. Only his L-series rocket, launched on March 26, 1937, reaching 2,438–2,743 meters (8,000–9,000 feet), exceeded this mark.

With the early success of the A-series, Goddard felt confident enough to invite Harry Guggenheim and the aviator Charles Lindbergh, both supporters and friends, to witness a flight. They arrived on September 22, 1935, but two attempts to initiate a launch failed, each a "fizzle." Goddard's supporters left, never witnessing a Goddard rocket in flight but also not giving up on the importance of his work.

In the fall of 1935, Lindbergh urged Goddard to piece together one complete A-series rocket and donate it to the Smithsonian. Goddard obliged in November. His records suggest that the rocket seen here was the very one that had failed to launch for his distinguished visitors on September 23. It also had the distinction of being the first liquid-fuel rocket to enter the collections of the Smithsonian Institution.

ABOVE: In Roswell, New Mexico, Robert Goddard prepares to launch (note the activation controls by his left hand) and then observe his rocket, mid-1930s.

"Vengeance Weapon 2": The V-2 Ballistic Missile
1940s

"[In London]… the reverberations from each [V-2] rocket explosion spread up to 20 miles so that millions …
felt a personal interest in every one that landed … clearly, in the event of another world war far more
deadly developments of these weapons would threaten mankind in every corner in every country."

—*Christian Science Monitor*, APRIL 28, 1945

ABOVE: An A-4 (or later known as V-2)
launches on a test flight from Peenemünde
in 1943.

PERHAPS no other artifact so elegantly captures the social, political, and military origins of spaceflight than the V-2. It was *not* built as a space rocket—Nazi Germany built it as a revolutionary weapons system, the world's first ballistic missile. As such, the V-2 was a harbinger of the most frightening armaments of the Cold War—intercontinental missiles that could deliver a nuclear warhead to the other side of the world in half an hour. Yet, it had its roots in an interwar movement to spark interest in spaceflight and, after the war, became a step toward exploring space. Its lineal descendant, the Soviet R-7 ICBM, put Sputnik into orbit in 1957.

In 1929–1930, the German army began to research rocketry's military potential, work that eventually led to the V-2. At that time, longer ranges were only possible using higher-energy liquid propellants. In the German-speaking world, Hermann Oberth first laid out the theoretical basis for liquid-propellant rocketry for spaceflight in 1923. His followers set up several small rocket groups, most notably in Berlin. Out of this group came the young engineering student and space enthusiast Wernher von Braun, who was hired by the German army in December 1932.

Owing to the Nazi seizure of power in 1933, the German army had increasingly more money for rearmament. Moreover, von Braun proved to be a brilliant engineering manager, one able to inspire his team with his vision of the future of rocket technology. In 1936 the army and the Luftwaffe (air force) decided to dramatically increase funding. They built a highly secret rocket center on the Baltic coast at Peenemünde, which opened in 1937. Despite many setbacks, on October 3, 1942, von Braun's group launched the first successful V-2 190 kilometers (110 miles) downrange; on the way it arced 90 kilometers (56 miles) high, the first human object to touch the edge of space.

This success, combined with Germany's rapidly deteriorating war situation, moved Adolf Hitler to order the weapon into mass production, drawing von Braun and his superiors even more deeply into the Nazi apparatus. Concentration camp workers helped assemble the missile, resulting in 10,000 to 20,000 prisoner deaths. After many delays and problems, the first V-2s were launched against London and Paris in September 1944; nearly 3,000 were fired by March 1945, causing about 5,000 deaths. As a weapons system, the V-2 proved to be a spectacular but highly inefficient way to drop a ton of high explosives—its only warhead—on enemy cities.

The real beneficiaries of Germany's investment in rocketry were the major Allied powers. The U.S., the USSR, Britain, and France all moved to grab technology and personnel because they saw the ballistic missile's military potential. The United States got Wernher von Braun and the core of the Peenemünde group, as well as parts for nearly a hundred V-2s. More than 70 were launched in the U.S., almost all in New Mexico, where they served as the first American rockets for exploring near space. The Soviets, meanwhile, began firing their own reconstructed V-2s and produced a copy, the R-1, which became the foundation for their strategic rocket forces.

The Museum's V-2 is actually a composite artifact, made from several captured rockets. Most of the fuselage comes from one the U.S. Air Force gave to the Smithsonian in 1949. Originally covered in spotted camouflage, it was repainted to resemble the October 3, 1942, vehicle before the new Museum building was opened in 1976.

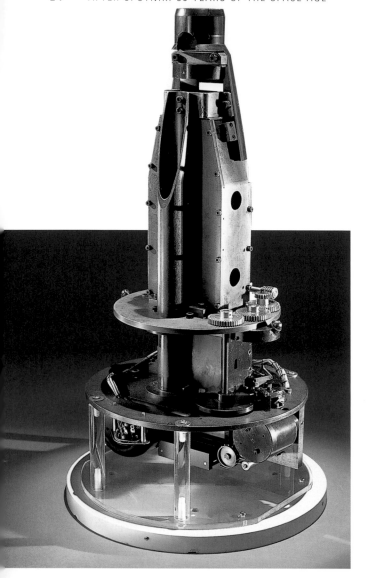

V-2 Spectrograph
1946–1947

"These rocket spectra are certainly fascinating. My first look at one gives me a sense that I was seeing something that no astronomer could expect to see unless he was good and went to heaven!"

—Astronomer Henry Norris Russell, 1947

No ultraviolet spectrographs ever looked like the one Richard Tousey and his Naval Research Laboratory team designed and built for flights on V-2 missiles.

Traditionally a boxlike laboratory instrument using prisms or a grating, spectrographs photographed the "fingerprints" of glowing substances, taking advantage of the fact that each substance exhibits a unique pattern of emission or absorption of light. But in the immediate aftermath of World War II, Tousey acquired a new "laboratory." V-2 missiles, captured in Europe and transported to the U.S., offered a unique opportunity to send redesigned "old" instruments into the Earth's upper atmosphere and the edge of space. Shaped like a bullet to fit in the nose of a V-2, spectrographs such as this one sought to view the Sun unobstructed by the Earth's atmosphere—a feat only possible with the advent of high-powered rockets.

A particular need motivated these postwar flights. The Navy wanted to know the structure of the high atmosphere, specifically the nature of the ozone layer. As a V-2 spectrograph flew toward space through the Earth's atmosphere, it gathered Sun spectra. Together these readings provided a profile of the atmospheric gases between the rocket and the Sun.

Tousey faced a major challenge retrieving his data. His spectrographs recorded data on photographic film—a medium vulnerable to destruction but, at that time, one that yielded the best data. On returning to Earth, V-2s crashed, and they did not land gently. In designing the spectrograph, Tousey made every effort to protect the film. During flight and after exposure, the instrument channeled film into a 1.3-centimeter (.5-inch)-thick-walled cylinder of steel.

The first flight was launched from White Sands, New Mexico, in June 1946. After reaching an altitude of 100 kilometers (62 miles), the missile returned to Earth at hypersonic speed, burying itself in the desolate Tularosa Basin. The Army Corps of Engineers dug for weeks to find the pieces. Tousey inspected the debris, looking for his film canister. He never found it.

What to do? Over the summer, someone suggested repositioning the spectrograph from the missile's nose to a tail fin. Engineers took an additional step. To slow the speed of the missile during reentry, they added charges to blow apart the vehicle, allowing the remnants (such as the fins) to tumble more slowly to the ground.

The next flight, on October 10, 1946, proved the new technique a success. V-2 #12 rose to 173 kilometers (107 miles), the spectrograph worked fine, and on reentry the bits and pieces of the V-2 scattered over a wide area between 25 kilometers (16 miles) and 30 kilometers (19 miles) from the launch pad. Searchers located the fin section on October 16, rushing the film canister back to Washington, D.C., to Tousey and his staff. Within days, the world saw its first glimpse of the ultraviolet spectrum of the Sun as seen from above the Earth's atmosphere, providing the Navy with information on the vertical distribution of ozone above White Sands. This flight did not change scientists' understanding of the Sun or provide information about the ozone layer, but it did demonstrate that laboratory-quality data could be taken during a rocket flight and retrieved for analysis.

The Museum has preserved two of the roughly one dozen V-2 spectrographs built on this design for the Navy. One may well be the original flown in October 1946 or soon after in March 1947. The other, illustrated here, was not flown.

ABOVE: NRL scientist Richard Tousey inspects debris, June 1946, from a U.S.-assembled V-2 that launched, reached the upper atmosphere, then crashed. The flight carried one of his specially designed ultraviolet spectrographs.

Minimum Orbital Unmanned Satellite of the Earth (MOUSE) Mock-up
1954

"… something tangible and graphic to get the idea across."

—S. FRED SINGER, 2006

HE SHOULD call it "House" rather than "Mouse," University of Chicago professors laughed when S. Fred Singer presented his idea for a small robotic scientific satellite at a meeting in 1953. In the years before Sputnik, many considered the idea of space satellites with skepticism and assumed that any possible satellite would have to be large. As Singer, then a physicist at the University of Maryland, recalled, "They didn't realize how much we could do in miniaturization." Indeed, few then realized how effectively technology might be packed into a vessel little larger than a ripe watermelon. Despite such doubts Singer advocated his idea with conviction.

Singer's thoughts crystallized in speeches he made between 1952 and May 1954, when he spoke at the Hayden Planetarium and announced his concept for a "Minimum Orbital Unmanned Satellite of the Earth," or MOUSE. Soon after, Singer proposed, with support from U.S. and European colleagues, that MOUSE become part of one of the twentieth-century's great scientific undertakings, the International Geophysical Year (IGY), planned for 1957–1958. MOUSE and other satellites, Singer reasoned, could provide a unique vantage for studying the Earth—the focus of the IGY. Singer and his colleagues immediately began to explore the numerous possible scientific tasks for an attitude-controlled spinning satellite. Suggestions included measuring the composition and physical structure of the outermost regions of the Earth's atmosphere, cloud cover, and the ultraviolet spectrum of the Sun, as well as the nature of solar-terrestrial relations and the composition and character of primary cosmic radiation.

After Singer's presentation at the Hayden, Israel M. Leavitt, director of Philadelphia's Fels Planetarium and a space enthusiast, wanted to popularize the concept. At Leavitt's urging, Singer designed and directed the construction of a model, because, as he recently recalled, one "had to do something tangible and graphic to get the idea

across." It went on display at the Fels Planetarium in late 1954.

The model probably was the first attempt to create a realistic satellite mock-up—and the Museum artifact is the original, donated by Singer in 1973. It weighs about 45 kilograms (100 pounds), contains real Anton Geiger counters for measuring cosmic ray intensity, and sensors for scanning the Earth's cloud cover. To support these instruments, Singer also included telemetry electronics for sending data back to Earth and a magnetic data storage device. For power, the satellite model incorporated experimental solar photocells, newly developed at Bell Laboratories by his scientific colleague John Pierce. To Singer, MOUSE represented a basic set of instruments capable of returning useful scientific information from orbit.

Singer's MOUSE concept received mixed reviews—Harvard astronomer Harlow Shapley hailed it as a major highlight of the year. But the attention brought to the idea also raised doubts about the scientific benefits of spaceflight and pointed to the tension between university-based scientists and the military in the Cold War. Shapley called Singer's proposal an "artificial tax-supported satellite that is designed to … sail around the earth, reporting as it goes on what this planet does to missiles and meteorites at an altitude of two or three hundred miles." But in the months after its public unveiling, MOUSE receded into the background. As the IGY gained momentum and its value to Cold War politics increased, the United States government and its leading scientific bodies pushed for an official satellite project. In July 1955, the Department of Defense and the National Academy of Sciences announced Project Vanguard, under development by the U.S. Navy, as the American contribution to IGY. Singer's novel proposal yielded to events, but it signaled the deep interest that existed in the possibilities of satellites—for science and other purposes—in the years before Sputnik.

Snark Guidance System
1950s

"'It's a Snark!' was the sound that first came to their ears, / And seemed almost too good to be true.
Then followed a torrent of laughter and cheers: / Then the ominous words 'It's a Boo—' / Then, silence."

—LEWIS CARROLL, *The Hunting of the Snark*, 1876

ABOVE: Workers at Cape Canaveral Auxiliary
Air Force Base, Florida, ready a Snark for launch.

AT THE end of World War II the victorious Allies held German scientists and engineers in high regard. Their advances in aeronautics and rocket technology often were at the leading edge of innovation. The unpiloted V-1 "buzz bomb," which terrorized London until the British mounted an effective defense, proved one of the most successful German weapons. In the late 1940s, as relations with the Soviet Union deteriorated, the United States Air Force sought to develop a similar weapon, the "Snark"—a robot bomber capable of carrying a nuclear warhead to targets across the Atlantic. But developing a long-distance missile posed a special problem: how to guide it accurately over thousands of miles to hit a designated target.

Under Air Force contract, the Northrop Aviation Company of Hawthorne, California, proposed two designs: the subsonic "Snark" and a supersonic "Boojum" (never developed), both named after mythical characters in the Lewis Carroll poem, *The Hunting of the Snark*. Though inspired by the German V-1, the Snark was larger—20 meters (67 feet) in length—and possessed a much longer range—up to 10,186 kilometers (5,500 nautical miles). This guidance system (donated to the Museum by Northrop in 1970) solved the problem of directing a missile automatically over intercontinental distances and represented the greatest technological breakthrough of the project.

The basic guidance was inertial, that is, it used a set of gyroscopes and accelerometers that fed data into a computer, which then calculated navigational position and velocity. What challenged the missile's designers was the complex profile of a mission, which had three distinct phases. Two solid-fuel rockets launched the missile into the air from a stationary platform (the Snark needed no runway). Four seconds later the boosters dropped away and a conventional turbojet engine took over, flying a mission that could last as long as 11 hours. When the Snark got near its destination, its nose cone and warhead separated and dropped to the target. An inertial system proved suitable for guiding the first and third phases, but during the long "cruise" phase the gyros tended to drift and required periodic corrections. Snark did this with a "star tracker," a device that used sightings of stars to provide additional information on a missile's position. Thus the Snark navigated as sailors had done for ages, only autonomously and with no human intervention.

During development, Snark encountered a number of problems, but the guidance system worked—engineers claimed the missile could deliver a warhead to within 2.4 kilometers (1.5 miles) of a target after a journey of thousands of miles. Air Force technicians joked about the "Snark-infested waters" off the coast of Florida, where they conducted tests, many of which resulted in missile crashes. In 1960, the Air Force briefly deployed the missiles from a base in northern Maine, but by then ballistic missiles such as the Atlas made them obsolete. The missiles were destroyed, save for components such as this guidance system.

The vacuum tubes visible around the sides of the guidance system give an indication of the primitive level of its technology, in comparison to the silicon chips of today. At the time, however, the guidance system was one of the most advanced pieces of aeronautics technology. The missile itself may not have been a success, but its guidance system laid the foundation for many long-range aircraft, cruise missile, and even spacecraft guidance systems of the following decades.

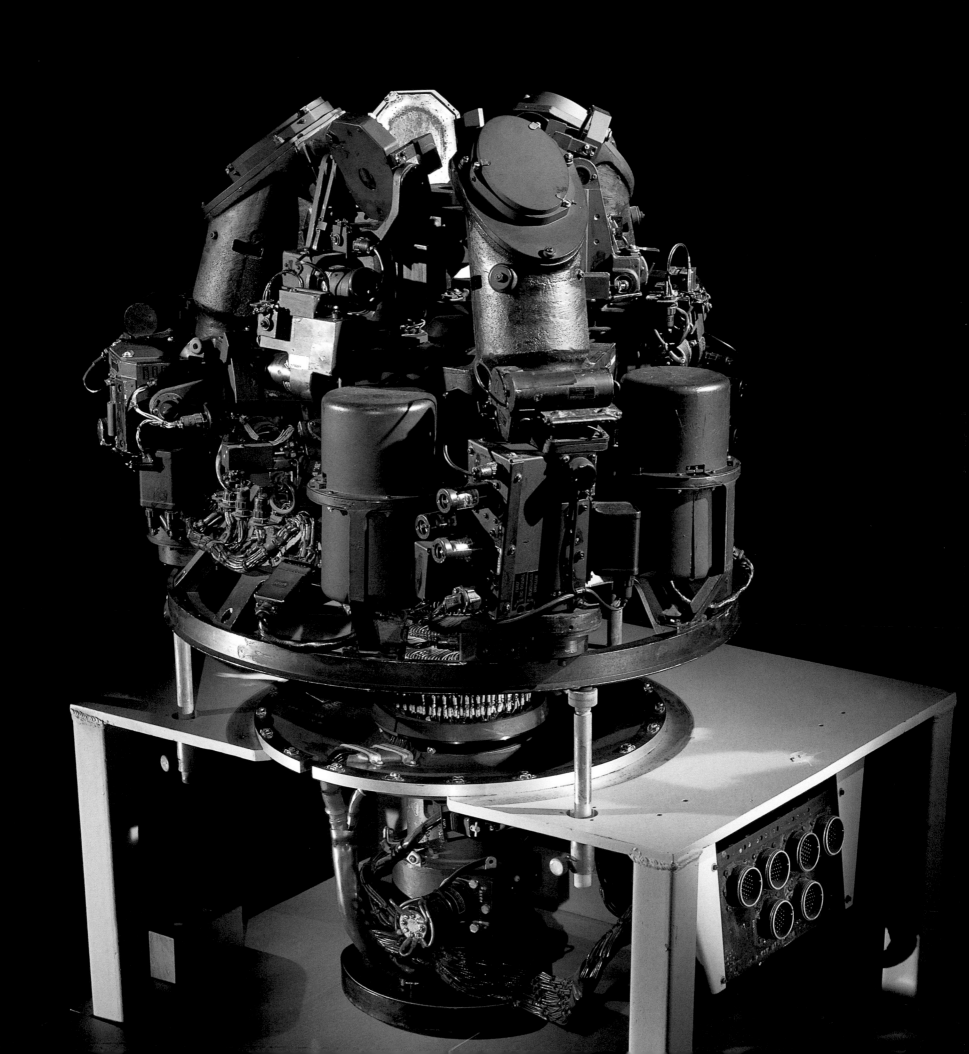

Goodyear Three-Stage METEOR Jr. Spacecraft Model
Mid-1950s

"Now, Mr. Romick, in a sequential engineering study, has shown how such ferry-stage rockets could become the initial building blocks for a space station."

—*Popular Science*, MAY 1956

YEARS before the Space Shuttle, before Sputnik demonstrated the reality of spaceflight, some engineers began to design winged spaceships—vehicles that might go into space, return, and go again. Heralded in newspapers and popular science magazines, METEOR was such a spaceship—fantastic in look and concept. In 1952, Darrell C. Romick, a Goodyear Aircraft engineer, dreamed up METEOR and then elaborated his designs over the next several years.

Born on an Iowa farm, Romick loved aviation so much he took his first job with a flying school in Pennsylvania during the Depression and learned to fly and rebuild entire aircraft. By 1940, he and a friend had designed a low-cost twin-engine aircraft called the Monarch. They test flew the plane, but potential customers were looking for other types of aircraft as World War II loomed. The experience was not wasted, however, as designing a complete mechanical-electrical flight system served Romick well when it came to conceiving the METEOR.

Romick joined Goodyear in 1946 and became the project engineer for the experimental MX-778, a 160-kilometer (100-mile) range missile. This stimulated Romick to study everything about missiles—and space travel—and he soon learned of concepts for artificial satellites and human-operated space stations that others were proposing.

"The one thing that was missing in these studies," Romick later recalled, was a "practical launch vehicle that you could run like an airliner—the transport of cargo on a regular basis." By 1948 he came up with a solution—a recoverable booster. His earliest sketch was a three-stage vehicle with oversized delta fins. Each stage fit into the other, could operate separately, and had a small crew for flying the ship back to a safe landing. The next year he convinced Goodyear of the idea's value and the company let him use their artists and other materials

to further develop it, but it remained Romick's private project.

The concept grew and evolved into the METEOR (Manned Earth-Satellite Terminal Evolving from Earth-To-Orbit Ferry Rockets). Romick wrote a series of papers on successively more complex versions of METEOR and presented them before the American Rocket Society.

By 1954, METEOR was designed as a ferry rocket to take sizeable payloads into 805-kilometer (500-mile) orbits and then return. The ferry stood 43 meters (142 feet) tall and could carry 35,000 kilograms (35 tons). Wernher von Braun was especially interested in METEOR. Romick, in turn, found inspiration in a von Braun idea—his famous "space wheel" space station. In its 1956 form, Romick envisioned an entire space "city" of some 20,000, people and a fleet of his now gigantic 87-meter (285-foot)-tall, 13-meter (42-foot)-wide METEOR ferries servicing the station. The first stage, with 51 engines, was to generate more than 191 million newtons (about 43 million pounds of thrust). Yet this, and the other stages, were all returnable. The Museum's model, donated by Goodyear, is a version of this concept, called METEOR Jr.

Romick's concept was so spectacular it was featured in *Time, Newsweek, The New York Times*, and on covers of *Mechanix Illustrated* and other magazines. Romick himself appeared on the original *Today* TV show and *What's My Line?* For a time, he was almost as well known across the country as Wernher von Braun.

But METEOR would have cost billions, and it was never built. Looking back, years later, Romick thought that although his idea was "too ambitious," it may have helped plant the seeds of the recoverable booster idea that eventually became the Space Shuttle.

Navaho Missile Rocket Engine
Late 1950s

"The Navaho cancellation was a tough blow to the men who worked so hard on the program, but they can take great satisfaction in the important help they have given to other defense projects—and taxpayers can also rejoice."

—Los Angeles Times, 1958

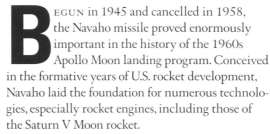

ABOVE: A Navaho missile launches on a test flight at Edwards Air Force Base in the 1950s. Its rocket engines only powered the first stage then dropped away.

TOP: A top-end view of the Navaho Rocket Engine.

BEGUN in 1945 and cancelled in 1958, the Navaho missile proved enormously important in the history of the 1960s Apollo Moon landing program. Conceived in the formative years of U.S. rocket development, Navaho laid the foundation for numerous technologies, especially rocket engines, including those of the Saturn V Moon rocket.

Near the end of World War II, North American Aviation Company, under contract to the Army Air Forces, undertook a study to develop an 800-kilometer (500-mile) range missile, later called Navaho. Known as a leading aircraft manufacturer, the company built superb planes—such as the P-51 Mustang and B-25 Mitchell bomber—but had no experience at all with rockets. As they began their effort, the German V-2 rocket of World War II reigned as the world's largest and most powerful missile, possessing a range of 320 kilometers (200 miles) and a liquid-fuel rocket engine of nearly 250,000 newtons (about 56,000 pounds) of thrust. North American used this as a starting point, and then systematically developed its own idea for a rocket.

In "Phase I," they studied V-2 documents closely and interrogated former V-2 technicians. In "Phase II," they built their own V-2 engines to learn how to make large liquid-fuel rocket engines. But "Phase III" of the Navaho program saw the creation of a completely new and revolutionary type of rocket engine. The German V-2 engine featured an hourglass shape and 16 injectors on top for spraying in the propellants. By contrast, North American designed a motor that was tubular in shape and more compact, with a simplified, central "flat plate" injector head.

The Navaho missile program expanded in scope and continually changed design under new contracts. By 1948, the plans called for an 800-kilometer (500-mile) range missile. Instead of a rocket, the new vehicle design called for two ramjets to power a cruise missile, with a liquid-propellant rocket booster. Next it evolved into a 1,600-kilometer (1,000-mile) missile, and eventually into a huge, dual-ramjet-powered 8,850-kilometer (5,500-mile) range intercontinental cruise missile with two powerful liquid-fuel boosters.

Engineers scaled up the rocket engine accordingly. By 1950, the engine produced more than 330,000 newtons (about 75,000 pounds) of thrust, but even this proved insufficient for the larger model of the Navaho. The engine did suit another rocket then under development, however, the 240 to 320-kilometer (150 to 200-mile) Redstone, which became the United States' first successful ballistic missile. This rocket was later modified and fitted with upper stages to become the launch vehicle for the country's first satellite, Explorer 1, in 1958. In 1961, another version of the Redstone, the Mercury-Redstone 3 (MR-3), launched the United States' first astronaut into space, Alan B. Shepard Jr.

Before the Navaho program's cancellation, North American upgraded the combustion chamber, increasing thrust by more than 50 percent. The next version of the boosters used an engine consisting of a pair of these improved combustion chambers, shown here, with a total thrust of more than 1 million newtons (about 240,000 pounds).

In 1957, Navaho finally underwent flight tests—none were a resounding success, primarily due to the missile's complexity. The military cancelled the program, partly for budgetary reasons, but also because the new ballistic missiles proved superior weapons. The years of effort were not wasted, however, as North American used the work on Navaho to develop each of their large-scale engines, which included those for the Atlas, Jupiter, and Thor missiles and all those for the mighty Saturn V.

The Rocketdyne Division of Rockwell International (the successor corporation to North American) donated this Navaho engine to the Museum in 1969.

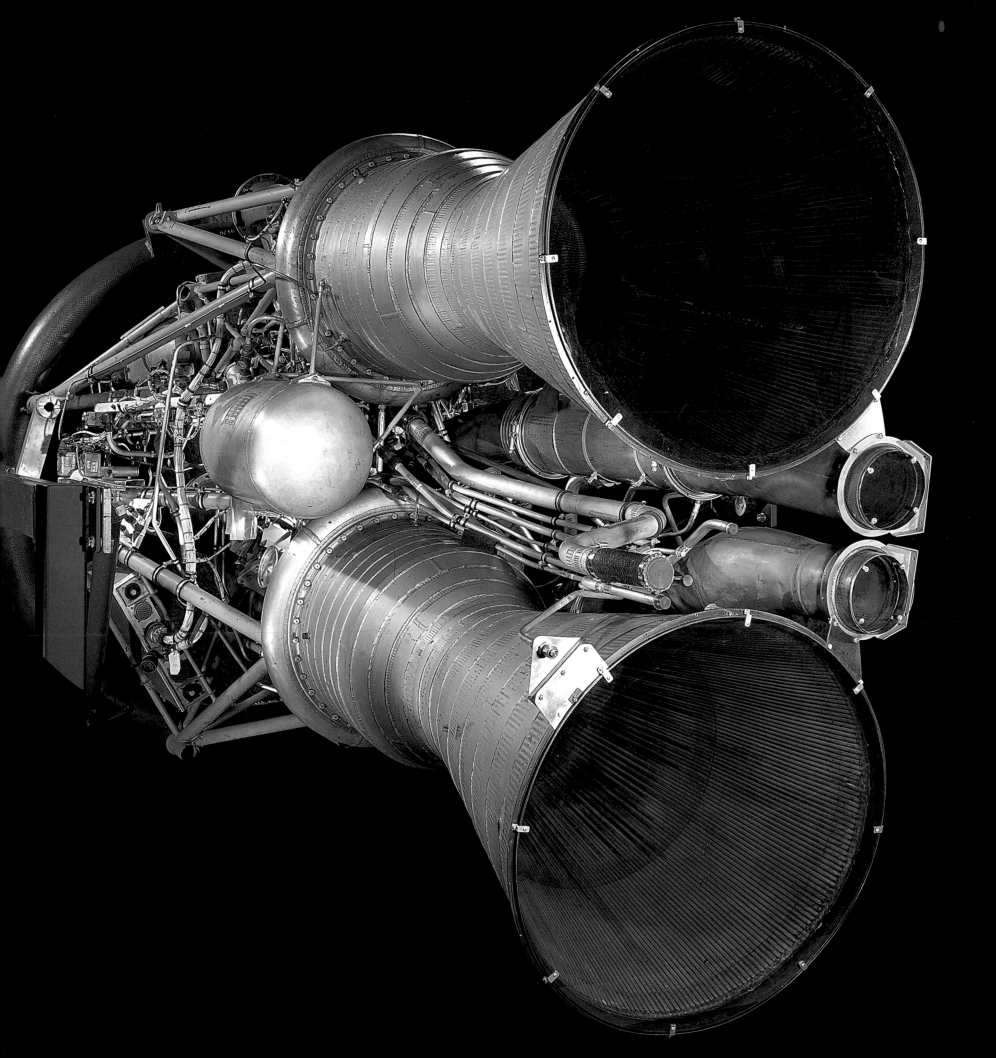

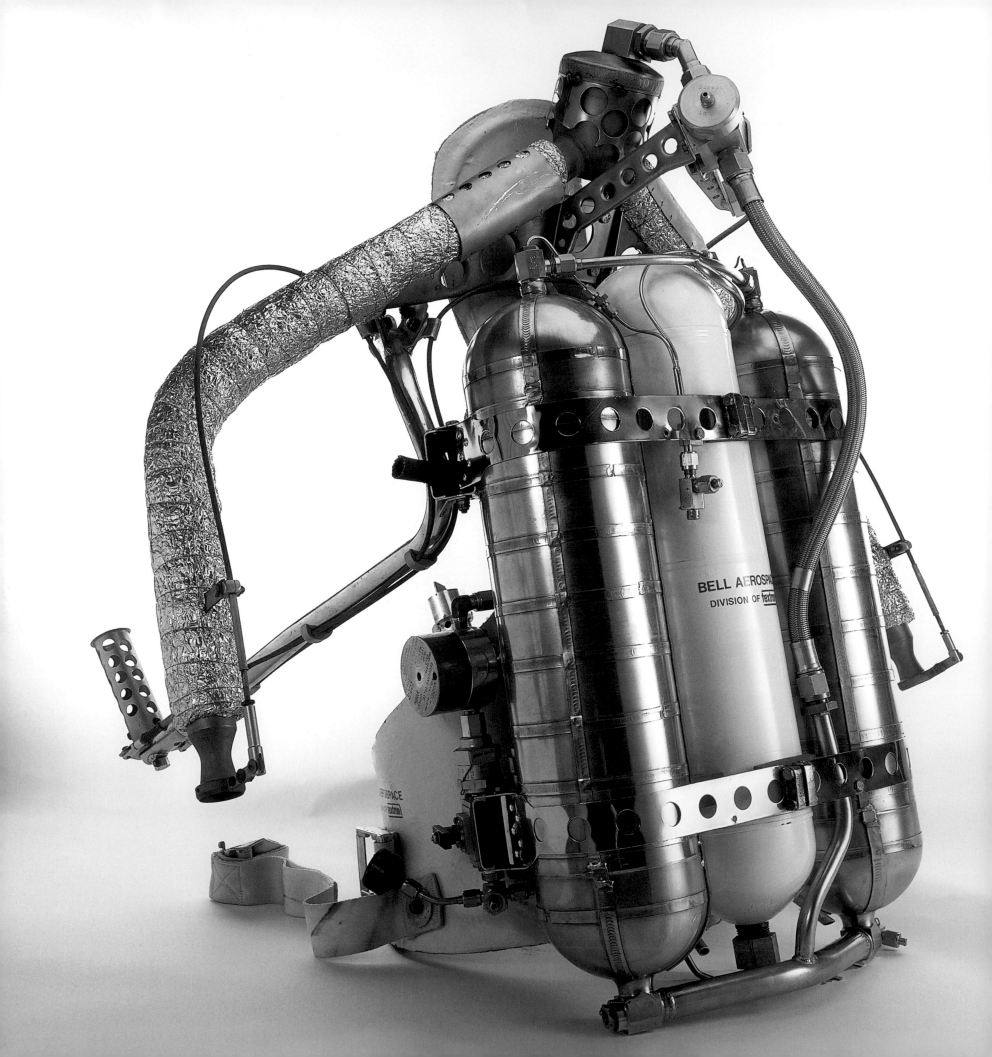

Bell Rocket Belt No. 2
Late 1950s

"Man's age-old dream of flying like a bird, free of any clumsy machinery, may be nearer than we think."

—*Popular Science*, SEPTEMBER 1958

THE ROCKET belt seems an exotic invention—something heralded in science fiction, a technology of the distant future. In fact, the idea *did* appear in science fiction. It first showed up in the *Buck Rogers* comic strip pages in 1929. Twenty years later the device gained further attention in the movies, in the action-packed serial *King of the Rocket Men*. In recent years, it reappeared in the feature film *The Rocketeer*. Such fictional examples, though, stand in contrast to the real-world effort to develop a rocket belt.

In the late 1950s, Wendell Moore, a young crew-cut engineer with the Bell Aerospace Company, was the first to build and test a rocket belt—the artifact seen here, donated to the Museum by Bell in 1972. It is not known whether he was influenced by the fictional Buck Rogers belt or by a movie version, but his ingenious rocket belt was operated and used differently from fictional examples.

In typical fictional representations, writers and artists portrayed the belt as a small rocket strapped to the back of a pilot. On command from the pilot, the belt, spewing hot gases from combustion, allowed a sudden leap over a very steep hill, perhaps, then a controlled, safe landing. In the *Buck Rogers* stories and films, Buck used the rocket belt to escape quickly from menacing aliens on another planet or, on Earth, to escape terrestrial villains.

Moore thought more about the rocket as an innovative device for earthly uses rather than its possible applications in space. He conceived of the device enabling soldiers, firefighters, or any rescue worker to leap across ravines, embankments, or other impassible barriers in order to complete a military or humanitarian lifesaving mission rapidly.

Moore realized that in order for the rocket belt to be safe for the user—and reusable—it had to operate with a noncombusting propellant to minimize risk from heat and possible explosion. He chose gaseous hydrogen peroxide, which, when used with a chemical catalyst, came out as powerful but noncombusting steam through the belt's two tiny downward-aiming nozzles. The belt attached with body straps and buckles to a form-fitting jacket or corset. The two outside tanks carried fuel while the third, smaller tank contained high-pressure nitrogen for forcing the fuel into the small catalyst bed that turned it into steam. Valves controlled pressure levels and activation.

Moore impressed the U.S. Army, as well as the Marines, with the concept and successful demonstrations of the device in the early 1960s, but they soon observed that the invention had problems. The rocket operated for a very short duration—about 20 seconds—a serious drawback. This highlighted another shortcoming. Once the pilot reached a destination, the fuel was exhausted or near depletion. How might the pilot then return, perhaps across a river or ravine? And because the rocket belt user traveled relatively slowly through the air, and the belt made a fairly loud whooshing sound as it left the ground, a pilot in combat would be noticeable and more vulnerable to enemy gunfire.

The military lost interest quickly, and the belt's short duration also proved a drawback for firefighters and other rescue personnel. The rocket belt, in both its fictional representation and experimental forms, was certainly exciting to watch in action. A failure in practical terms, it took on a new and more glamorous career—as entertainment at world's fairs, half-time shows, and in the movies.

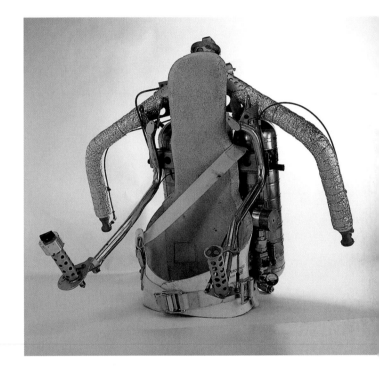

ABOVE: Gripping the hand controls, a pilot elevates in a vertical line as he tests a Bell rocket belt.

Mark II, Model "R" Full Pressure Suit
Mid-1950s

"From the day that Wiley Post brought his idea for such a garment to the Goodrich Company, engineers and Aero Medical men in many countries have been striving for a device to protect the air crewman's life when he was exposed to the rarefied atmosphere of high altitudes."

—T.M. WALKER, *The Project Engineer*, MAY 1956

THE FIRST spacesuits had a long history—they derived from aviation's high-altitude flight suits. The Mark II, Model "R" Full Pressure Suit proved crucial in the transition from high-altitude flying to spaceflight. Developed in the mid-1950s by B.F. Goodrich, a rubber and synthetic products company, and the U.S. Navy to protect its pilots in high-performance aircraft, the Mark II and related suits served as prototypes for astronaut spacesuits.

The Mark II originated in the 1930s, when aviators competing for altitude, distance, and speed records encountered two crucial problems at high altitudes: lack of oxygen and extreme cold. To combat these conditions, mechanic and noted aviator Wiley Post conceived the idea of a full pressure suit to protect pilots. As he prepared to compete in a long-distance air race by flying at altitudes of 9,000–12,000 meters (30,000–40,000 feet) in an unpressurized aircraft, Post designed three suits, the third proving effective. Working with B.F. Goodrich, Post turned this initial research into the first working protective full pressure suit in the United States.

Dramatic advances in aircraft construction, speed, and altitude made during the 1940s and 1950s raised new challenges for high-altitude, high-performance flight, and two avenues of development ensued: pressurization of aircraft cockpits or cabins and, building on Wiley Post's efforts, improved pressurized suits for pilots. The Mark II was developed during these years. B.F. Goodrich continued as an important manufacturer of pressure suits and obtained the Navy contract for the series of "Mark" pressure suits.

The Mark II, as developed by the Navy and B.F. Goodrich, performed two basic functions. As with Wiley Post's suits, the Mark II protected the pilot from the cold, but primarily it provided him with oxygen. Pressurization of the suit was essential to force the oxygen into the lungs. At high altitudes, without the aid of a protective suit, a pilot's brain becomes oxygen-starved, resulting in blackout in a very short time, placing pilot and aircraft in extreme danger.

This Mark II, Model "R" suit was last in the series of developmental suits used to specify a design before the production of flight versions. The advanced features of this suit included detachable gloves and allowed the wearer to walk erect and sit-stand-sit while pressurized. This suit had zippered sizing adjustments that made for easy helmet attachment and removal. During testing, though, the developmental suit revealed a serious drawback. If a pilot had to make an emergency landing at sea, the suit could fill with water—a potentially deadly circumstance for Navy pilots, who often flew over water.

The Mark II was one of a series of pressure suits developed by the Navy and B.F. Goodrich through the 1950s, culminating with the Mark V. This line of development proved crucial for the United States' human spaceflight program in the first years after Sputnik. The Mark IV suit was adopted by NASA and modified to create the Mercury spacesuit. The Naval Air Material Center in the Navy Bureau of Aeronautics transferred this suit to the Museum in 1956.

Impacts and Influences

Promoting Space Exploration before Sputnik

The *Collier's* Space Series, 1952–1954

LESS THAN a decade before *Sputnik's* launch, the American public scarcely believed in the imminence of spaceflight. For many, science fiction characters like Buck Rogers and Flash Gordon had so fictionalized and fantasized the subject that it seemed preposterous. When the Gallup organization ran a poll in late 1949 on what scientific and technological advances to expect by the end of the century, only 15 percent of Americans believed that "men in rockets will be able to reach the Moon within the next 50 years." By contrast, 63 percent thought there would be atomic-powered trains and airplanes.

The *Collier's* magazine space series, which ran between March 1952 and April 1954, helped change the mindset of the American public with regard to the feasibility of spaceflight. The magazine's editors, excited by reports of a spaceflight symposium held in New York City on Columbus Day 1951, sent writer Cornelius Ryan to an Air Force conference in San Antonio. There Ryan met Wernher von Braun, the ex-Nazi rocket engineer, as well as Fred Whipple and Joseph Kaplan, two leading scientists who believed in space travel. Enthused by their ideas, Ryan began organizing a special issue of the magazine, bringing in the leading space artist, Chesley Bonestell, plus two other talented illustrators, Rolf Klep and Fred Freeman, for maximum visual impact. Whipple, Kaplan, and other space advocates, most notably the writer Willy Ley, contributed to that first issue, dated March 22, 1952, but von Braun's ideas dominated. He proposed sweeping, romantic visions of giant space stations and multi-stage, winged space planes that appealed to Ryan's—and the public's—imagination.

Reflecting the fear of the spread of communism that prevailed during the Korean War years, von Braun's proposed space station included an orbiting reconnaissance platform and nuclear missile launch base for achieving "space superiority" over the Soviet Union—an idea the magazine urgently endorsed in an opening editorial, "What Are We Waiting For?" Yet, in the long run, von Braun's military space ideas would eventually be eclipsed in the public mind by his vision of human flights in space.

Encouraged by the enthusiastic reader reaction, *Collier's* published seven more issues on the possibilities of space travel over the next two years. These issues anticipated the first human trip to the Moon, the training and outfitting of astronauts, procedures for emergency escape, and Wernher von Braun's imagined 10-ship expedition to Mars, based on his technical treatise *The Mars Project*. Robotic spacecraft essentially did not exist in von Braun's vision of space exploration, which he modeled on the polar and aviation expeditions of the earlier twentieth century. Von Braun's article about the "Baby Space Station," his name for a monkey-carrying satellite, only came about because Ryan—and the magazine's readers—pressed him for some indication of what might precede human spaceflight. Above all, von Braun was fascinated by the prospect of traveling in space—personally, if possible.

Public skepticism—even ridicule—did not end with the *Collier's* series, but the endorsement of space travel by respected scientists and engineers did convince many Americans that spaceflight might soon become reality after all. The series also launched an important spin-off: in 1954, the Walt Disney Company hired von Braun and Willy Ley to develop spaceflight programs for its new ABC television series. The first two broadcasts in 1955 reached an even bigger audience than *Collier's*. Thanks in large part to the magazine and the TV series, the percentage of Americans who believed in a human lunar trip within 50 years jumped from 15 to 38 percent by 1955. Events close in time to the 1955 broadcasts reinforced those beliefs: in July, President Dwight D. Eisenhower announced the U.S. intention to launch an Earth-circling satellite as part of the scientific International Geophysical Year, set to begin in 1957.

Chesley Bonestell. *Lunar Landscape.* Oil on canvas mural
1957

"'The moon, gentlemen, has been carefully studied,' continued Barbicane. 'Selenographic charts have been constructed with a perfection which equals if it does not even surpass, that of our terrestrial maps. Photography has given us proofs of the incomparable beauty of our satellite.'"

—JULES VERNE, *From the Earth to the Moon*, 1865

"NO SPACESHIP reservations are needed for a startlingly realistic visit to the moon," announced a press release issued by the Boston Museum of Science on March 29, 1957. The day before, George Gardner Jr., president of the museum, and Floyd Blair, representing the Charles Hayden Foundation, had unveiled Chesley Bonestell's mural, *Lunar Landscape*, on the lobby wall of the museum's Hayden Planetarium. Measuring 12 meters (40 feet) long by 3 meters (10 feet) tall, the dramatic lunar panorama was the masterwork of an artist who helped set the stage for the space age.

Born in 1888, Chesley Bonestell grew up on San Francisco's Nob Hill and survived the 1906 earthquake to emerge as a leading American architectural designer. He left his mark on the Chrysler Building, the U.S. Supreme Court, and the Golden Gate Bridge, then moved on to Hollywood, where he created background paintings for such films as *The Hunchback of Notre Dame* (1939) and *Citizen Kane* (1941).

Hollywood paid the bills, but Bonestell's real interest lay elsewhere. Always fascinated by astronomy, he began combining the best available science with his own artistry to produce paintings of the surfaces of other worlds. Over the next several years, his work reached millions through popular magazines, books, and even a classic spaceflight film, *Destination Moon* (1950).

Bonestell was at the peak of his career, in 1956, when the Boston Museum of Science commissioned *Lunar Landscape*. As in the case of all of his paintings, the artist planned the mural in meticulous detail. He positioned the viewer on a spot 396 meters (1,300 feet) up the south wall of an imaginary lunar crater ("similar to Albateguius, but smaller"), located 7 degrees from the Moon's North Pole and 5 degrees to the left of the center of the lunar disc. He went so far as to specify that it was three o'clock, Boston time, on a late June afternoon and calculated the position of the planets and stars accordingly (Jupiter over the central peaks, Antares below and to the right of the Earth).

Nineteenth-century artists had produced huge canvases to introduce stay-at-home Americans to the scenic wonders of the West. Chesley Bonestell invited viewers to imagine what it would be like to stand on the surface of another world. And if his predecessors had taken artistic license to emphasize the grandeur of the western landscape, Bonestell portrayed a dramatically lit Moonscape, with sharp peaks, jagged canyons, and precipitous crater walls.

The Soviets orbited Sputnik on October 4, 1957, a little more than six months after the unveiling of *Lunar Landscape*. Over the next few years, images returned by the first spacecraft to visit the Moon revealed a very different, and far less dramatic, world than the artist had envisioned. Recognizing that the mural was not an accurate depiction of the Moon, Boston Museum of Science officials removed it from the wall in 1970 and presented it to the Smithsonian Museum six years later.

"I tried to make it as dramatic as I could," the artist explained, while admitting that the Moon looked "nothing like" his masterpiece. He continued to insist, however, that a mural with the soft, rolling hills of the real Moon "wouldn't have looked very interesting." As artist Ron Miller suggests, Chesley Bonestell gave us the Moon "as it should have been," the Moon that inspired our romantic desire to stand on a spot where the blue-green Earth floats above the rugged lunar highlands. Today, *Lunar Landscape* remains as both a reminder of an era when human beings could only dream of space travel, and a casualty of a dream realized.

Sputnik and First Steps into Space, 1957–1961

I N T H E Y E A R S before Sputnik, the United States and the Union of Soviet Socialist Republics (USSR) speculated about, planned for, and initiated programs to achieve spaceflight. Each anticipated a Sputnik-like event in 1957 or 1958. Yet when on October 4, 1957, around 10:30 p.m. Moscow time, the Soviet Union launched Sputnik, the first human-made object to reach space and orbit the Earth, the event triggered a wide-ranging response. The artificial "moon" sent a pulse through western culture that reconfigured the geopolitics of the Cold War, redirected U.S. national policies, and birthed a new era in human history, the "space age."

That Sputnik might evoke such a response was not obvious in advance. Sputnik itself seemed the essence of simplicity. A brightly polished orb, 58 centimeters (23 inches) in diameter, with four swept-back antennae, it was equipped only to do one thing: emit a radio "beep" at two separate frequencies. In the USSR, on October 5, Sputnik received only minor notice in the state-run newspapers, perhaps reflecting the uncertainty of Soviet leaders on the significance of the achievement.

But elsewhere, especially in the U.S., Sputnik drew near instant attention as a stunning event, historic in its implications. Military installations, telecommunications companies, and radio amateurs picked up Sputnik's "beep" on its first orbits—which, with the time difference, occurred in late afternoon October 4 on the eastern seaboard of the United States. Before the day was out, the major radio and television stations interrupted their programming to bring Sputnik's audio signature—the soundtrack of the "space age"—into millions of cars and homes. In the evening, families across the eastern half of the country (in the path of Sputnik that day) looked heavenward in attempt to spot the Soviet "moon." Sputnik was a phenomenon, conveying a mix of dread, awe, and, as one British newspaper

stated, "flabbergasted admiration." On October 5, in contrast to USSR press coverage, U.S. and European newspapers blanketed their front pages with everything Sputnik: details of orbits and tracking techniques; broad assessments of implications; quotes from political leaders, the military, corporate executives, scientists, and men and women on the street; and, of course, plenty of speculation on Soviet technology and intentions. Seeing the reaction of their Cold War adversaries, the Soviets now hailed the achievement. Preexisting themes of the Cold War took on new intensity: the importance of scientific and technical advance and the confrontation between communism and capitalism. And a new powerful theme took hold: awe at being witness to a new epoch in history, one in which humans might begin to reach beyond the Earth. When, in early November, the Soviets launched Sputnik 2, carrying Laika the dog into orbit, this mix of Cold War fears and transcendent anticipation intensified. As the U.S. space effort languished in late 1957 and early 1958, space and spaceflight became the dominant symbols of the United States' ability (or inability) to compete with the Soviet enemy and lead free peoples into a new future.

Sputnik tapped into the preoccupations, fears, and hopes of 1957 in unique ways. In the hard give-and-take of the Cold War, Sputnik's "beep" was only part of the story. The R-7, the rocket that launched Sputnik into space, signaled that the Soviets had perfected a rocket sufficiently powerful to carry a nuclear warhead thousands of miles to the United States. Advances in bombers and nuclear bombs (with stockpiles on each side in the thousands) already had created what one nuclear strategist called a "delicate balance of terror" between the Cold War superpowers. The possibility of missile attack, accomplished over transcontinental distances in less than 30 minutes, raised the potential shock and devastation of nuclear warfare to a new level.

OPPOSITE: After his capsule's splashdown on its return from space, May 5, 1961, astronaut Alan Shepard and his craft are hoisted from the water off the coast of Florida by a Marine helicopter.

In the strange world of the Cold War, the R-7 and Sputnik stood not only as specific technical accomplishments, but as potent symbols. They served as telltale markers of the relative capabilities of the two superpower societies. The two dramatic Soviet successes in October and November, followed by a much-publicized U.S. flop—the catastrophic explosion of Vanguard on the launch pad in December—raised the question of whether state-driven communism might, indeed, possess an edge over democratic capitalism in the crucial areas of scientific and technical advance.

ONE senator quipped that "while we [the U.S.] have been thinking about more attractive fins on our automobiles, they have been thinking about building a missile to get outside the earth's orbit." The R-7 symbolized the unique status of the rocket in the post-Sputnik Cold War: as the ultimate terror weapon and as the essential tool of the space age—whether for making space a battlefield or for peaceful exploration. Worries of a "missile gap" quickly emerged, claiming that U.S. missteps in managing dozens of separate and largely uncoordinated programs in the military services left the country vulnerable to the more advanced Soviets. The ever-present Wernher von Braun amplified such concerns: "We don't have a real powerful rocket engine today simply because none of our present crash missile programs needs it…. But in order to beat the Russians in the race for outer space we absolutely need it—and the development of such an engine requires several years."

Clare Booth Luce—playwright, journalist, congresswoman, and wife of *Life* magazine publisher Henry Luce—lamented that Sputnik's success represented "a baleful intercontinental, outer space raspberry." The *Christian Science Monitor* editorialized that while the Soviets pursued their goals with discipline, the consumer-oriented, individualistic United States had turned toward "vast expenditure on creature comforts and a preoccupation with a soft materialism." Senator Lyndon Baines Johnson pronounced that if the United States did not "control" space the Soviets would. In this new arena, the U.S. had to be "strong enough so it could prevent domination of the world by a nation with evil intent." Johnson began to organize a series of hearings that brought a "who's who" of the scientific, technical, business, and political elite in front of the nation, all calling for an invigorated effort in spaceflight. The testimony ranged from the minutely practical to the near whimsical (renowned physicist Edward Teller offered, "Shall I tell you why I want to go to the Moon?… I don't really know…. I am just curious.").

After Sputnik 2, President Eisenhower tried to reassure the nation in November through a televised Oval Office talk. To highlight U.S. technical accomplishment, the president proudly displayed an inert lump of artificial rock—a Jupiter-C warhead—that had arced into space and back to Earth in August. The warhead demonstrated a key technology for allowing space-reaching objects to return to Earth and endure the extreme heat of reentry. But in contrast to Laika the orbiting dog, the lumpish warhead only seemed to emphasize the Soviet advantage.

Sputnik opened a deep fissure in American self-assurance that raised fundamental questions about American democracy's capacity to confront Cold War stresses, the effectiveness of the nation's political leaders and institutions, and the character of the populace. In his state of the union address in early 1958, President Eisenhower conveyed something of this uneasiness:

> But what makes the Soviet threat unique in history is its all-inclusiveness. Every human activity is pressed into service as a weapon of expansion. Trade, economic development, military power, arts, science, education, the whole world of ideas—all are harnessed to this same chariot of expansion. The Soviets are, in short, waging total cold war.

The response to Sputnik in the months and years that followed revealed the ways in which traditional American values and ideals were recast to meet the Sputnik challenge.

Two responses stood out, each a way of "doing something" in the face of the soul-searching worries catalyzed by Sputnik, each resulting in a new, shiny bureaucracy. One response revolved around the question of centralizing the direction and control of government programs to stimulate new science and technology. This issue struck at the heart of American identity. Creativity and innovation occurred, Americans long believed, through the free-ranging interest of individuals and institutions to pursue their muse. But the Soviet challenge seemed to point in the opposite direction. Newspaper editorial pages called for "if not a Manhattan District project [which developed the atomic bomb in World war II], then at least a centralized control over programs now scattered among universities, industry, the services." In short, many argued that to compete with the Soviets we needed to become more like them in managing science and technology. The result, in early 1958, was the Advanced Research Projects Agency, a new entity within the Department of Defense that is still in existence. ARPA made acceptable a new value in American life: that concentrated government direction of technological innovation on a broad front was proper.

The other response to Sputnik was the establishment, in fall 1958, of the National Aeronautics and Space Administration, long the most visible expression of United States space activity. But NASA's founding revealed, too, deep questions on American values in meeting the Sputnik challenge. Space had clear, potential military uses—from using satellites as "eyes" in space to spy on adversaries to creating orbiting platforms that might launch nuclear weapons to the Earth below. But spaceflight quickly acquired powerful meanings that highlighted its importance as an undertaking on behalf of all humanity. One British newspaper noted, regarding Sputnik, that it was one of the "biggest things that has happened in the realm of natural science for centuries … records and firsts may be national but progress is always international." Another claimed that the prospect of "venturing out among the stars" made "this mid-century decade … the prologue to one of the greatest adventures men have ever undertaken—the exploration of interplanetary space."

BUT the contrast between these views and military possibilities loomed large. One editorial starkly laid out the options: "Shall our leap out into space be used for the benefit of mankind, in fabulous new knowledge and relationships, or shall it be used to seek to enslave human society under planetary bombardment directed from space platforms or satellites?" NASA represented a very careful attempt to dampen such extreme Cold War fears and align a significant part of the U.S. space effort with the transcendent hopes for spaceflight. But as NASA took shape, public and open in its mission, national security space efforts also coalesced. By 1960, the United States possessed three separate space efforts: NASA; highly-secret programs to gather intelligence from Earth orbit; and a program (some parts of which also were secret) that focused on support of military operations. These decisions made in the wake of Sputnik still structure the United States space effort today. NASA stood front and center in the public eye, as the others receded into the background.

The rich creative effort that came out of all three initiatives produced in a few short years a wide array of new technologies. Satellites of various types soon headed to the launch pad: photographic reconnaissance, electronic eavesdropping, missile launch detection, weather, communications, and science. Capsules for carrying animals and humans went from drawing board to reality. Probes reached for the Moon. The sub-technologies essential for spaceflight improved, including navigation, guidance, and communications. Launch complexes and ground and tracking stations were built. Spaceflight, almost suddenly, was about *real* not *imagined* things and events. Statistics and "scorecard" accounts of a succession of space firsts became the measure of an intense U.S.-USSR competition.

The space age and spaceflight created its own lingo, fresh images, and a powerful sense of new things "just about to happen." As money began to flow, communities all across the United States had a direct stake in space. From east to west coast, America's major corporations built the first generation of massive research and assembly facilities dedicated to space technology. University departments in nearly every state took on government space contracts. NASA began to create its system of centers that reach across the continent; the military created its own overlapping complex of government, industry, and university teams. The launch complexes at Cape Canaveral in Florida and at Vandenberg in California pulsed with new excitement.

Sputnik promptly raised the question of spaceflight's meaning for humanity. When, in April 1959, NASA selected the first batch of astronauts, the Mercury 7, this deep question finally moved from abstraction to flesh-and-blood presence. Spaceflight now took on a human face—seven specific faces, the clean, competent look of test and fighter pilots. They were a vanguard for all humanity *and* the "American Davids who volunteered to slay the Soviet Goliath." America could not resist the story; newspapers, magazines, and the burgeoning world of television filled the public appetite. *Life* magazine, with NASA's permission, entered into a contract with the astronauts to bring their stories, at home and work, into the American mainstream. Despite the broad range of the emerging space effort, the astronauts became the center of public attention, the focus of spaceflight's future and of the competition with the Soviets.

By January 1961, less than four years after Sputnik, as President John F. Kennedy succeeded President Dwight D. Eisenhower, the basic elements of a U.S. space capability had been organized and built. But what were the country's goals in space, particularly in its public manifestation at NASA? In April, when cosmonaut Yuri Gagarin became the first human into space, orbiting the Earth, and then weeks later, when astronaut Alan Shepard arced into space, the question loomed ever larger. By the end of May, President Kennedy offered an answer: a bold, clear objective to best the Soviet Union by an unprecedented journey to the Moon—to send, land, and return humans to Earth in a grand act of exploration.

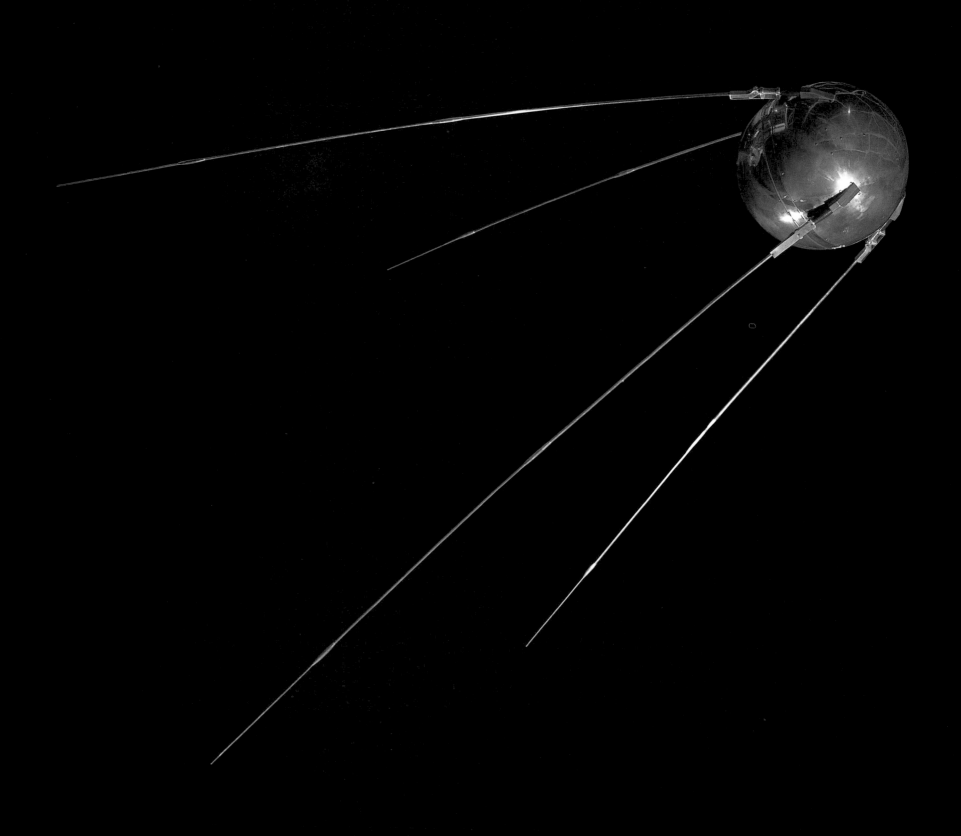

Sputnik Model
1960s

"The first artificial earth satellite in the world has now been created.... Its flight will be observed in the rays of the rising and setting sun with the aid of the simplest optical instruments such as binoculars and spyglasses."

—Soviet News Agency TASS, 1957

SPUTNIK, launched by the Union of Soviet Socialist Republics (USSR) on October 4, 1957, marked a simple, yet profound event in history: the placement of the first human-made satellite into Earth orbit. Since that October date, Sputnik has possessed a dual identity, as an icon of technological achievement and as the starting point for a Cold War Space Race.

As significant icon, known throughout the world, Sputnik the object seems small and unimpressive. About the size of a beach ball, at 58 centimeters (23 inches) in diameter, the satellite operated on a single watt of power provided by three silver-zinc batteries. These batteries powered the satellite's most famous feature, a 0.4-second "beep" produced on two radio frequencies, heard by amateur radio operators throughout the world. The sphere weighed 83 kilograms (183 pounds) on Earth and had an exterior of aluminum alloy 2 millimeters (.1 inch) thick, polished to a high sheen, which facilitated visual tracking via telescope. On the satellite's exterior, four antennae, positioned symmetrically, swept back at 35-degree angles. On Earth, the design conveyed an image of flying through space in a single, purposeful direction. In reality, the satellite did not glide through space—it tumbled.

Sputnik functioned for 21 days, emitting its "beep," until its batteries depleted. It continued in orbit for more than two additional months, burning up on reentry into the Earth's atmosphere on January 4, 1958.

Sputnik was the brainchild of two men: Sergei Korolev, a rocket engineer who managed the development of the first intercontinental ballistic missile (ICBM) and the first decade of Soviet spaceflight, and Premier Nikita Khrushchev. Sputnik, though, was not Korolev and Khrushchev's first choice for a satellite. The original plan called for a satellite carrying scientific instruments to investigate conditions above the Earth's atmosphere to coincide with the International Geophysical Year (IGY). But design of this satellite did not match the rapid development of the USSR's newest weapon, the R-7, the world's first successful ICBM.

Sputnik coverage actually predated its launch, and may have predated the physical existence of the object itself. During the summer of 1957, a series of articles and announcements about plans for a Soviet satellite launch during the IGY appeared in the widely read journal *Radio*. In an announcement on July 7, 1957, the Soviet Academy of Sciences made an open call for international assistance in tracking a satellite by amateur radio enthusiasts, to prepare for "the reception of signals of satellites launched in the USSR." These announcements worked—at Sputnik's launch, amateurs worldwide played a crucial role in documenting the Soviet achievement.

In the years after Sputnik, Soviet officials distributed display models of the satellite throughout the world—including the Sputnik model that hangs in the Museum, which was sent by the Soviet Academy of Sciences in 1976. This model contains no instruments or working parts, nor any indication of its provenance. The Museum displays the last surviving piece of Sputnik—the arming pin, a small square of metal with a pull-ring that separated the batteries from transmitters, removed just prior to launch.

ABOVE: This 1957 Soviet photograph shows a Sputnik partially disassembled. It is unknown if this is the actual Sputnik launched October 1957. Note the oblong fixtures (to the left) that secured the satellite's distinctive antennas.

Impacts and Influences

Sputnik's Reverberation through Popular Culture

Louis Prima's "Beep! Beep!," 1957

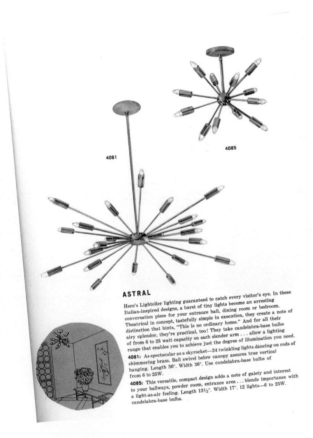

ASTRAL

Here's Lightolier lighting guaranteed to catch every visitor's eye. In these Italian-inspired designs, a burst of tiny lights become an arresting conversation piece for your entrance hall, dining room or bedroom. Theatrical in concept, tastefully simple in execution, they create a note of distinction that hints, "This is no ordinary home." And for all their airy splendor, they're practical, too! They take candelabra-base bulbs of from 6 to 25 watt capacity on each slender arm . . . allow a lighting range that enables you to achieve just the degree of illumination you need.

4081: As spectacular as a skyrocket—24 twinkling lights dancing on rods of shimmering brass. Ball swivel below canopy assures true vertical hanging. Length 36". Width 36". Use candelabra-base bulbs of from 6 to 25W.

4085: This versatile, compact design adds a note of gaiety and interest to your hallways, powder room, entrance area . . . blends importance with a light-as-air feeling. Length 13½". Width 17". 12 lights—6 to 25W. candelabra-base bulbs.

ABOVE: The Lightolier Company's space-age-inspired lamps, mid-1960s.

"Never before had so small and so harmless an object created such consternation."

—DANIEL J. BOORSTIN, HISTORIAN

W HEN THE Soviet satellites Sputnik and Sputnik 2 orbited the Earth in October and November 1957, they created a sensation. As the public debated their political or strategic significance, their distinctive shapes and sounds became a part of American popular culture. American restaurateurs and bartenders quickly capitalized on the Sputnik phenomenon. An Atlanta restaurant introduced a Sputnik burger featuring a hamburger and a small dog (a cocktail sausage) topped with "Tzarist Russian dressing" and a satellite olive. Bartenders promoted Sputnik cocktails made with Russian vodka, an echo of the popular joke that an American Sputnik cocktail would be "one part vodka, three parts sour grapes."

Sputnik's distinctive silhouette, as seen in the Smithsonian's model, also influenced interior design. "Starburst" patterns featuring long, thin rods decorated light fixtures and mirrors. Contemporary lamps showed silver globes extending from long, straight rods or seeming to orbit on curved arms. Satellite shapes decorated clothing and Christmas ornaments, and manufacturers of toys, candies, and bubble gum freely used the Sputnik name and image.

The satellite's name also entered American language as "-nik" became a popular suffix. Laika, the dog that flew on Sputnik 2, became "Mutnik." After the Vanguard rocket collapsed, leaving the American satellite TV3 chirping inertly on the ground, the press dubbed the attempt "Flopnik," "Stayputnik," or "Kaputnik." The 1950s word *Beatnik* also came from Sputnik. Journalist Herb Caen coined the term by "Russifying" the name of Jack Kerouac's antimaterialist literary movement, the "Beat generation."

Sputnik's distinctive beeping was as familiar to Americans as its characteristic shape and name. NBC radio declared the "beep" "forevermore separates the old from the new," and musicians were quick to adopt it. Al Barkle and the Tri-Tones' 1957 "Sputnik" began with Barkle imitating the familiar beeps. Satellites orbited in and out of love songs like Teresa Brewer's 1958 "Satellite" ("My love takes the path of a satellite / In an orbit around your heart") and in rockabilly lyrics that alternated "rock it" with "rocket"—remember Skip Stanley's "Satellite Baby" or teen guitarist/songwriter Carl Mann's "Satellite No. 2". Bluesman Roosevelt Sykes mooned to his "Sputnik Baby" in 1957, "Listen Mr. Khrushchev, I heard a lot of talk/Satellites and missiles and Eisenhower's fault/ But now you better listen to what I've got to say/ … I've got a satellite baby that can rock it all night through." And in 1958, Ray Anderson and the Homefolks worried about "Sputniks and mutniks flying everywhere/ … Are they atomic? / Those funny missiles have got me scared."

Sputnik songs launched careers and boosted established artists. Inspired by newspaper accounts, Jerry Engler wrote "Sputnik (Satellite Girl)" during his lunch break at the Eastman-Kodak Company. When a label signed the song, promoting it as "out of this world," its success launched Engler and the Four Ekkos onto tours with stars including Buddy Holly and Sam Cooke. For swing singer Louis Prima, a Sputnik song became part of his late-1950s Vegas stage show, which used singer and partner Keely Smith as the straight woman for his wild antics and broad humor. Recorded in 1957 on Capitol Records, Louis Prima's comical single compares a wayward girlfriend to the Soviet satellite: "My baby's going on a trip to the Moon/ And she won't be back soon./She doesn't write me, and I can't sleep/ All I hear from her is [beep, beep, beep sound effects]."

Sputnik Music Box
Late 1950s–early 1960s

*"Arise ye workers from your slumbers / Arise ye prisoners of want / For reason in revolt now thunders
And at last ends the age of cant. / Away with all your superstitions / Servile masses arise, arise
We'll change henceforth the old tradition / And spurn the dust to win the prize."*

—*The Internationale*, EUGENE POTTIER, 1871

POLITICS and technology propelled the Cold War between the Soviet Union and the United States. In that competition, each superpower used propaganda and symbols to confront each other and to enlist support from peoples around the world. This was especially true in the Space Race—perceived by both sides as a symbolic confrontation between the two ideologies. This Sputnik music box exemplifies how each country marshaled everyday items into the global conflict. This small music box, only 15 centimeters (6 inches) tall, portrays the first artificial satellite of the Earth, Sputnik. A stylized rendition of the satellite flies above a round base that represents the globe. On the base is inscribed the date October 4, 1957 (Sputnik's launch date), and the letters CCCP (Cyrillic for USSR).

When wound, the music box plays a few short phrases from the *Internationale*, communism's most prominent anthem, followed by beeps—the same sounds emitted by Sputnik as it orbited the Earth. These beeps, detected around the world by amateur radio operators in the fall of 1957, became the signature sound of the space age. Produced in the 1960s largely for Western consumption at state-run hard-currency stores, this small souvenir captures the essence of Soviet efforts to equate their ideology and political system with mastery of space technology. The beeps of the Sputnik music box became the chorus to communism's *Internationale*, a song that called for workers to rise against their oppressors.

The Museum collected this object to represent the far-ranging cultural and psychological impact of Sputnik. A private collector who purchased the music box while on a trip to the Soviet Union in 1964 donated it to the Museum in 1985.

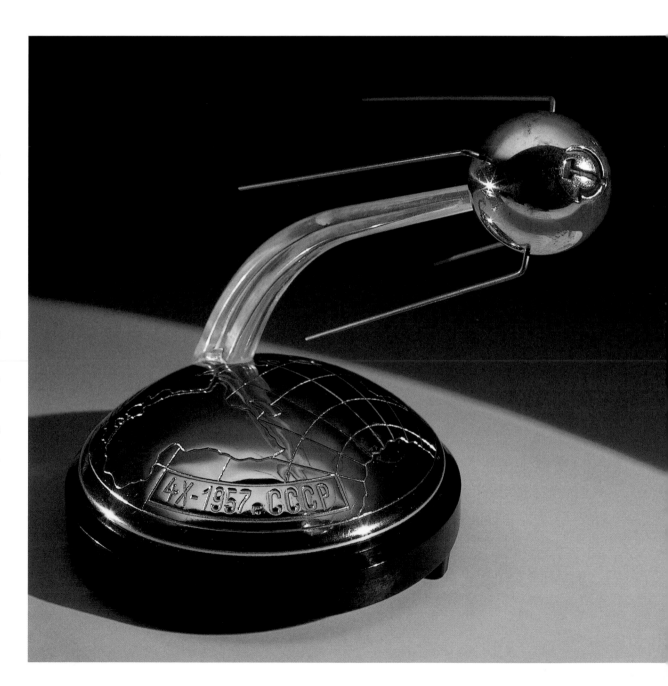

Jupiter-C Nose Cone
1957

"One difficult obstacle to producing a useful long-range weapon is that of bringing a missile back from outer space without its burning up.... Our scientists and engineers have solved that problem. This object here in my office is the nose cone of an experimental missile [that] has been hundreds of miles into the outer space and back."

—President Dwight D. Eisenhower, November 7, 1957

ABOVE: Eisenhower addresses the nation from the White House on the challenge of the Soviet Sputniks, using the Jupiter-C nose cone as evidence of U.S. achievements relating to space.

With U.S. prestige suffering greatly in the wake of the Soviet launch of the first satellite to orbit the Earth, the Eisenhower administration took a number of steps to reassure America. On November 7, 1957, one month after Sputnik and days after Sputnik 2, the president sought to bolster the American public with a televised address from the White House, entitled "Science and Security." In it, he recited a long list of scientific achievements relating to space. One was the one-third-scale Jupiter-C nose cone launched and retrieved after a suborbital flight on August 8, 1957—the first U.S. reentry vehicle to be recovered from space. Although the press covered the event in August, the November address was the first time that the administration confirmed the accomplishment. As Eisenhower talked, the nose cone sat on prominent display next to his desk. One year later, the Army donated it to the Smithsonian in a highly publicized ceremony.

Launched atop a long-range ballistic missile, the primary purpose of the Jupiter-C nose cone was to carry a nuclear weapon to the edge of space and then to a target back on Earth. The Army's August flight test achieved an important reentry vehicle milestone by solving a key problem: protecting the nuclear warhead inside the nose cone from the tremendous heat generated during reentry into the Earth's atmosphere. In the mid-1950s, both the Army and Air Force were designing long-range

missiles. Initially, each service adopted a different solution to the reentry problem. The Army favored the ablation technique in which heat-shielding material on the nose cone eroded or melted, while the Air Force preferred the heat sink approach in which metals on the nose cone absorbed the heat. The Jupiter-C nose cone proved the suitability of an ablative design.

Beginning in 1955, the Army initiated testing of the ablation concept at its Redstone Arsenal, using rocket engines on stands to heat wood soaked in water or phenolic resin compounds. The results encouraged the Army to conduct two flight tests, in September 1956 and May 1957, using one-third-scale nose cones designed for Jupiter intermediate-range ballistic missiles. The service did not recover either of these reentry vehicles.

The next test, in August 1957, achieved the milestone. An Army Jupiter-C rocket launched from Cape Canaveral placed the reentry vehicle, coated with a ceramic ablative material, in space. It reached an altitude of 435 kilometers (270 miles) and a temperature of 1,100°C (2,000°F). U.S. Navy ships recovered the nose cone more than 1,850 kilometers (1,150 miles) downrange. The event proved extremely significant in the development of U.S. space technology. Except for the very earliest ones, all U.S. intercontinental ballistic missile reentry vehicles, as well as the Apollo-era capsules that carried astronauts, used ablative heat shielding.

Vanguard TV3 Satellite
1957

"U.S. Moon Rocket Wrecked"

—*Chicago Daily Tribune* HEADLINE, DECEMBER 7, 1957

UNDER this dramatic but misleading headline, the *Chicago Daily Tribune* reported the launch failure of a Vanguard rocket carrying America's first satellite, the Vanguard TV3. Lyndon Baines Johnson, then Democratic Senator from Texas, declared it the "most humiliating failure in America's history." Project Vanguard did not aim for the Moon—rather it was the United States' first effort to place a satellite in orbit. In the wake of the launch of the Soviet Sputnik on October 4, Vanguard took on new, highly charged significance—as the American answer to the Communist challenge. As expressed by Johnson and others, Vanguard's failure seemed to signal further national shortcomings.

The deep symbolism of the failure led to the satellite's donation to the Smithsonian. As launch neared on December 6, the slim 22-meter (72-foot) Vanguard rocket stood on a Cape Canaveral pad, ready to orbit the TV3 satellite, a 16-centimeter (6.4-inch) aluminum sphere that looked, with its folded antennae, like a robotic daddy longlegs spider. The rocket started to rise on a cushion of fire and smoke, only to settle back after a few pitiful feet, crushing into the pad, and consuming itself in flames. The nose cone shroud split off from the disintegrating rocket. The satellite disgorged onto the tarmac, beeping as it fell. It was recovered after this spectacular explosion and eventually presented to John P. Hagen, the director of the Vanguard program. In the early 1970s, Hagen lent and then donated the crumpled satellite to the Museum, which dramatically exhibited it in the *Apollo to the Moon* gallery in 1976. The satellite sat in the outstretched hand of a statue of Uncle Sam as a symbol of how Vanguard's failure shaped American political resolve in the early space program.

Before the fall of 1957, TV3 seemed but one part of the growing interest in space exploration.

In 1955, President Eisenhower announced that the United States planned to launch a small Earth-orbiting satellite as part of the country's participation in the International Geophysical Year (IGY), a coordinated, multination scientific undertaking to study the Earth, from mid-1957 to mid-1958. The military services competed for the project, but the Navy's proposal—Vanguard—won out, based on the perception that it derived from peaceful civilian scientific technologies fostered at the Naval Research Laboratory (NRL).

The laboratory assumed responsibility for the project, with funding from the National Science Foundation. As part of the effort, NRL, based in Washington, D.C., built three prototype satellites for a series of test flights. Each satellite in the Test Vehicle (TV) series included a simple payload: seven mercury cell batteries, in a hermetically sealed container, two tracking radio transmitters, a temperature-sensitive crystal, and six rectangular clusters of solar cells on the surface of the sphere.

The Navy finally successfully launched a Vanguard satellite, TV4, in March 1958. Identical to TV3, the satellite achieved a stable orbit, and its radio continued to transmit until 1965. Tracking data from the satellite helped to improve knowledge of the shape of the Earth. In late 1958, the Vanguard program was transferred to the newly created NASA. The program drew to a close with the launch of Vanguard 3 in 1959.

Journalists, pundits, and historians have long debated the meaning of the Vanguard failure. Was it a "legitimate test failure," an "inexcusable example of over publicity," as a *Christian Science Monitor* reporter wondered immediately after the launch, or did it represent the United States' "most humiliating failure"? The TV3, dented and bent, brings to life these questions and the extreme pressures of the Cold War in the late 1950s.

ABOVE: On December 6, 1957, the TV3 satellite sat atop a Vanguard rocket as it catastrophically exploded. As a rhyme with Sputnik, the press dubbed the failed Vanguard "flopnik."

Explorer 1
1958

"Don't give me any of this probability crap, Hibbs. Is the thing up there or not?"

—MAJ. GEN. JOHN B. MEDARIS, 1958

ABOVE: William Pickering, James Van Allen, and Wernher von Braun (left to right) celebrate the successful launch of Explorer 1, the first U.S. satellite, at a press conference in January 1958.

O N JANUARY 31, 1958, the successful launch of Explorer 1 was critical. Months before, the Soviets had orbited Sputnik to great acclaim. The United States tried but failed to match this riveting accomplishment when, in December, Vanguard TV3 exploded on the launch pad. The burden of a success now fell to Maj. Gen. John B. Medaris, commander of the Army's Redstone Arsenal of the Army Ballistics Missile Agency. At Cape Canaveral, he watched the Jupiter-C rocket lift off flawlessly, carrying Explorer 1, and wanted to know, "Is the thing up there or not?"

The question reverberated across the country. At the National Academy of Sciences in Washington, a room full of press waited for news with the "brains" behind Explorer 1—William H. Pickering, head of the Jet Propulsion Laboratory (JPL), famed rocket engineer Wernher von Braun, and State University of Iowa physicist James Van Allen. As midnight passed, Medaris hounded JPL scientist Al Hibbs and his satellite tracking staff for confirmation that Explorer 1 had achieved orbit. Hibbs seemed sure it had, but no confirmation was possible until he detected an actual radio signal, a signal that now was 30 minutes past their estimate for contact. As the minutes ticked by, Hibbs at last received the signal. The U.S. had—finally—joined the space age.

Bullet-shaped, some 17 centimeters (6.5 inches) in diameter and about 1 meter (3 feet) long, the satellite housed three micrometeorite detectors, temperature sensors, and a special cosmic-ray chamber developed at the State University of Iowa. The Museum's Explorer 1 spacecraft originally was a fully instrumented flight backup, transferred in 1962 from NASA's Jet Propulsion Laboratory. Van Allen and his graduate student George Ludwig cannily designed the cosmic-ray detection system, with a Geiger counter and electronics, to fit inside either the 51-centimeter (20-inch) Project Vanguard sphere or the bullet-shaped Explorer payload.

Once in orbit, Explorer 1 instruments transmitted scientific data from its elliptical orbit for 105 days, varying between 356 kilometers (221 miles) and 2,475 kilometers (1,538 miles) above the Earth's surface, inclined 33 degrees to the Equator. Because of the rushed preparations, Explorer 1 did not possess any means to store observations on board. Ground stations, located primarily in the U.S., only communicated or received data when the satellite was in direct "sight"—Medaris's frustration on launch night resulted from this limitation. These intermittent transmissions resulted in the capture of only a small fraction of Explorer 1's scientific harvest and made interpreting the data tentative at best, delaying the understanding of Explorer 1's most significant discovery.

Explorer 1 seemed to detect varying amounts of cosmic radiation at different altitudes, with an apparent and puzzling absence of radiation when the satellite was above some 960 kilometers (600 miles). Only after the successful launch of Explorer 3 in March 1958 (Explorer 2 failed) did Van Allen and his team understand the data. Explorer 3 had a tiny—by today's standards—wire recorder for storing data when the vehicle was not in line with a ground station. As the satellite traveled in its highly elliptical orbit, collecting data over several weeks, Van Allen's team realized that the Geiger counter's "silence" at certain altitudes above 960 kilometers (600 miles) was actually the result of radiation saturation from belts of high-energy particles trapped by the Earth's magnetic field—the Van Allen belts.

The "belts" quickly entered into popular lore, generating excitement but also misunderstanding. One reporter dubbed the phenomenon a "hot band of peril" and suggested that "death [was] lurking ... above the Earth." While potentially harmful to electronics in spacecraft, the Van Allen belts held no significant danger for astronauts during the brief intervals required to pass through these regions.

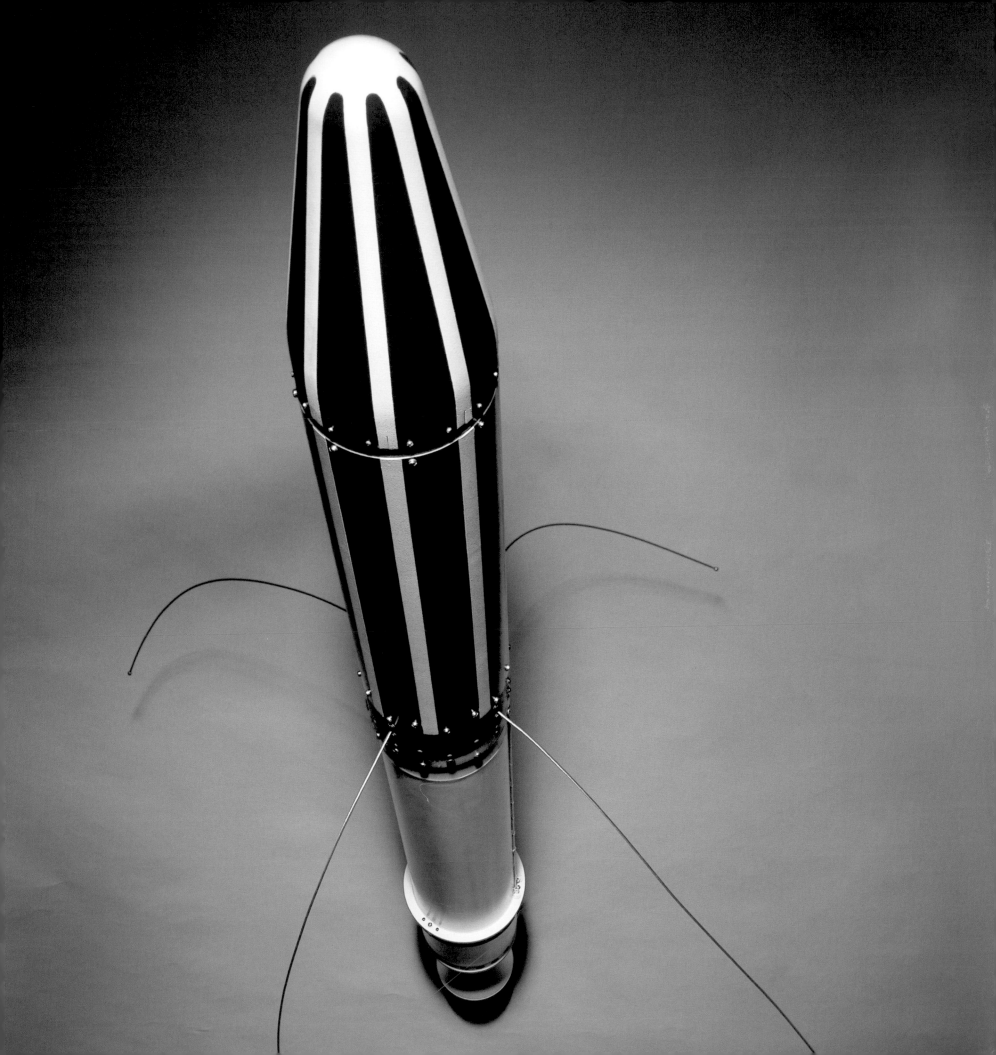

HYAC Panoramic Camera
1958

ABOVE: The protective housing obscures most of the actual camera. Visible is the camera lens (the silver-colored "bell"). During operation, it swung in an arc, allowing the camera to take a succession of high resolution photographs of the Earth below.

"Several balloons carrying equipment which included automatic cameras for aerial photography … have been brought down in the airspace of the Soviet Union lately.… The Soviet government again protests to the U.S. government against the launching of American balloons over the territory of the USSR."

—Soviet diplomatic note to the United States, September 3, 1958

Under international law it is illegal for one nation to fly through the airspace of another without permission. Violating a nation's airspace to conduct reconnaissance is thus risky and provocative. In the Cold War during the 1950s, though, the United States did exactly that by flying camera-equipped aircraft (including the famous U-2) and balloons over the USSR. This HYAC (for "high acuity") camera was the same type used on balloons the U.S. directed over the Soviet Union in a 1958 spy mission.

President Dwight D. Eisenhower approved the first Air Force espionage balloon project, codenamed GENETRIX, in December 1955. To take advantage of atmospheric wind streams, balloons flew over the Soviet Union from west to east at an altitude of 16,750 meters (55,000 feet), carrying specially designed film cameras. More than 500 balloons were launched from Western Europe early the next year, but only 46 were recovered. Of these, only 34 brought back any usable imagery. The Soviets or their allies shot down many of the balloons and others simply descended prematurely in Eastern Europe or the Soviet Union. The violation of Soviet airspace was obvious and provoked a wave of criticism. In February 1956, the USSR displayed for the international press recovered GENETRIX cameras, equipment, and photographs. Because of the tremendous negative publicity, President Eisenhower soon terminated the project.

This did not deter the Air Force, however. It soon discovered that during six weeks each year there was an east-to-west jet stream over the Soviet Union at an altitude of 33,500 meters (110,000 feet). Officials believed that balloons at that height were invulnerable to attack and probably not even detectable. The Air Force accordingly proposed a new espionage balloon project, designated WS-416L, which President Eisenhower reluctantly approved in June 1958.

Several balloons were launched from an aircraft carrier in the Bering Sea the following month. But a human error in setting the timing devices caused the gondolas with the cameras to separate prematurely from the gasbags. The gondolas fell to the ground in Poland, not in Western Europe as planned. The USSR and its Warsaw Pact allies lodged strong diplomatic protests (including the one quoted above), once again embarrassing the United States on the world stage. Nearly two years later the U.S. and the Soviets reenacted a similar drama of protest and embarrassment after the Soviets shot down Francis Gary Powers's U-2 aircraft in May 1960, an event that finally ended U.S. photographic balloon and aircraft overflights of their adversary.

The HYAC panoramic camera was built by Itek specifically for the WS-416L project. It featured a long-focal-length lens that swung in an arc, producing high-resolution strip photographs on 70mm film that covered long sections of terrain—a design radically different from the GENETRIX camera. The HYAC missions and camera played an important role in the development of space-based photographic reconnaissance. The failed balloon missions reinforced the view that Earth-orbiting satellites were the only feasible way to spy on the Soviet Union. More important, the HYAC design provided the basis for the KH-1 camera (also built by Itek) used in early CORONA photoreconnaissance satellites.

Lockheed Missile and Space Corporation, who built the spacecraft for the CORONA program, donated this HYAC camera to the Museum in 1983.

SCORE (Signal Communications by Orbital Relay Equipment)
1958

"This is the President of the United States speaking. Through the marvels of scientific advance, my voice is coming to you from a satellite traveling in outer space. My message is a simple one: Through this unique means I convey to you and all mankind, America's wish for peace on Earth and goodwill toward men everywhere."

—President Dwight D. Eisenhower, December 1958

IN THE months after Sputnik, the public notion of spaceflight embraced two radically different views. One emphasized the new endeavor's continuity with the Cold War arms race of the 1950s—an extension of the push to build missiles and nuclear weapons. The other envisioned spaceflight as a rare moment in human history— an opportunity to celebrate the "family of man." Orbiting communications satellites, with their potential ability to link peoples across the vast stretches of the globe, surmounting national boundaries and differences, seemed a perfect expression of this space age idealism. SCORE, the first communications satellite, launched on December 18, 1958, aboard an Atlas missile, embodied both views.

In December 1957, Atlas, the United States' first intercontinental ballistic missile (ICBM), launched successfully for the first time, making it possible to deliver a nuclear weapon from the U.S. to the territory of the USSR. Over the ensuing year, the Air Force conducted 14 more test launches, readying the missile as an ICBM and as a launch vehicle for the Mercury human space program, then in the early stages of planning. For both purposes, the Air Force sought to demonstrate the missile's lifting prowess and planned to place an entire Atlas into orbit. This provided the basis for project SCORE, as an orbiting Atlas made a suitable platform for testing space communications.

The U.S. Army Signal Research and Development Laboratory in New Jersey began to develop a communications payload for the Atlas. An important design consideration was the expected orbit—so low that communications required a technique known as "store-and-forward." The satellite passed over a ground station receiving a signal, stored it on an onboard tape recorder, then replayed it as the satellite flew toward another ground station. The recorder also made it possible to store a message on the satellite before launch and then broadcast it anywhere as the satellite traveled around the Earth.

Work on SCORE progressed in strict secrecy. As the December 18 launch neared, Eisenhower was persuaded to record the message above. On the day of liftoff, the president hosted a diplomatic dinner at the White House. A U.S. naval aide whispered the news of the launch into the president's ear. Eisenhower stopped the festivities saying, "Ladies and gentlemen. I have something interesting to announce. I have just been advised that a satellite is in orbit and that its weight is nearly 9,000 pounds." According to *Time* magazine, the crowd burst into applause, including the representative from Communist Poland.

In highlighting the weight, Eisenhower was pointing to the United States' ability to send nuclear weapons over intercontinental distances. The next day, SCORE broadcast his message of goodwill around the world. The press noted both accomplishments, calling the undertaking the "talking Atlas," perfectly condensing the project's dual message of nuclear war and universal peace.

SCORE operated for 12 days, before the onboard batteries failed, proving the feasibility of space-based communications. Ground stations located in Georgia, Texas, Arizona, and California participated in the tests. SCORE's technical accomplishments provided the basis for future communications satellites.

This artifact is a backup payload, originally transferred in 1965 from the U.S. Army Signal Corps.

ABOVE: As the "talking Atlas" orbits the Earth and broadcasts President Eisenhower's recorded message of good will, Air Force cadets gather around a radio receiver and try to tune in.

Redstone Missile
Late 1950s–early 1960s

"This is the missile that provided the first stage for the Jupiter-C that pushed three Explorer satellites into orbit and is now being considered for advanced man-in-space projects."

—*Baltimore Sun*, SEPTEMBER 18, 1958

THE Redstone missile is one of the most important in the history of U.S. rocketry and spaceflight. Its history traces the development of U.S. missiles after World War II, revealing the advances, sudden changes, and dead ends as engineers grappled with developing a new, complicated technology in the postwar years.

In November 1944, as the Germans bombarded London and other European cities with V-2s—the world's first large-scale liquid-fuel rocket—the U.S. Army realized it lagged seriously in missile development. To advance this new technology the service initiated Project Hermes (after the Greek god). The General Electric Company won the contract and planned to use the German V-2 design to develop a family of missiles under the project.

One concept was the Hermes C-1, a proposed three-stage rocket glider. In early 1951, as military planners looked to place atomic bombs on missiles as well as on aircraft, the design changed. The bomb's heavy weight caused engineers to reconfigure the C-1 to a single-stage ballistic missile (not a glider) with a range of 400 kilometers (250 miles).

At the same time, the Army discovered that the 333,000 newtons (75,000 pounds) thrust XLR-43 rocket engine then under development for the Air Force's Navaho missile fit the new C-1 perfectly. The timing proved fortuitous—the Navaho design also had changed and the Air Force transferred the engine to the Army. As the Army further developed the missile, it underwent a series of name changes: Major, Ursa (as in the star Ursa Major), and then finally Redstone—after the Army arsenal developing the missile. Wernher von Braun and his team of engineers, the German builders of the V-2 who were captured by the U.S. after the war, led the effort to create a working missile.

Redstone made its first flight in 1953 from Cape Canaveral, Florida. The engine fired properly, but the missile's control system malfunctioned, causing it to veer off course and land in shallow water. Despite this failure, the test achieved a crucial objective: for the first time a U.S. missile showed that a warhead could separate from the main body of the missile. This design innovation reduced drag and increased the warhead's range. With this innovation, missiles became an even more formidable weapon.

Redstone subsequently became the United States' first operational ballistic missile in 1958 and first to use an all-inertial guidance system. Redstone performed so successfully that those who worked with it called it "old reliable." Several "old reliables" were fitted with upper stages to become the Jupiter-C that tested reentry nose cones (or warheads) for other missile systems.

Von Braun had such confidence in this version of the Redstone that he pushed for a Jupiter-C (slightly modified to place an object in orbit) to launch the United States' first satellite, Explorer 1, on January 31, 1958. This came after the embarrassing failure of a Vanguard launch to achieve the same feat in December 1957. With the Soviet October 1957 success of Sputnik creating more and more concern in the U.S., the modified Jupiter-C came through.

Three years later, on May 5, 1961, a Redstone helped achieve another milestone: topped with a space capsule, the rocket lifted into space the first American, Alan B. Shepard, on a suborbital mission called Mercury-Redstone 3 (MR-3). The rocket carried a second American, Virgil I. Grissom, to space on June 21 in the MR-4. The Army finally retired Redstone in 1964. The Army's Ballistic Missile Agency transferred this rocket to the Museum in 1978.

ABOVE: Designed initially as a military missile, Redstone was adapted as a launch vehicle for the first Mercury flights. Here a Redstone stands ready to launch astronaut Alan Shepard, first American into space.

OPPOSITE: Detail of the Museum's Redstone, showing engine and fuel tank components.

Missile Mobile
1958

"Designed to be an ideal decoration for a boy's room or recreation room, the mobile can be hung from the ceiling or from a wall projection."

—*Playthings*, MARCH 1959

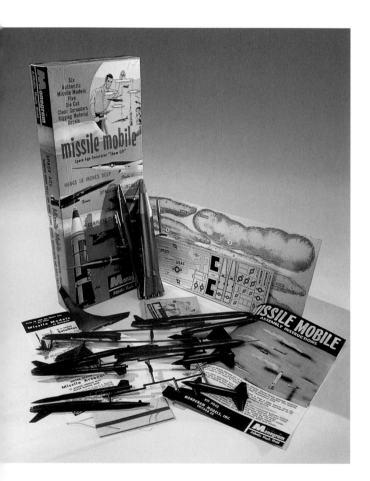

IF YOU were a teenager in America in 1958 you couldn't help but notice all the excited talk on TV and radio of the "space age" and "missile race." Specific missiles also made headline news—both failures and successes—missiles such as the Atlas and Snark, the United States' first intercontinental "birds." Others, like the sleek, delta-finned Bomarc, soon became one of our country's strongest weapons of defense against the dreaded enemy, the Soviet Union.

Naturally, model-makers like Monogram quickly capitalized on the publicity and appeal of these awesome machines to youngsters and began producing the first missile models. Monogram Models, Inc., founded in the 1940s by Jack Besser and one of the largest model producers in the U.S., originally made balsa wood warships and planes. By 1954, they moved into plastic kits that were far easier and less costly to mass-produce—and less difficult for the kids to assemble, with their interlocking parts. And the colors looked more real—mimicking the shiny metal fuselages and the distinctive emblems of actual missiles.

Earlier in 1958, Monogram entered the market with their first missile kit, called the "U.S. Missile Arsenal," which featured some 31 separate missiles plus a booklet by the famous German-born space and rocket writer Willy Ley, still widely respected for his classic popular history *Rockets, Missiles, and Space Travel*. The Arsenal kit apparently sold well—in a few months Besser's team repackaged six of the best-looking models into a new kit called the "Missile Mobile."

The "Missile Mobile" featured the Regulus II, Bomarc, Rascal, Matador, and Snark, and the Atlas, which Monogram altered slightly in this new offering. All were at a scale of 1:128, unusual for models at that time. The mobile box also included "cloud spreaders" (blue plastic cutouts of clouds from which to hang the missiles), "rigging material" (string to hang the "birds" on the clouds), and the inevitable decals.

Besser, an excellent marketer selling a unique product, achieved another success with the mobile. This led to other Monogram missile kits, such as "Five Missiles," "Ten Missiles," "Eight Missiles," "Three Missiles," "U.S. Space Missiles," and individual missiles like the "Little John XM-47 with Jeep." "U.S. Space Missiles" and "US/USSR Strategic Missiles" came later. Monogram also made human spacecraft models up to the Space Shuttle, while another big hit was the "Factual Futuristic" series, speculative, eye-catching space vehicle concepts from an imagined future designed by Ley and featuring his likeness and autograph on the kit boxes.

Now, many years after Monogram introduced the "Mobile"—and as those late-1950s kids verge on granddad age—the kit is prized even more. On E-Bay, such nostalgia has a price of $100–$150—compared to $2 at a hobby shop in 1958. G. Harry Stine, a pioneer in model rocketry in the U.S., donated this set to the Museum in 1973.

Atlas Missile or Rocket Skin Fragment
Late 1950s–early 1960s

"I think you should produce the Atlas at the maximum logical practical rate, because you are going to get it first. It is the only ICBM weapons system that has really fired up to now and it is a good weapon. It is almost a proven weapon."

—GEN. THOMAS S. POWER, COMMANDER, STRATEGIC AIR COMMAND, 1959

MANY people in the United States believed a "missile gap" existed between the two superpowers after the USSR successfully launched Sputnik. They argued that the nation was in great peril from the reportedly large and growing numbers of Soviet intercontinental ballistic missiles (ICBMs). Pressure increased to develop and deploy U.S. ICBMs to thwart the Soviet advantage—a disparity that Americans learned a few years later never really existed.

In the wake of Sputnik, the U.S. Air Force Atlas was one of two (the other was Titan I) ICBMs nearly ready for service. Development of Atlas began shortly after World War II. Work accelerated in the mid-1950s as all U.S. missile programs received a high national priority.

Atlas started with a unique design. In an effort to minimize the rocket's structural weight and maximize the amount of fuel carried, the "skin," rather than internal struts, carried most of the vehicle's stresses. When filled with fuel, the Atlas was much like an inflated "steel balloon"—its nickname among project engineers. To counteract stresses from weight and during flight, engineers designed a honeycomb material for the skin.

The Atlas design ran counter to the conservative engineering approach used by Wernher von Braun and his U.S. Army "Rocket Team." Von Braun thought the Atlas too flimsy. Reservations melted away, however, when Atlas personnel pressurized one of the boosters and dared one of von Braun's engineers to knock a hole in it with a sledge hammer. The blow left the booster unharmed, but on recoil the hammer nearly clubbed the engineer.

The Atlas D became the first operational U.S. ICBM in 1959, designed to carry a single nuclear warhead as far as 9,650 kilometers (6,000 miles).

But this missile (and the Atlas E and F deployed in the early 1960s) required fueling above ground before launch, making it extremely vulnerable to a Soviet attack. The Air Force withdrew all from service by 1965 and replaced them with Minuteman I, a more accurate ICBM, launched from hardened underground silos.

In contrast to its short service as an ICBM, the Atlas established a lengthy history as a space launch vehicle. Initially used in this capacity in December 1958, an Atlas B launched into Earth orbit project SCORE, the world's first communications satellite. Mercury-Atlas rockets launched John Glenn, the first American to orbit the Earth, on his historic February 1962 flight and all three subsequent Project Mercury astronauts on their orbital flights.

Beginning in the 1960s, Agena and Centaur upper stages were added and NASA and the Air Force used them to place a variety of civilian and military payloads in space. Over the last 15 years, Atlas IIs, IIIs, and Vs (built by Lockheed Martin) have launched a wide range of spacecraft.

The Museum's Atlas fragment reveals the honeycomb innovation for the skin. Although the Museum possesses no record of this fragment's history, the artifact speaks to the hurried, experimental character of rocket development in the late 1950s and early 1960s. By today's standards, launches were frequent and many ended in failure, on the pad or once in flight. The fragment may have come from a failed launch, or been the result of typical "breakup" as a missile or launch vehicle reentered the Earth's atmosphere after arcing into space. The Museum acquired this Atlas fragment, measuring roughly 1.2 x 1.8 meters (4 x 6 feet), in 1971 from the Smithsonian Astrophysical Observatory.

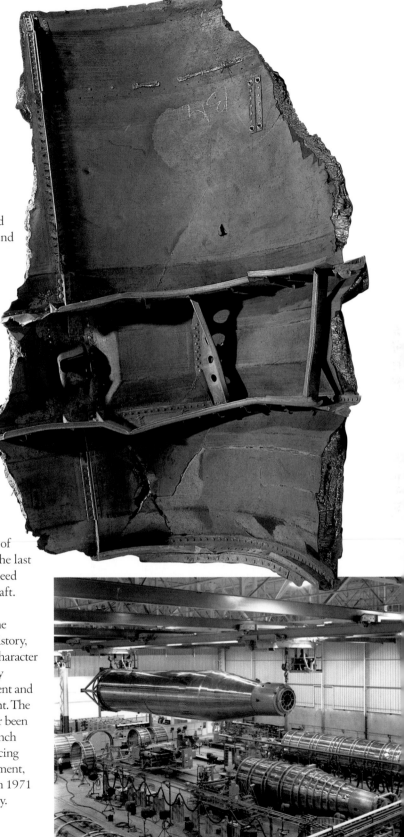

ABOVE: At a Convair factory in San Diego workers assemble Atlas missiles. Note the ribbing on the missiles on the floor (lower right), just as on the Museum's fragment.

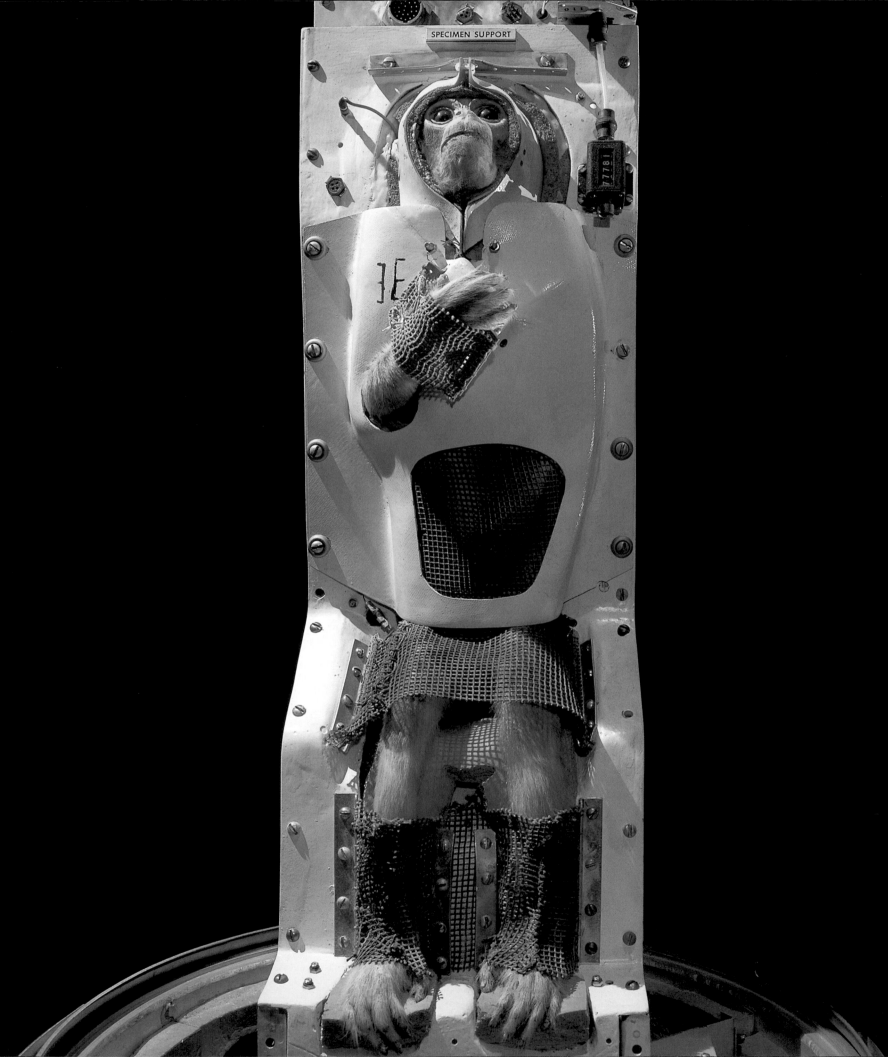

Able Cradle and Biocapsule
1959

"This rhesus monkey … was one of two American pioneers to travel 500 miles into outer space, thus blazing the trail for human beings who will follow, and widening the horizons of scientific knowledge."

—THE AMERICAN SOCIETY FOR THE PREVENTION OF CRUELTY TO ANIMALS'
    CERTIFICATE OF MERIT, AWARDED POSTHUMOUSLY TO ABLE, 1960

IN THE wake of Sputnik, the Soviet Union, and, then, the United States, sent animals into space. The flights served as precursors to the possibility of human space exploration, investigating the biomedical effects of the harsh environment of space. Among the questions to be answered were whether living organisms could survive long periods of weightlessness and other possible hazards in space.

In November 1957, Sputnik 2 carried Laika, a dog, the first animal to orbit the Earth. She stayed alive for 10 days, according to the Soviet scientists monitoring her condition. The initial U.S. efforts to launch and retrieve animals from space during this period failed.

The first successful U.S. experiment involved Able, an American-born female rhesus monkey, and Baker, a South American female squirrel monkey. Just days before the flight, they were selected from a colony of eight that the Army's Medical Research and Development Command had trained, conditioned, and monitored for possible use in space.

To prepare for flight, Army and Navy medical investigators first strapped Able and Baker into cradles and then placed them inside biocapsules, each with self-contained systems for heating and cooling, for exchanging the air every 30 seconds, and for disposing of carbon dioxide and moisture. Instruments obtained data concerning their medical condition for transmittal back to Earth.

On May 28, 1959, a Jupiter rocket carried the two monkeys in its nose cone, launching from Cape Canaveral on a flight of over 1,930 kilometers (1,200 miles) down the Atlantic Missile Range. The nose cone reached a maximum altitude of 482 kilometers (300 miles) and a top speed of 16,000 kilometers per hour (10,000 miles per hour) during reentry into the Earth's atmosphere. Able and Baker experienced six minutes of weightlessness in the flight downrange and 30 to 40 times their own weight when returning to Earth. Ablative material on the vehicle protected them from the very high temperatures generated during reentry. Approximately 45 minutes after launch, the spacecraft landed in the ocean and Navy ships quickly recovered it.

Both monkeys survived the trip in good shape. Unfortunately, Able died a few days later during an operation to remove electrodes. Taxidermists preserved her body, which is on display in the cradle. Baker died in 1984 at age 27. The U.S. Army Ballistic Missile Agency transferred the cradle and biocapsule to the Museum in 1960.

Able and Baker received extensive press coverage and captured the attention of many. The American Society for the Prevention of Cruelty to Animals presented its Medal of Honor and Certificate of Merit to each, the first time that a major humane organization gave such an award to animals involved in scientific research. Along with the other animals that flew in space, Able and Baker made an important contribution to America's space program and helped pave the way for sending humans into this new frontier.

LIFE — WHAT THE MONKEYS' RIDE TELLS US AND PLANS FOR MAN IN SPACE — BIG RIDDLE FOR THE U.S. FAMILY: WHERE DOES THE MONEY GO? — AMERICA'S SPACE TRAVELERS: ABLE AND BAKER — JUNE 15, 1959

TOP: Army and Navy personnel securely fix Able into her cradle before placing her into the biocapsule, in preparation for her trip into space on May 28, 1959.

TIROS (Television Infrared Observation Satellite)
1960

"This country's weather satellite, Tiros I, has been in orbit about two days now, but already its launching opens a new phase of man's activity in outer space. That phase is the practical utilization of our species' new capabilities above the atmosphere to serve workaday needs on this planet itself."

—*The New York Times*, APRIL 3, 1960

MARK TWAIN reputedly said that everybody talks about the weather but nobody does anything about it. TIROS I, the world's first weather satellite, launched on April 1, 1960, generated plenty of talk and public interest—on its role as a harbinger of space technology's potential benefits to humanity, on the images it produced, and on its place in the continuing Cold War struggle with the USSR. Soon after it achieved orbit, TIROS sent back its first images, sweeping views of the planet and its weather patterns. The story and images made front-page newspaper headlines across the country. Within days of the launch, RCA, the builder of the satellite (first under contract to the Army and then NASA), trucked an engineering version to Washington, D.C., to give eager lawmakers a close-up view.

TIROS came at a unique juncture in the still young space age. "Useful" (such as communications and weather) satellites, whether civilian or military, had not yet made their mark. Before TIROS, the only communications satellite to fly was SCORE, launched in December 1958. The public linked TIROS to the military effort to create a reconnaissance satellite, which also sought to take images of Earth from space. This effort, the Discoverer program, frequently made the news in the year before TIROS, but had not yet achieved a full success. Press accounts speculated that TIROS might provide images useful to the military and thereby taint the weather program's symbolism as an undertaking for all humanity.

But the scientific information TIROS harvested quickly made these public concerns secondary. Weather forecasters and scientists saw directly for the first time the massive scale of our planet's weather systems. The satellite transmitted thousands of images of cloud patterns and other phenomena to ground stations during its three-month life. A total of nine TIROS were launched from 1960 to 1965, laying the groundwork for a system of satellites to monitor the Earth's weather and atmosphere constantly. In conjunction with a companion program, Geostationary Operational Environmental Satellites, TIROS is still in existence today and provides the images from space for forecasting that we see every day on television news.

The early history of TIROS had its Cold War twists. In the early 1950s, RCA participated in secret engineering studies to develop a reconnaissance satellite, using television technology. As this effort developed, engineers realized that space-borne cameras, rather than TVs, would better provide the highly detailed images needed for reconnaissance. RCA then channeled its research into TIROS—and into a parallel, distinct, and secret military program of weather satellites that served as an adjunct to photographic spy satellites. Information on cloud cover over the Soviet Union helped to guide the reconnaissance effort. As RCA worked on both programs, technological innovations moved back and forth (but primarily in the military to civilian direction). In the 1970s, government officials began to merge the two programs.

The cylindrically shaped TIROS rotated like a top to maintain stability in orbit. The satellite's main instruments were narrow and wide-angle television cameras that looked out through the base. Beginning with TIROS II, satellite designers added instruments for observing the Earth in infrared wavelengths. This enabled the satellite to "see" at night and through cloud cover, providing additional and near-continuous information on atmospheric behavior.

This backup for TIROS I and II satellites was transferred from NASA to the Museum in 1965.

ABOVE: This image from TIROS I, launched April 1960, was the first television view of Earth from a satellite. This and other TIROS images provided the first broad view of the planet's weather dynamics.

SolRad/GRAB
1960s

"Eisenhower insisted that we be able to turn the system off if he said it was over.... Eisenhower only let us turn it on 23 times. He was extremely concerned about overflight. He never let us turn it on on successive passes."

—Reid D. Mayo, former Naval Research Laboratory engineer, 1998

B Y THE 1950s, the Naval Research Laboratory (NRL) in Washington, D.C., had developed expertise in electronics, radio, atmospheric research, and other fields. This knowledge was not only used in a wide range of open scientific experiments, but also in highly classified projects conducted completely out of the public eye. Many of these involved the collection and analysis of foreign communications and radar signals.

It was critical during this period to learn the exact numbers, locations, and technical characteristics of the various Soviet air defense radar systems, as U.S. war plans relied heavily on the use of Strategic Air Command bombers to deliver nuclear weapons on the USSR. Because most such radar signals simply travel into space, the worldwide network of U.S. ground stations, ships, and aircraft could not intercept them.

NRL scientists had successfully installed radar detectors on submarine periscopes, permitting the crews to learn when ships and aircraft were searching for them. They began investigating whether similar equipment could be placed in orbit. Their calculations were encouraging and they proposed building a spacecraft to do exactly that.

President Eisenhower approved full development of the Galactic Radiation and Background (GRAB) electronic intelligence satellite in August 1959. However, a cover story was needed to explain the vehicle and the worldwide network of ground stations to service it. GRAB's developers approached their NRL colleagues engaged in solar research, and they agreed to incorporate a solar radiation (SolRad) measurement experiment. The GRAB equipment and mission was not shared with them—these

individuals only discovered the satellite's true purpose when the program was declassified in 1998.

The satellite was a 51-centimeter (20-inch) diameter metal sphere. Its antennas collected each pulse of a radar signal, and the transponder sent a corresponding signal to mobile ground stations in the United States and its allies for recording on tapes. These were forwarded to the NRL, National Security Agency, and Strategic Air Command for analysis.

The launch of SolRad/GRAB-1 took place on June 22, 1960, and was America's and the world's, first successful reconnaissance satellite. The GRAB equipment was activated on July 5 and operated during the next three months when the president directed. Four more SolRad/GRABS were launched from November 1960 to April 1962, but only the one in June 1961 reached orbit. Its electronic intelligence package was also turned on only with presidential approval.

GRAB's data proved invaluable. It produced a large quantity of information on the location and technical characteristics of Soviet air defense radars. This allowed the Strategic Air Command to develop electronic countermeasures and plot the safest and most effective routes for its bombers in the event of war. GRAB also proved that the USSR had an anti-ballistic-missile radar.

The successor to GRAB was another electronic intelligence satellite built by the NRL and code named POPPY. These satellites were successfully launched from 1962 to 1971. Such programs have been highly classified. The U.S. government only revealed GRAB to the public in 1998. The NRL transferred this satellite, a backup for the first GRAB, to the Museum in 2002.

ABOVE: A GRAB satellite undergoes final checkout on the launch pad. To help maintain secrecy, it "piggybacks" atop a non-secret Transit 3A satellite (beneath the grating, not visible). This launch in November 1960 failed.

The Agena-B
1960–1966

"A satellite engine hailed as promising the United States greatly increased agility in space exploration was unveiled today … [it is] the first to be developed which will allow a heavy military satellite to change orbit."

—*Los Angeles Times*, JANUARY 28, 1960

ABOVE: An Atlas missile lifts off with an Agena-B as a second stage.

THE AGENA was the most used and most successful of all the United States' upper stages. An "upper stage" usually refers to the third or fourth stage of a multistage rocket. Its purpose is to provide greater orbital altitude for a payload (such as a satellite or spacecraft) by taking advantage of the heights already achieved by a rocket's lower stages. But Agena had an additional capability—it could serve as a combined rocket and spacecraft, carrying a payload to orbit and then providing power and other "housekeeping" functions.

The Agena program began in 1956, when the Air Force granted a contract to the Lockheed Missile Systems Division for the development of a military reconnaissance satellite system. At that time, the United States' rockets lacked sufficient power to place heavy payloads into space and especially into polar orbits required for "recon" satellites. Therefore, the Air Force tasked Lockheed with developing a sophisticated upper stage to lift these payloads to low-Earth orbit and serve as a satellite operating the "recon" equipment.

In 1958, Lockheed named their program the Agena in keeping with a tradition of using celestial names. (Agena is the popular name of the star Beta Centauri.) The company built at least three versions of the Agena, each designed to satisfy specific military requirements. Agena's main lift came from a Bell Aerospace Company rocket-powered engine derived from the B-58 Hustler bomber program, improved to provide a total thrust of 66,000 newtons (15,000 pounds).

The first Agena model, known as the Agena-A, used a Thor as the lower stage booster rocket. This type orbited Discoverer reconnaissance satellites (starting with Discoverer 1 in 1959), placing the first U.S. satellites into polar orbit. This orbit had special value: as satellites orbited over the poles, the Earth revolved below, allowing instruments (such as reconnaissance cameras) to "see" the entire planet over a period of time. Agena thus provided a crucial capability, allowing the U.S. to conduct systematic reconnaissance during a crucial time in the Cold War.

Then came the Agena-B. It differed from the "A" model in several ways. It used a Bell engine that allowed the engine to stop combustion and then restart—a useful feature that allowed Agena to adjust its orbit from elliptical to circular—which was of particular value in conducting reconnaissance. Also, designers integrated the propellant tanks with the vehicle body and lengthened the craft to carry more fuel. The Agena-B used a Thor booster but later switched to the Atlas.

Agena-B's first launch in October 1960 carried Discoverer 16, but the rocket failed to orbit. It went on to launch many other "spysats," including the Midas "early warning satellite" (to detect enemy ballistic missile launches) and Ferret (electronic surveillance). Its usefulness soon made Agena a general-purpose booster and satellite.

In this role, from 1961, it launched the first OSCAR amateur radio satellite (as a "piggyback" with a Discoverer satellite), Nimbus weather satellites, the Alouette (Canada's first satellite), the Orbiting Geophysical Observatory scientific satellite, and Echo 2, a passive "balloon" communications satellite.

But there were planetary missions, too. The Atlas/Agena-B sent Rangers 3 through 9 to the Moon and, its most successful payload, Mariner 2, to Venus, Mariner 2 being the first spacecraft to encounter another planet. Agena-B last flew in 1966 and gave way to Agena-D (Agena-C never materialized), the most widely used of the Agenas.

The Museum received its Agena-B from the U.S. Air Force in 1965.

Impacts and Influences

Making Science Fiction Real in the New Space Age

Men Into Space, 1959–1960

"The window to outer space is open and the potentials of discoveries ... are so tremendous that it would be a tragic fallacy not to use this opportunity."

—WERNHER VON BRAUN, WITH EDWARD R. MURROW, ON CBS'S *Small World*, APRIL 1960

IN 38 black-and-white, half-hour episodes aired on CBS from September 1959 to September 1960, *Men Into Space* depicted the adventures of Colonel Ed McCauley, fictional head of the American space program. Aimed at adults, and with the cooperation of the U.S. Air Force, "technical advisor" Wernher von Braun, and space artist Chesley Bonestell, this short-lived television series tried to give a realistic picture of future spaceflight in the budding space age.

Men Into Space dramatized scientifically based space possibilities. McCauley used multi-stage rockets and full pressure suits. Void of atmosphere, space was silent. Rotation produced artificial gravity. Rather than fighting aliens or other fantastic threats, McCauley battled saboteurs, budget cuts, defective equipment, and other real-world problems. The show even filmed at real aerospace sites, including Edwards Air Force Base, California, and Cape Canaveral, Florida.

Filming realistic-looking spaceflight was technically complicated. Ziv Television Programs, Inc., spent over $100,000 to build the *Men Into Space* sets, which included the Moon, various planets, launch pads, space platforms, and several rocket-ship interiors. To depict weightlessness, three giant cranes held set pieces in mid-air while actors rigged into a complex grid of cables and pulleys floated around them. Scenes without dialogue were shot in slow motion to make the action seem smoother and less staged.

Despite the effort, the show received mixed reviews. The Chicago *Daily Tribune*'s television writer described Bill Lundigan's McCauley as "a spaceman whose exploits put Buck Rogers to shame" but expressed doubt that adults cared to watch space adventures. To entertain its viewers, the show walked a fine line. Realistic plots, the show's forte, proved dull, but taking factual liberties to enhance drama drew criticism. As Marvin Miles, the *Los Angeles Times* space-aviation editor, wrote, "Even the youngsters spotted the glaring technical errors." In some ways, the show's problems echoed contemporary questions about the plausibility of human spaceflight. Although engineers such as von Braun pushed for missions to begin soon, critics derided the "space idiots" who promoted human rocket flights.

By autumn 1960, CBS had cancelled the show. In addition to requiring complicated sets, *Men Into Space* suffered from airing opposite ABC's long-running series *The Adventures of Ozzie and Harriet.* (The monochromatic show also competed with NBC's color broadcast of *The Price is Right.*) By the time Murray Leinster published a *Men Into Space* novel in October 1960, the program was already off the air.

Within a year, space reality on television trumped any space fiction. Alan Shepard became the first American in space with his successful May 5, 1961, suborbital flight in the *Freedom 7* Mercury capsule, now held by the Museum. And with all of 15 minutes of flight time under America's belt, President John F. Kennedy, before a joint session of Congress on May 25, 1961, committed the United States to achieving a human mission to the Moon within the decade. The major television networks covered the remaining five Mercury missions and the 10 Gemini flights. By the end of the 1960s, half a billion people (or one-seventh of the world's population) watched astronauts Neil Armstrong and Buzz Aldrin walk on the Moon.

Yuri Gagarin 10-Kopek Stamp
1961

"Your glorious deed will be remembered through the ages as an example of courage, bravery, and heroism for the sake of mankind performed by you. The flight opens a new page in the history of mankind."

—NIKITA KHRUSHCHEV'S CONGRATULATORY MESSAGE TO YURI GAGARIN, APRIL 12, 1961

ONE WAY the Soviet Union tried to popularize major initiatives and programs was with government-issued stamps. Stamps reached nearly every person in every household and thereby served as a powerful means of advertising official programs and celebrating national achievements. This Soviet practice continued a tradition of the tsarist period, when Russian stamps featured projects in exploration and the personalities associated with them. In the first decades of the Soviet era, beginning in 1917, stamps often featured major development projects or prominent scientists and artists. At the dawn of the space age, the Ministry of Communications (which included telephone, telegraph, and the post) chose stamps with themes that honored Soviet activities in space.

This stamp celebrates the first flight of a human into space. On April 12, 1961, Yuri Gagarin made a single orbit of the Earth on board the Vostok spacecraft. The Soviets carefully managed public announcements of his flight, including issuance of the stamp in 1961. After Gagarin returned from space, he immediately began a goodwill tour that spread the word of his flight and Soviet successes throughout the world. His spacecraft and its capabilities, however, remained shrouded in mystery. After the flight, the Soviet government chose to reveal a few technical details, as part of its active publicizing and promotion of Gagarin's mission.

The illustration of the spacecraft on the stamp bore little resemblance to that of the Vostok capsule. Gagarin's spacecraft entered orbit as a cylinder with a cone on one end and returned to Earth as a sphere. The illustration on the stamp adheres to the imagery of spaceflight of the pre-Sputnik era. Titled "Man from the Land of the Soviets in Space," this stamp illustrates the message that the Soviet government crafted about the first human spaceflight. The cosmonaut looks outward and upward as the rocket lifts over the Kremlin, clearly demonstrating the political origins of his flight. The date of Gagarin's flight is written in the plumes. The highly stylized versions of the rocket and spacecraft resemble the popular image of a rocket ship from the 1930s, directly connecting popular images of spaceflight with Gagarin's successful Vostok mission.

A military attaché at the Soviet Embassy in Washington, D.C., donated this and other USSR stamps to the Museum in the late 1970s.

Mercury-4 Earth Path Indicator
1961

"Our job was to load [Mercury capsule MA-4] up for a canned man mission. That was the term we gave to the 'crewman simulator' that would ride in the astronaut's couch. Every conceivable piece of data was to be recorded: noise levels in the spacecraft, vibration, radiation levels, temperature."

—Guenter Wendt and Russell Still, 2001

Looking at the Mercury capsule from a current perspective, one gets an impression that it was a very primitive piece of technology: cramped, equipped with only rudimentary controls, and outfitted with relatively unsophisticated switches and dials. Yet Project Mercury was a major leap into the unknown. The fastest aircraft of the day, even the rocket-powered X-15 never approached the velocities and altitudes of the Mercury spacecraft. Some of the data gathered by high-speed and high-altitude aircraft proved useful to Project Mercury. But in most respects Mercury literally flew into uncharted territory—propelled by rocket engines the capsule reached a velocity of over 28,324 kilometers per hour (17,600 miles per hour, or 25,700 feet per second) and an altitude of 160 kilometers (100 miles), above most of the Earth's atmosphere.

The Earth Path Indicator, an instrument mounted on the center panel of the craft, illustrates this paradox well. It had one job: to tell the astronaut, as he orbited, his approximate position over the Earth. But at this early stage of spaceflight, the development of such instruments was vital to Project Mercury's purpose of putting a human being into Earth orbit and returning him safely.

The Earth Path Indicator was simple in design. It featured a small globe about 10 centimeters (4 inches) in diameter, driven by a mechanical clockwork. Once a Mercury capsule separated from its Atlas booster rocket, the astronaut wound up the clockwork and set the basic parameters of the orbit: its period (the time it took to complete one revolution of the Earth), inclination (its angle as it crossed the Equator), and the capsule's current position. Ground stations, which tracked the launch by radar, determined those numbers as soon as an orbit was achieved, and ground controllers radioed them up to the astronaut. Mercury capsules did not have the capability of modifying their orbit once in space (other than to return to Earth), thus, once set, the Indicator needed no further adjustments.

This Earth Path Indicator, transferred to the Museum from NASA in 1972, was flown on the "Mercury-Atlas 4" mission. This orbital flight carried no human or other living passenger but, rather, a heavily instrumented robot—the "crewman simulator" referred to by engineer Gunter Wendt above. The purpose was to gather further data on conditions for an animal or human flying inside the capsule. That mission flew successfully on September 13, 1961, and paved the way for an orbital flight that carried a chimpanzee (Enos) two months later, and finally John Glenn's historic flight in February 1962. Glenn's flight showed that ground-based tracking stations, plus the astronaut's keen eyesight, were more than adequate to locate the spacecraft's position. Subsequent Mercury flights did not carry an indicator. Mercury's descendant, the two-person Gemini spacecraft, navigated with digital electronics, not mechanical clockwork.

Alan Shepard's Mercury Spacecraft Training Diagram
1961

"What a beautiful view."

—ALAN SHEPARD, 1961

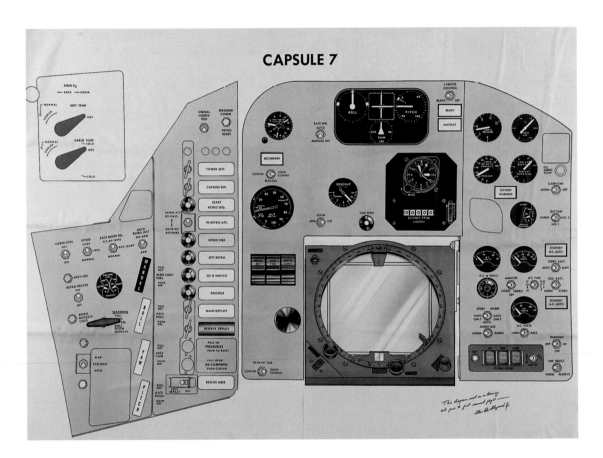

BEFORE Alan Shepard flew his historic 1961 suborbital mission on *Freedom 7*, he had to learn how to operate his spacecraft. This simple photographic chart, which is 75 centimeters (29.5 inches) high and a bit over a meter (40 inches) wide, helped him memorize the layout of his instrument panel. The very simplicity of this training aid, and of the instruments it represents, provides a window on the earliest days of U.S. human spaceflight. The rather crude aid highlighted a central feature of *Freedom 7*'s cockpit instrumentation: it was sparse, even compared to later Project Mercury orbital flights. Still, this simplicity can be misleading, as the Mercury capsule was a complicated piece of machinery requiring equally elaborate training.

This diagram is labeled "Capsule 7" because Shepard's was the seventh spacecraft in the series. It shows the four main panels in front of the astronaut. The most noticeable feature (represented in bottom center) is the large circular screen used to display images from a periscope that looked outside the capsule—this provided Shepard's main view. In the original Mercury design, two small portholes also provided an exterior view, but a very limited one. The astronauts fought against the porthole concept and won a rectangular window in front of the astronaut's head—an intervention later made famous by Tom Wolfe's book *The Right Stuff*. Shepard, however, flew the last of the porthole-equipped capsules, so that during his brief time in suborbital transit through space, the periscope provided his only effective view of the Earth from a hundred miles up. But the view was "beautiful," leading him to describe cloud cover on the East Coast and the islands and reefs of the Bahamas.

Other instruments in the center panel provided basic orientation. A clock, along with a time counter in seconds, provided the time and the mission elapsed time—on a 15-minute mission he did not need to count hours, or even minutes, from launch. At the top of the panel there are three needles giving the orientation of the capsule in roll,

pitch, and yaw. At upper left on the center panel is a gauge giving acceleration and deceleration forces in g's—multiples of the Earth's gravitation pull. Shepard and all the early astronauts endured extremely heavy forces: 8 g's on the way up and 11 coming back down. Finally, for the parachute descent of his capsule into the ocean, Shepard had an altimeter and a rate of descent gauge.

The far right panel provided fundamental information on the cabin atmosphere and cooling and the capsule's batteries and electrical power, plus controls for the radios. Immediately to the left of the main panel are lights for various important functions such as separation of the capsule from the booster, retrofire, and parachute deployment. They would light green when things worked; red

when they did not. To the left are switches for manually overriding these critical functions, on which Shepard's life depended. Finally, switches and the gauge on the left provided control over cabin lights and the attitude control system. Shepard had two independent attitude control systems and two sets of control jets. A control stick in his right hand, not shown in the diagram, provided the mechanism for changing the attitude of his capsule in space by commanding the firing of the jets.

After his flight, Alan Shepard hung on to this poster-board diagram as a souvenir before giving it to the Smithsonian in the fall of 1965. On doing so, he signed it, making it an even more valuable artifact of the earliest days of humans in space.

Freedom 7 Mercury Capsule
1961

"You don't get into it, you put it on."

—MERCURY ASTRONAUTS, EARLY 1960S

ABOVE: Astronaut Alan Shepard sits in his Mercury capsule during an "all up" flight simulation test, with Shepard in his complete spacesuit and the capsule mated to the Redstone booster.

O N MAY 5, 1961, Cmdr. Alan B. Shepard became the first American in space in a small spacecraft named *Freedom 7*. He was hurled 187.6 kilometers (116.5 miles.) high by a Redstone rocket and landed 486 kilometers (302 miles) downrange after a 15-minute, 22-second suborbital hop. On the way, he spent about five minutes in weightlessness and endured heavy g forces during the rocket boost and during reentry into the Earth's atmosphere. Shepard's feat followed that of Yuri Gagarin by 23 days. The Soviet Air Force major's historic one-orbit flight in Vostok 1 on April 12 had made him the first human in space.

The Mercury capsule was very much a product of the post-Sputnik Space Race. In the rush to put the first man into space (women were not seriously considered at first by either side), engineers discarded plans to develop space planes with wings and landing gear in favor of "quick and dirty" solutions derived from ballistic missiles and their reentry vehicles. The United States faced an additional handicap: because of a belated start in ballistic missile development and more advanced, lighter, nuclear warheads, it possessed smaller boosters. The Mercury capsule could not weigh much more than 1,400 kilograms (3,080 pounds), assuming launch on the Atlas ICBM, the first large rocket available. Mercury was thus very compact, just big enough to carry a person and the basic equipment needed for environmental support, attitude control, and recovery. Not surprisingly, astronauts joked, "You don't get into it, you put it on."

In order to get a human into orbit, the principal objective of Project Mercury, NASA decided to use the Redstone ballistic missile to launch astronauts on practice missions. This decision allowed NASA to accelerate the flight schedule and better accomplish all of Mercury's goals. Called Mercury-Redstone 3 (MR-3), Shepard's mission followed two unpiloted tests, one of which carried the chimpanzee "Ham." Because of problems with Ham's launch, NASA undertook an extra booster test (without a production-line Mercury spacecraft), delaying Shepard's flight, which was originally scheduled to launch before Gagarin's historic mission. As a result, Shepard lost his chance to be the first human in space.

A key feature of the Mercury capsule is its large, gently curved heat shield. Facing forward during flight, it served its purpose during the fiery minutes of reentry, holding off a large fraction of the heat generated by friction with the atmosphere via the shock wave created in front of it. The rest of the heat was absorbed or removed either by a beryllium "heat sink" shield on suborbital missions like Shepard's, or by an "ablative" fiberglass/resin shield that charred and eroded away during reentry from orbit. A distinctive feature of *Freedom 7*, compared to all later Mercury capsules flown by astronauts, is the small, round porthole (there is another on the opposite side). The astronauts fought for, and got, a larger window directly over their heads, but Shepard flew the last of the original Mercury capsules. (The large, rectangular opening in the photo at right is the hatch, with door removed.)

He named his spacecraft for the values the United States represented in the Cold War and for the seven Mercury astronauts. Following his example, all the other astronauts put a "7" in the name of their capsules, too.

NASA transferred *Freedom 7* to the Museum in 1962, making it the first human spacecraft to enter the collection. It was immediately placed on exhibit in the Smithsonian's Arts and Industries building.

Space Race: The Heroic Years, 1961–1972

ON THE morning of July 16, 1969, astronauts Buzz Aldrin, Neil Armstrong, and Michael Collins sat strapped into their Apollo command module *Columbia*, a small, conical enclosure sitting atop a mammoth Saturn V, the largest rocket ever built, taller than a football field is long. The three-stage rocket and its cargo—lunar module, service module, and *Columbia*—was a marvel. In the annals of American technological accomplishment stretching back to the founding of the republic, this assemblage, with its Moon-bound astronauts, perhaps ranked highest. In ingenuity, innovation, and daring it reflected the best of the American spirit. It embodied, too, a strong belief that technology was not just about the practical (such as making a better automobile) but a means to larger ends. In spaceflight, those ends embraced Cold War objectives of besting the Soviets and showing the strength and will of a democratic society in its confrontation with communism as well as cultural leanings to fulfill human destiny by exploring beyond Earth.

Thousands gathered that July morning at Kennedy Space Center, including hundreds of journalists from all over the world, within sight of launch complex 39-A. Millions in the United States watched on television, as did millions more in other countries, courtesy of the first network of communications satellites, which had become operational just weeks before. It was, in the lingo of the time, a "happening." About one in seven persons in the world witnessed the Apollo 11 mission over the next eight days as the astronauts traversed from launch pad, to first steps on the Moon, to splashdown in the Pacific Ocean on July 24.

The firing up of the Saturn V at 9:32 a.m., EDT, concentrated the multiple meanings and emotional intensity of the epic event— one that began with one of the most dramatic decisions in the history of American politics. In May 1961, President John F. Kennedy exhorted the nation to achieve "the goal, before this decade is out, of landing a man on the moon and returning him safely to Earth." Over the 1960s, in tense competition with the rival Soviet space program, six Mercury human spaceflights, 10 Gemini, and four Apollo (7, 8, 9, and 10) prepared the way for the Apollo 11 liftoff. The launch, occurring "before this decade is out," vindicated Kennedy's expansive belief in American vitality and national purpose. During the 1960s, this confident view of the American character was tested—by the assassinations of President Kennedy, civil rights icon Martin Luther King, Jr., and Robert F. Kennedy; by problems of race, urban blight, and rural poverty; by the Vietnam War; and by a tragedy within the space program itself, the death by fire of astronauts Roger Chafee, Gus Grissom, and Ed White in a 1967 ground test. But the Moon landing and the astronauts' successful return to Earth offered a contrast, a clear note of grandeur, idealism, and "can do" accomplishment. Greeting Aldrin, Armstrong, and Collins on the USS *Hornet*, President Nixon declared the interval from launch to return "the greatest week in the history of the world since the Creation," claiming that "as a result of what you have done, the world's never been closer together."

The goal of landing humans on the Moon initiated the largest peacetime mobilization in U.S. history. In a few years, buoyed by the expenditure of at least $25 billion (well over $130 billion in 2005 inflation-adjusted dollars), NASA organized more than 400,000 workers around the country, many within its system of technical centers, and even more in private industry and universities. Building on a network of expertise developed for national defense during and after World War II, the bulk of the

OPPOSITE: Astronaut James B. Irwin, lunar module pilot, salutes the U.S. flag during Apollo 15 at the mission's Hadley-Apennine landing site.

work occurred in a crescent (called by some the "gun belt") that arced from New England, through the southern states, into California (with San Diego and Los Angeles crucial hubs), and then north into Washington state. In 1962, *The New York Times*, noting the surging effect of Apollo, reported that its journalistic colleague, the *Los Angeles Times*, ran 20 pages of help-wanted advertisements from aerospace firms in a recent edition. But the classifieds "weren't for Rosie the Riveter ... [but] Egbert the Engineer—servo-systems, reliability, thermodynamics, propulsion, and a score of other categories of engineers." "Rosie" represented the past, in which raw labor and production won World War II; "Egbert" captured the new forces of the Cold War and the space age in which science, technology, and "brainpower" comprised the crucial assets of a transformed industrial base and the driving rationale of government policy. Animated by New Deal–style values, this policy sought to use Moon dollars to spur science and technology-based regional economic development in places like Oklahoma and West Virginia, to bring "Egbert" to those expanses of America that had been bypassed for years in favor of "belt" states.

THIS immense enterprise followed a specific, Kennedy-inspired formula for space exploration: a clear, bold goal; strong presidential leadership; human (rather than machine) exploration as the preferred mode of spaceflight; and national commitment (and even mobilization). Language mattered, too. President Kennedy came into office with the campaign theme of the New Frontier, a metaphor that easily resonated with a centuries-long tradition of Western expeditions around the world and a fascination within the United States for its own westward settlement by European immigrants. Placing astronauts as explorer-emissaries at the forefront of the venture into space helped connect the massive national effort to these traditions. This mix of elements proved potent—the formula confronted the Soviet Cold War challenge, addressed national social and economic issues, and anchored space exploration in deep ideas of American culture.

Through the 1960s, this mix caught the attention and, for the most part, the support of the American public. Via television, magazines, newspapers and a burgeoning consumer culture, astronaut and space age phenomena, serious and frivolous, reached into nearly every community and home. The Kennedy formula and the Space Race were tailor-made for the relatively new and rapidly expanding medium of television. Launches were a spectacle of roaring sound and flame. Flights occurred on a well-planned schedule that allowed networks to delve into each with detail, highlighting the astronauts, their families, launch, recovery, new technologies and feats, the Cold War stakes, and the larger meaning of reaching for the stars for the country and for humanity. Each was an epic drama that linked the public with America's past, present, and future. This public context, not coincidentally, gave birth to the Smithsonian's National Air and Space Museum in founding legislation in 1966 (an update to 1946 legislation that had created a National Air Museum) and gave impetus to the building of a dedicated facility on the National Mall that opened in 1976, part of the bicentennial celebrations in honor of the republic's founding. Around the time of the legislation, NASA entered into an arrangement that ensured that the Museum would be the primary repository for preserving and sharing with the American public the treasures of the U.S. space program—a role the Museum continues to this day.

As the Apollo missions commenced, they prompted reflections on the transformative possibilities of spaceflight that were heartfelt and immediate. Typical was this newspaper editorial assessment during Apollo 8: "None of us knows what the final destiny of man may be ... we in the United States are convinced that the power to leave the earth—to travel where we will in space—and to return at will marks the opening of a brilliant new stage in man's evolution." The possibility of space stations and human colonies on the Moon and Mars seemed just a few rocket launches away. This outward look dovetailed with an emerging concern for the future of the Earth. Space-based images of the "blue planet," seen through the very human eyes of the astronauts, helped invigorate views of the Earth as a precious home, with its delicate ecological balances in need of protection. But lurking near these "dawn of a new age" assessments were concerns. Did the overwhelming large systems of technology required for spaceflight diminish the role of the human, make the individual a mere cog in a relentless techno-social machine? (The word *cyborg*, a human-machine hybrid, was coined during space research in the early 1960s.) Both outlooks intertwined in the 1960s and found expression in such popular culture icons as the classic film *2001: A Space Odyssey*.

As natural as it seemed in the Apollo years, President Kennedy's emphasis on human exploration and frontier motifs was not foreordained, but grew out of the challenges of the Cold War in the late 1950s and early 1960s. As Kennedy took office, spaceflight already had become *the* measuring stick in the Cold War struggle between democratic capitalism and communism. As the first ballistic missiles became operational, the consequences of a nuclear conflict grew too devastating to contemplate. Space "spectaculars," in contrast, offered a peaceful means for each nation to display its prowess and, by implication, the merits of its vision of society. The stakes were high. After World War II, dozens of former European colonies in Africa and Asia gained independence and became separate nations. For the U.S. and USSR, spaceflight accomplishment—especially human accomplishment—became a crucial way to win the "hearts and minds" of these newly free peoples and align them with their respective politics.

In early 1961, the United States possessed an impressive range of space projects—in military and intelligence areas, and in the newly formed arena of civilian space, overseen by NASA. As significant as these were, they lacked a clear, riveting message on the ability of a free society to marshal science and technology to create a better future for peoples around the world. If the U.S. was to win the wide-ranging contest with the USSR, it needed a high-profile, open endeavor that displayed the full scope of ingenuity and commitment of a democratic society—something that went beyond spy, weather observation, science, and communication satellites and, even, single historic feats such as the United States' first human spaceflight by astronaut Alan Shepard on May 5, 1961.

ROUND the time of Shepard's feat, the Kennedy administration vigorously pursued the question of the future direction of spaceflight. Cosmonaut Yuri Gagarin achieved the distinction of becoming the first human into space, just weeks ahead of Shepard and near the time of a foreign relations fiasco, a failed U.S. Central Intelligence Agency–backed attempt to overthrow Fidel Castro, the Soviet-supported leader of Cuba. Both events raised the geopolitical stakes in the Cold War. In early May, two heavyweights of the administration, James E. Webb, NASA administrator, and Robert F. McNamara, Secretary of Defense, prepared a secret, lengthy memorandum on the state of space activities and options for besting the Soviets in the years ahead. Their analysis and recommendations boiled down to a simple logic. In the areas of civilian and military satellite activities, the U.S. was equal with or ahead of the USSR. But in "prestige" capabilities, in readiness for human exploration, the U.S. stood at a disadvantage in the "battle along the fluid front of the cold war." To win, the response seemed clear: dramatically elevate the goals for human exploration (without diminishing emphasis on other space activities). The memo then pointed space exploration in a clear direction:

> It is our belief that manned exploration to the vicinity of and on the surface of the moon represents a major area in … international competition.… The orbiting of machines is not the same as the orbiting or landing of man. It is man, not merely machines, in space that capture the imagination of the world.

Weeks later, with that argument in hand, Kennedy redirected the U.S. effort on a course for the Moon.

Just over eight years later, on July 16, 1969, Aldrin, Armstrong, and Collins sat in *Columbia* at launch complex 39-A. While not obvious at the time, that moment and the eight days to follow, as the Apollo 11 mission traveled from launch pad to the Moon and back, marked the high tide of human space exploration—at least as measured by the Kennedy vision. As NASA embarked on six more Apollo lunar missions, ending in 1972, space exploration as a bold endeavor began to ebb. In the ensuing decades, this 1960s vision cast a long shadow, setting expectations for space exploration that never reached fulfillment. Changes in international and American politics made human space exploration less of a priority and public interest waned. Advocates invoked the Kennedy language but politicians and the public chose not to make those visions substantial.

Nevertheless, in the 1960s, behind the wall-to-wall attention paid to human space exploration and its impressive accomplishments, the less heralded aspects of spaceflight made strides. After Sputnik, in the 1950s, the U.S. launched, military and civilian combined, a total of 19 satellites. In the 1960s, those numbers exploded. The military launched more than 450 satellites over the period and NASA close to 300. This magnitude and range of accomplishment laid the foundation for an alternate (but less well-articulated) concept of spaceflight rooted in Earth-oriented objectives—many of them practical—for intelligence, military operations, science, weather prediction, and communications. In the 1970s, 1980s, and 1990s, these vied to define spaceflight and its implications.

OSCAR 1 (Orbiting Satellites Carrying Amateur Radio)
1961

"The project chairman was high in his praise for the widespread cooperation furnished the project and stressed the fact that OSCAR is a fully private project—not government in any sense. 'It proves that in the United States, at least, the private citizen has a voice and is accorded a hearing—and cooperation.'"

—*Los Angeles Times,* DECEMBER 8, 1961

ABOVE: Lance Ginner, one of the founders of Project OSCAR, stands in his backyard in San Francisco holding the first OSCAR satellite before its launch in December 1961.

FROM ITS inception, the space age seemed the province of the "big boys"—government and industrial contractors organizing large teams of taxpayer-funded experts to tackle the complex problems of spaceflight. But amateurs interested in radio communications ("hams," in their own slang) looked for ways to join the new field.

U.S. radio amateurs grew into an active community in the early twentieth century, soon after the invention of radio, and originated many technological innovations in the ensuing years. After the launch of Sputnik, hams worldwide played a critical role in tracking the satellite. Sputnik's own radio "beep" alerted U.S. amateurs to the possibility of space communications, but they wanted to do more than listen—they wanted to build their own satellite. From 1959 to 1960, a small group located in the San Francisco area conceived Project OSCAR, an effort to place in orbit an amateur-designed and -built satellite.

Much of the effort embodied vintage amateur enthusiasm and ingenuity. Led by hams Don Stoner and Fred Hicks, the project built the OSCAR 1 satellite for a mere $63. But technology was the easy part. Project OSCAR, an official undertaking of the American Amateur Radio League (AARL), the amateurs' national association, did not possess its own rocket. It needed to hitch a ride on a government launch vehicle, but that required, first, the permissions of the U.S. State Department and the Federal Communications Commission, and then, the cooperation of the U.S. Air Force. The project acquired go-aheads from all three, perhaps helped by the facts that Hicks worked for defense contractor Lockheed and Stoner served as the head of NASA's Jet Propulsion Laboratory Space Instrumentation

Section and as a national representative of the AARL.

On December 12, 1961, the Air Force "piggybacked" OSCAR 1 onto the launch of Discoverer 36. This payload, ironically, was a mission to place a CORONA spy satellite into orbit. "Discoverer" served as the cover name for the highly secret CORONA program. As the spy satellite climbed toward its orbit over the Earth's poles, it dropped OSCAR off at a lower altitude. OSCAR worked beautifully. Its communications package consisted of a simple beacon that broadcast out in international Morse code a simple message of "Hi Hi." This simplicity had value, allowing amateurs all over the world to identify the signal and provide global data on the properties of radio reception from space. In advance of the mission, Stoner published an article on OSCAR in AARL's magazine, noting that the satellite could "supply the scientific world with volumes of communication data gathered by observing and using the OSCAR satellite." The satellite operated for nearly 20 days. During that time, radio operators in 28 different countries reported more than 5,000 receptions of its beacon.

OSCAR 1 marked the beginning of the OSCAR program that continues to this day, under the auspices of the Radio Amateur Satellite Corporation (known as AMSAT), which operates approximately 20 OSCAR satellites—as of the year 2000. The program has sparked innovations in spacecraft design and manufacture and enabled radio enthusiasts to participate in satellite communications.

The Museum's artifact is a full-size scale model of the spacecraft built by the members of Project OSCAR and was acquired in August 1963.

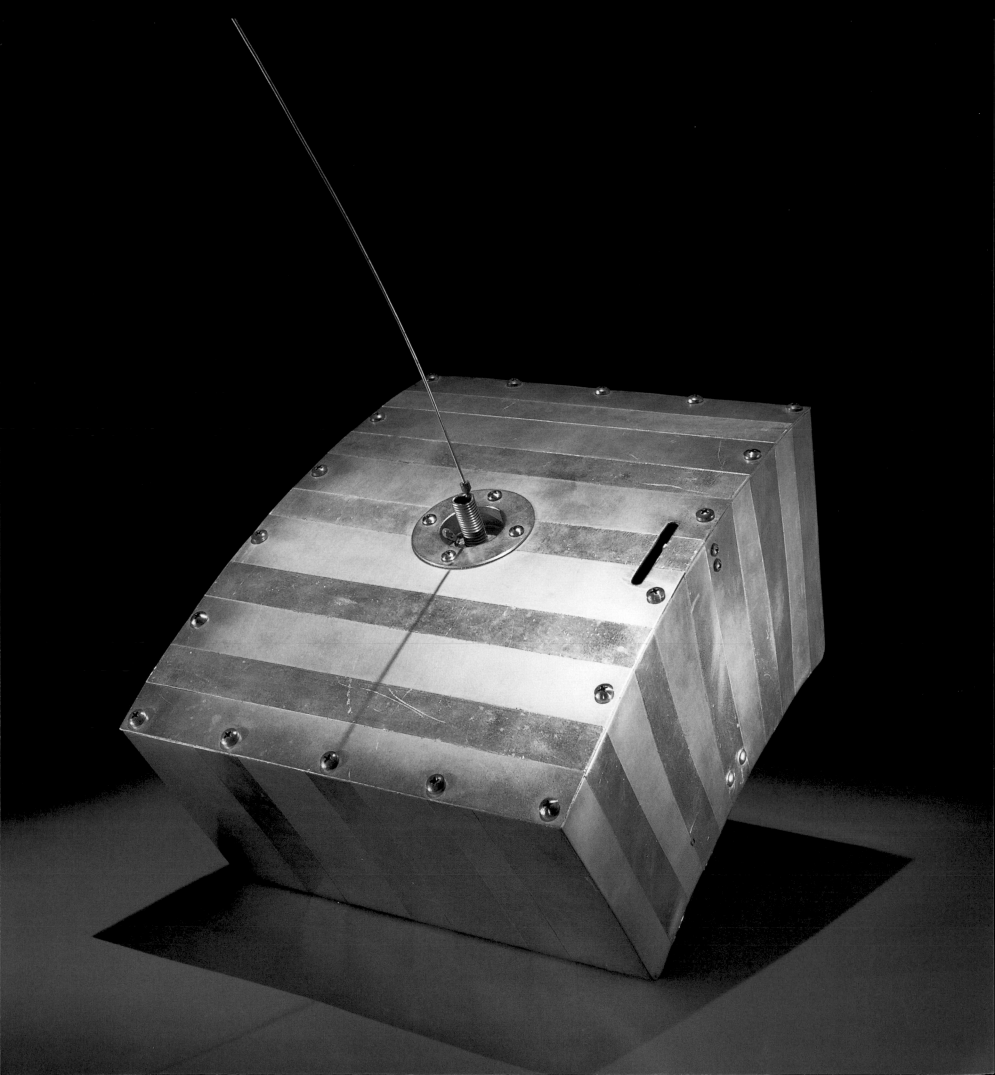

John Glenn's Spacesuit
1962

"Godspeed, John Glenn."

—SCOTT CARPENTER, FEBRUARY 20, 1962

DURING the 1950s, as high-performance aircraft achieved greater speeds and higher altitudes, pilots required specialized suits to handle the g forces and inadequate oxygen encountered in such flight. The U.S. Navy and U.S. Air Force independently pursued a range of research and development programs to design such suits, known as full pressure suits. This work served as the foundation for the development of the first spacesuits, and the Mark IV, developed by the Navy and the B.F. Goodrich Company, proved the best adapted to this new purpose. NASA chose the Mark IV, with modifications, for the Mercury program.

The Mercury spacesuit was—quoting the B.F. Goodrich news release—"a dazzling aluminum-coated nylon-and-rubber creation." In appearance, the suit was eye-catching. The aluminized silver coating made it look "space age"—certainly in comparison with the olive green Mark IV. This silver exterior had a purpose; it served as a thermal barrier and radiation shield. In its unpressurized mode, astronauts could move with comfort, but once the suit was pressurized, arm and leg movement was restricted. Engineers designed the Mercury spacesuit to be unpressurized during the mission. Its primary purpose was to provide an emergency backup system in case of sudden cabin decompression. If this occurred, the suit automatically inflated, keeping the astronaut alive during the remainder of a flight and reentry. In its pressurized mode, the relative immobility of the suit was not a drawback, as Mercury astronauts performed only a limited range of physical maneuvers during a flight.

John Glenn wore this Mercury spacesuit during his *Friendship 7* mission on February 20, 1962. NASA postponed the flight repeatedly due to bad weather, before sending Glenn into space as the first American to orbit the Earth. The mission lasted almost five hours, during which he orbited the Earth three times and traveled over 120,700 kilometers (75,000 miles). In the heat of the Space Race with the Soviets, the flight captured worldwide attention and on his return the public made Glenn an American hero. Images of Glenn in his spacesuit became part of this heroic persona. Saul Pett, a reporter with the Associated Press, captured the "down home" heroism personified by Glenn: "In the saddle of success, he rode loose and easy, and everyone found something to like."

NASA transferred the suit to the Museum in 1967.

LEFT: Astronaut John Glenn, wearing his full-pressure suit and helmet, practices entry into the *Friendship 7* capsule during preflight testing.

Friendship 7 Ansco Camera
1962

*"When I needed both hands, I just let go of the camera
and it floated there in front of me."*

—JOHN H. GLENN JR., 1999

HEN JOHN GLENN became the first American to orbit the Earth on February 20, 1962, he also became the first American to photograph the Earth and the stars from space. In his cramped capsule, *Friendship 7*, he carried two commercial, 35mm-film cameras specially modified for use while wearing spacesuit gloves. One, made by Leica, had a fixed 50mm telephoto lens that he used to take pioneering daylight photographs of clouds and surface features from a little more than a hundred miles up. The other, shown here, was a camera that Minolta made for the Ansco Corporation, extensively altered so that Glenn could carry out the first human-operated astronomical experiment in space.

The Ansco camera is upside down from what would be its normal orientation so that NASA technicians could attach a pistol grip handle for ease of use by Glenn. Because he could not put the regular viewfinder to his eye with his visor down, technicians also affixed a reticle on the bottom (now the top) of the camera so that he could line it up on the stars in the brilliant constellation of Orion, his target. Glenn's task was to take several pictures of the spectra of stars in Orion in the ultraviolet part of the spectrum, between 2000 and 3000 angstroms. On the ground, this light is blocked from view by absorption in the Earth's atmosphere. To turn the starlight into spectra, the camera was equipped with a quartz lens and prism. In the middle of his first orbit, in darkness over the Pacific but not long before sunrise, Glenn took six spectrographic photos of Orion, one of which turned out to be fairly clear and usable.

With the other camera, the Leica, Glenn took the first still photographs of the Earth. (Soviet cosmonaut Gherman Titov had already taken movie footage of orbital sunrise and sunset during his one-day mission in August 1961.) Glenn's Earth photographs revealed the value and beauty of full-color photography of weather patterns, the oceans, and the land, paving the way for even more extensive experiments on future human spaceflight missions. The astronaut found it easy to operate the two cameras, with zero gravity providing an extra benefit: he could simply hang the one he was using in space until he needed it again.

As Glenn moved into his second and third orbits, problems with his spacecraft preoccupied him more and more, giving him less time to take pictures. The automatic attitude control system faltered, so he had to take over manual control of the capsule's orientation in space. A signal that his heat shield might be loose also added to the concern of ground controllers and Glenn. To compensate, Mercury Control instructed him to reenter the atmosphere without jettisoning the retrorockets after firing—the straps of the retro pack might hold the heat shield in place. The warning signal proved false, though, and after a spectacular but frightening reentry, *Friendship 7* parachuted into the Atlantic, carrying with it some of the first photographs taken by humans in space.

After Glenn's historic flight, in 1963 NASA transferred to the Museum *Friendship 7* and other materiel from the mission, including the Ansco camera.

Revel Vostok Model
1969

"Vostok Details Are Depicted in a Soviet Photo"

—*Washington Post* HEADLINE, DECEMBER 23, 1967

MADE BY Revel, a major American manufacturer of toy and hobbyist models, this model presented a likeness of the Soviet Vostok, the spacecraft used by Yuri Gagarin in his historic 1961 flight as the first human to orbit the Earth. This child's toy may not seem significant in the history of the Space Race between the USSR and U.S., yet it conveys an important story about secrecy and technology during the 1960s.

The mere availability of the Vostok model, released in 1969, marked a notable change in the practices of strict secrecy with which the Soviets usually cloaked their space efforts. Nearly eight years after Gagarin's flight, the USSR reversed its policy of near absolute secrecy and released technical details and photographs of the spacecraft through its state-controlled news service, TASS. At last, the world saw a genuine example of the spacecraft that carried the first human into space. Model-makers were among those who anticipated this release of information. Now, they could create models of the Vostok with the same amount of detail as those of the Mercury spacecraft.

After his flight around the Earth, Soviet cosmonaut Yuri Gagarin immediately embarked on a world tour. After astronaut John Glenn's 1962 flight, America's first Earth-orbiting mission, Glenn *and* his spacecraft toured the world. NASA and State Department officials affectionately dubbed the tour "The Fourth Orbit of *Friendship 7*." The spacecraft was officially "orbited" throughout the world to demonstrate American technical expertise and the openness of the U.S. political system.

The Soviets never sent Vostok on a world tour, revealing a fundamental difference in the way each country used the flights to promote their respective political values. The Soviets kept Vostok home to protect a secret—one that if revealed might undermine Gagarin's distinction as the first to execute a spaceflight. The International Aeronautics Federation (the official body for establishing aviation and spaceflight records) stipulated that pilots and craft had to land together to count as a spaceflight. Only a few close to the Soviet space program knew that Gagarin's flight did not meet that requirement. The flight had involved a risky maneuver: as Gagarin's capsule descended through the atmosphere, he needed to eject from his craft at 6,090 meters (20,000 feet) and parachute to Earth. In the rush to get to space, the Soviets had not perfected their reentry landing techniques and the spherical Vostok returned to Earth at a speed too fast to land a human. Even after the second man in space, Gherman Titov, described the parachute and ejection system shortly after his own flight in August 1961, Soviet officials did not discuss Gagarin's flight and his spacecraft's capability. Titov's revelations led to an American challenge to his claim for a spaceflight duration record.

Throughout the 1960s, the Soviets did not publish any detailed photographs of a Vostok, nor did they publicly display a flown spacecraft until after 1965, when they supplanted this spaceflight design with the Soyuz spacecraft. In 1967, immediately prior to the first launch of a Soyuz spacecraft, the USSR finally released details about the Vostok—apparently to emphasize the improved capabilities (such as greater maneuverability) of the Soyuz.

When it came out in 1969, this Revel model reflected the most current and complete knowledge in the West of the Soviet Vostok. During the Cold War, model-makers prided themselves on accuracy and labored to provide a high level of detail. A private collector purchased this model and donated his large collection to the Museum.

Soviet Green Cabbage Soup
1960s

"The woman replied softly, with tears running down her sunken cheeks, 'My Vasya is dead—and, of course, my own death will come soon because my very head has been taken from me while I am still alive. But the shchi *must not go to waste. After all, it is salted.'"*

—IVAN TURGENEV, "SHCHI," 1878

ALL traditional Russian meals begin with a soup course. In Russian, the word *shchi* (pronounced *shee*) carries an ancient meaning that predates the cultivation of cabbage in Russia. It means liquid food. In the story above, famed Russian writer Ivan Turgenev draws on the deep association between the soup and Russian culture.

Over the last thousand years, *shchi* has come to refer to the traditional Russian sour cabbage soup—a soup that must be cooked in advance and fermented at least overnight. Depending on locale or region, cooks use additional ingredients to vary the basic recipe. Depending on social status, *shchi* can be sumptuous or quite simple. "Rich" *shchi* might include a wide range of ingredients such as meat, mushrooms, and tomatoes. "Poor" *shchi* consists of little more than sour cabbage, water, and salt. In 1812, Russian soldiers spread sour cabbage cuisine throughout the Eurasian continent when they brought their tastes to Paris. During the Space Race, *shchi* went into orbit around the Earth.

This is a tube of green sour cabbage soup prepared for use in spaceflight during the 1960s. The plastic-lined metal tube contains a single portion of *shchi*. A cosmonaut could empty the contents of the soup tube into his or her mouth without the benefit of a spoon. Thickened to a paste—and thus in texture differing from the traditional Russian opening course—this tube of food nonetheless brought the taste of a traditional Russian meal into orbit several hundred miles above the Earth. As with their U.S. counterparts, Russian technicians had to create food that was safe as well as appealing to the cosmonauts. On short flights, this was not important, but during long-duration missions, cosmonauts had to maintain their health even though their sense of smell diminished and their appetite waned.

This is an unflown sample that an American collector donated to the Museum during the early 1970s.

TOP: Americans Thomas Stafford (left) and Deke Slayton prepare to sample some Soviet space food, the first astronauts to do so, during the U.S.-USSR Apollo-Soyuz mission in 1975.

Missile Defense Alarm System Series III Infrared Sensor
1963–1966

"If successfully developed, this specific application of satellites [early warning] would represent a logical extension of the type of capability the United States has established in the Distant Early Warning Line and the Ballistic Missile Early Warning System ... which exert a stabilizing influence by helping to discourage surprise attack."

—DEPARTMENT OF STATE PRESS RELEASE, 1961

B Y THE late 1950s, U.S. policymakers believed that Soviet intercontinental ballistic missiles (ICBMs) equipped with nuclear warheads threatened this country's long-range missiles and bombers. This concern was based on a chilling reality: a Soviet ICBM required a mere 30 minutes from launch to reach a target in the U.S. Officials felt it was imperative to get the earliest warning of a possible attack to ensure the survival of U.S. strategic forces and of the ability to counterattack.

This emphasis on protecting missiles and bombers reflected prevailing nuclear strategy. In this view, a capability by each side to inflict massive damage in retaliation discouraged first strikes, as did mutual monitoring of military activities. This balance encouraged stability in the relationship between the two superpowers. The value placed on monitoring led to extensive efforts to provide warning in the event of attack.

One new system deployed in 1961–1963 to meet the threat of missile attack was the Ballistic Missile Early Warning System, a network of radars in Greenland, Alaska, and England. Under the best circumstances, this system provided 15 minutes warning of an ICBM attack.

A space-based system promised to detect enemy launches more quickly, however, yielding crucial additional minutes of warning. This led to a system ultimately known as the Missile Defense Alarm System (MIDAS) satellite. Infrared sensors on the spacecraft detected the exhaust of ICBMs ascending through the atmosphere, doubling the warning time to near 30 minutes. The technical challenges were daunting—especially in building

sensors capable of discriminating between the Earth's natural background radiation and an ICBM's hot plume.

The first flight tests ended in failure. From 1960 to 1962, MIDAS 1 and MIDAS 2, carrying, respectively, the Series I and Series II Infrared Sensors, encountered a variety of launch and electrical power problems. But launches of the Series III sensor did succeed.

Designed to view about 86 million square kilometers (25 million square nautical miles) of the Earth every 10 seconds from its polar orbit at roughly 3,200 kilometers (2,000 miles) altitude, the Series III permitted as many as nine "looks" at an ICBM between the moment shortly after its launch to burnout above the atmosphere. This number of "looks" theoretically could identify the missile's direction of travel. During 1962 and 1963, the Air Force sent the Series III aloft four times, but only the MIDAS 7 test in May 1963 was successful. Operating for six weeks, the satellite detected all the ballistic missile launches in its field of view and relayed the data to the control center on the ground. The Museum's Series III was built during this period but never flew.

The last three MIDAS with improved sensors were launched in 1966, but only two reached the desired orbit. Each operated successfully for almost one year, and together they detected all 139 Soviet and U.S. ballistic missile launches within their field of view.

Although beset with many problems, the MIDAS program proved the feasibility and worth of early warning from space and laid the groundwork for the more capable and reliable Defense Support Program satellites, which were launched beginning in 1970. MIDAS was an important step in the development of a more stable strategic relationship between the U.S. and the USSR.

SA-2 Launch Site Model
1965

"Intelligence, like knowledge, is power. Photographic intelligence, because of its currency, reliability, and relevancy, is an especially effective tool when it is applied to matters concerning strategic balances of power and knowledge of a crisis."

—DINO A. BRUGIONI, FORMER SENIOR MANAGER,
    NATIONAL PHOTOGRAPHIC INTERPRETATION CENTER, 1991

ONE OF the United States intelligence community's most difficult challenges is effectively communicating key information to civilian and military policymakers. These individuals often do not have the time or expertise to review lengthy reports or large numbers of photographs. Data must be presented to them in a succinct and understandable manner.

The Central Intelligence Agency's National Photographic Interpretation Center (NPIC), created in 1961, had the primary responsibility to analyze imagery from aircraft, drones, and satellites and to inform officials of the results. To accomplish the latter, it prepared a wide range of intelligence reports of differing lengths, classification levels, and technical complexity, depending on the recipients. NPIC also frequently disseminated copies of actual photographs to illustrate important points. But photographs of enemy activities or facilities could be difficult to interpret. So NPIC occasionally made three-dimensional models to assist policymakers in visualizing the information contained in photographs. According to records available today, NPIC made at least 800 models in the 1960s and 1970s, chiefly of Soviet missiles and their launch complexes, airfields and aircraft, shipyards and naval vessels, command and control centers, and other critical military targets.

The Museum's model and photograph depict a Soviet SA-2 surface-to-air missile complex at La Coloma, Cuba, during the Cuban Missile Crisis. These deadly missiles were well-known to the United States—the Soviets used one to shoot down Gary Powers's U-2 over the USSR in May 1960. High-flying U-2 aircraft initially detected this and other SA-2 complexes in August 1962. After nuclear-capable offensive missiles were discovered several weeks later, the Air Force and Navy began flying low-level reconnaissance aircraft missions over the island nation. This image came from a mission flown at the height of the crisis—and the only one that used color film. Arthur Lundahl, director of NPIC and a colleague of Dino Brugioni (quoted above), included the photograph in a briefing of President Kennedy and other key policymakers during the tense standoff. Reportedly, it particularly impressed the president because it clearly revealed the presence of the SA-2s.

As a result, in 1965, NPIC used this photograph to build a three-dimensional model for an exhibit on the Cuban Missile Crisis at the Central Intelligence Agency's headquarters. When the CIA dismantled the exhibit, it transferred the model and other objects from the display to the Museum.

Project Mercury Mission Control Tracking Icon
1962–1965

"The map contained a series of circles, bull's-eyes centered on the worldwide network of tracking stations. A toylike spacecraft model, suspended by wires, moved across the map to trace the orbit."

—GENE KRANZ, *Failure Is Not an Option*, 2000

ABOVE: During Project Mercury the control center was located at Cape Canaveral (NASA later moved Mission Control to Houston). Note the large map of the Earth with spacecraft trajectory paths; this icon tracked along those paths.

THIS TOYLIKE 15-centimeter (6-inch) piece of red plastic reminds us of a time when human spaceflight was new—when NASA's freshly established Mission Control learned by experience. This icon, more particularly, tells a story of the early years of computer technology. Computers and computer-driven displays still were special and hard-to-acquire devices. Computers delivered their results not in graphical form but in the form of long tables of numbers, printed on fan-folded paper. This icon represented that data physically—giving the position of the spacecraft in flight over a wall map of the globe.

Shaped in the form of a Mercury capsule, the icon was developed for Project Mercury. At the Mercury control center at Cape Canaveral, it was attached to wires and moved along a world map according to tracking data received by the Center. It was used for all Mercury orbital flights in 1962 and 1963, and the first few Gemini flights in 1965. With the transfer of Mission Control from the Cape to Houston for the Gemini IV mission in April 1965, NASA retired the icon and replaced it with more "modern" computerized displays.

America's human spaceflight program ventured into the unknown in countless ways. Spacecraft, with their astronaut crews, traveled into new territory. But the "new" occurred on the ground, too, including the invention of the concept of ground control—the management of a spaceflight from a specialized facility on Earth. Flights of ballistic missiles, beginning with the German V-2 during World War II, were controlled

from a blockhouse, usually built of reinforced concrete and located close to the launch site. But those flights lasted only minutes at most, with a range of about 320 kilometers (200 miles). The controllers stationed in those bunkers had a lot to do prior to a launch, but once in flight, their duties diminished. The flights of Project Mercury might last many hours, and traverse the entire globe. If a problem arose during a mission, both the astronaut on board and the controllers on the ground had to make life-and-death decisions in a matter of seconds. Project Mercury pioneered in developing such procedures, which have been in use ever since.

Immediately after a launch, the controllers first determined the spacecraft's orbit, checking its course and stability, which would confirm whether the spacecraft would continue safely around the Earth and not enter the atmosphere prematurely. Tracking stations downrange gathered radar and telemetry data. A complex of mainframe computers located in the Washington, D.C., suburbs processed that data and sent the results by teletype to the Cape. The icon showed the position of the capsule as determined by those calculations. Strictly speaking, the visual display was not necessary as the controllers were trained to read the tables of numbers. But in practical terms the visual display proved vital for the whole ground control team to comprehend the overall job they were doing.

In 1965, when Mission Control transferred to Houston, the control room at the Cape was demolished. One of the veterans of that era retrieved this icon from the debris, saved it, and gave it to the Museum in 2002.

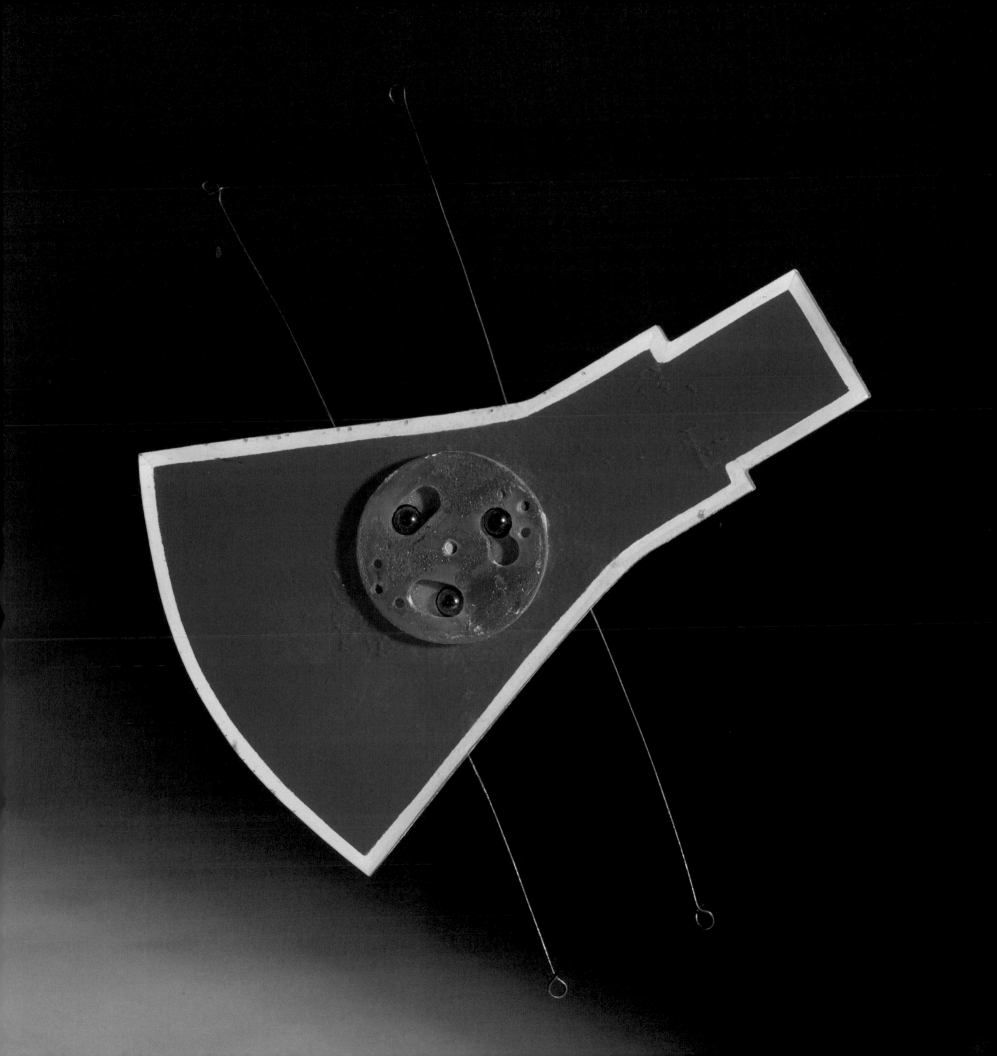

Space-themed Lunchbox and Thermos® Bottle
Mid-1960s

"If we are to win the battle for men's minds, the dramatic achievements in space which occurred in recent weeks should have made clear to all of us, as did the Sputnik in 1957, the impact of this new frontier of human adventure."

—President John F. Kennedy to a Special Joint Session of Congress, May 25, 1961

THIS LUNCHBOX set has a special place in the Museum's history. During the 1960s, it served as the light-hearted token of a community of scholars (known as the "Lunch Box Forum") interested in aviation and space. In making this playful lunchbox and Thermos bottle their symbol, these scholars highlighted how significantly—and how quickly—the images and realities of spaceflight changed in the wake of Sputnik.

Even before the Museum had its own building, it hosted the "Lunch Box Forum," an informal series of weekly talks about flight or astronautics. As early as 1967, these brown bag lunches in the Arts and Industries Building included lectures by invited guests and Museum staff. The forum's speakers included Grover Loening talking about working with the Wright brothers, pilot Jacqueline Cochran recounting her career, and aeromedical experimenter Dr. John Paul Stapp discussing his research. Smithsonian staff also described their ongoing projects. Early on, one of the speakers, Dr. James B. Edson of NASA, donated the space-themed kit that became the series' symbol; placing the lunchbox on the table signaled that the Forum had officially begun.

Although Dr. Edson gave the lunchbox and Thermos bottle as a set, they actually came from two different kits made by the same manufacturer three years apart. The colorful 1960 "Astronaut" dome lunchbox by King Seeley Thermos featured multistage rocket ships, wheeled space stations, and a human lunar base—images that closely resembled Wernher von Braun and Chesley Bonestell's 1950s *Collier's* magazine illustrations. John Polgreen, also known for his work on the children's *The Golden Book of Space Flight,* created the fanciful art. Despite the resemblance to von Braun and Bonestell's

concepts, however, KST never encountered legal problems; it did three years later, however, with its 1963 "Orbit" kit, the source of Edson's Thermos.

King Seeley Thermos had to withdraw "Orbit" from the market because some of its art had been copied without permission. One side of the rectangular metal box showed an Atlas rocket launching John Glenn into orbit on February 20, 1962. The other side featured a cutaway view of Glenn in the *Friendship 7* capsule. The accompanying insulated bottle featured realistic images from NASA's Project Mercury (including an Atlas-Mercury launch vehicle and Mercury capsules orbiting the Earth, reentering the atmosphere, and being recovered from the water). Since the Glenn cutaway graphics had been used without permission, however, KST eventually pulled the offending lunchbox from stores. Withdrawing the kit left many extra Thermos bottles. "Orbit" featured two different types: a standard half-pint size and a tall 283-gram (10-ounce) size. Fortunately, the 227-gram (8-ounce) Thermos bottles fit other KST lunchboxes, including the 1960 dome "Astronaut." Dr. Edson's gift included that combination: a fantastical "Astronaut" lunchbox with a Project Mercury "Orbit" Thermos.

Between 1960's "Astronaut" and 1963's "Orbit," the space age took hold. Yuri Gagarin became the first human in space; Alan Shepard, the first American. President Kennedy aimed NASA toward the Moon. John Glenn piloted the first American Earth orbits, and Valentina Tereshkova became the first woman in space. As Smithsonian curators began collecting the objects that chronicled this transformation, the talisman that sat on their weekly lunch table reflected the same shift: from speculation to reality.

Mariner 2 Mock-up
1977

"Before Mariner II lost its sing-song voice, it produced 13 million data words of computer space lyrics to accompany the music of the spheres."

—HAROLD J. WHEELOCK, COMPOSER, 1963

As WITH other space exploration, planetary exploration took shape as a race between the United States and the Soviet Union. As a first objective, both countries looked to send a spacecraft close to Venus. The Soviets launched a spacecraft (Venera 1) first, but the U.S. achieved the first success: Mariner 2 became the first spacecraft to reach the planet. The Mariner 2 mission revealed the tension between science and long-standing popular beliefs in describing Venus in the early space age.

Scientists in both the United States and the Soviet Union recognized the importance of Venus in understanding the solar system. But the planet, widely recognized as the evening and the morning star, had long enchanted humans—and all the more so after astronomers realized that a mysterious cloak of clouds shrouded the surface from view. In planetary terms and in the public mind, Venus stood as a near twin to Earth, almost identical in size, mass, and gravitation, but its impenetrable veil of clouds stimulated speculation on what might lie beneath.

Incomplete scientific knowledge gave rise to myriad speculations about Venus, including the idea that life existed there. In the first half of the twentieth century a popular theory held that as the Sun cooled over the millennia, the inner planets of the solar system, in turn, provided a haven for life. In this cycle, Earth now harbored life, Mars once had been habitable, and Venus just was starting to evolve. In this view, beneath the clouds, Venus might possess a warm, watery world, suitable to aquatic and amphibious life. As late as 1963, a publication from the Jet Propulsion Laboratory about the Mariner 2 mission noted, "It was reasoned that if the oceans of Venus still exist, then the Venusian clouds may be composed of water droplets," and "If Venus were covered by water, it was suggested that it might be inhabited by Venusian equivalents of Earth's Cambrian period of 500 million years ago, and the same steamy atmosphere could be a possibility." These ideas, still prevalent in the early 1960s, first took hold during the earliest years of the twentieth century. While most scientists questioned them, it took the data returned from Mariner 2 to push them out of popular and even NASA publications.

Mariner 2 launched successfully on August 27, 1962 (Mariner 1 failed at launch the previous year). A 204-kilogram (450-pound) vehicle, it carried six scientific instruments, a two-way radio, a solar-power system, and assorted electronic and mechanical devices. A ground control team of roughly 75 worked from NASA's Jet Propulsion Laboratory in Pasadena, California.

Mariner 2 made its closest approach to Venus on December 14, 1962, after a 108-day journey from Earth, coming within 34,839 kilometers (21,648 miles) of the planet. Its instruments provided information on the nature of Venus's atmosphere and surface as well as interplanetary space. A microwave radiometer measured the temperatures of both the planet's surface and its upper atmosphere, and found that the surface of Venus was indeed very hot— around 427°C (800°F). A magnetometer revealed that the planet had no appreciable magnetosphere. A high-energy radiation experiment discerned that radiation levels in interplanetary space were nonlethal—that an astronaut could survive a four-month interplanetary voyage to Venus. NASA lost contact with Mariner 2 on January 2, 1963, and it is now in orbit around the Sun.

This object is a full-scale engineering mock-up, constructed from flight spares by engineers at the Jet Propulsion Laboratory in 1977, then transferred to the Museum.

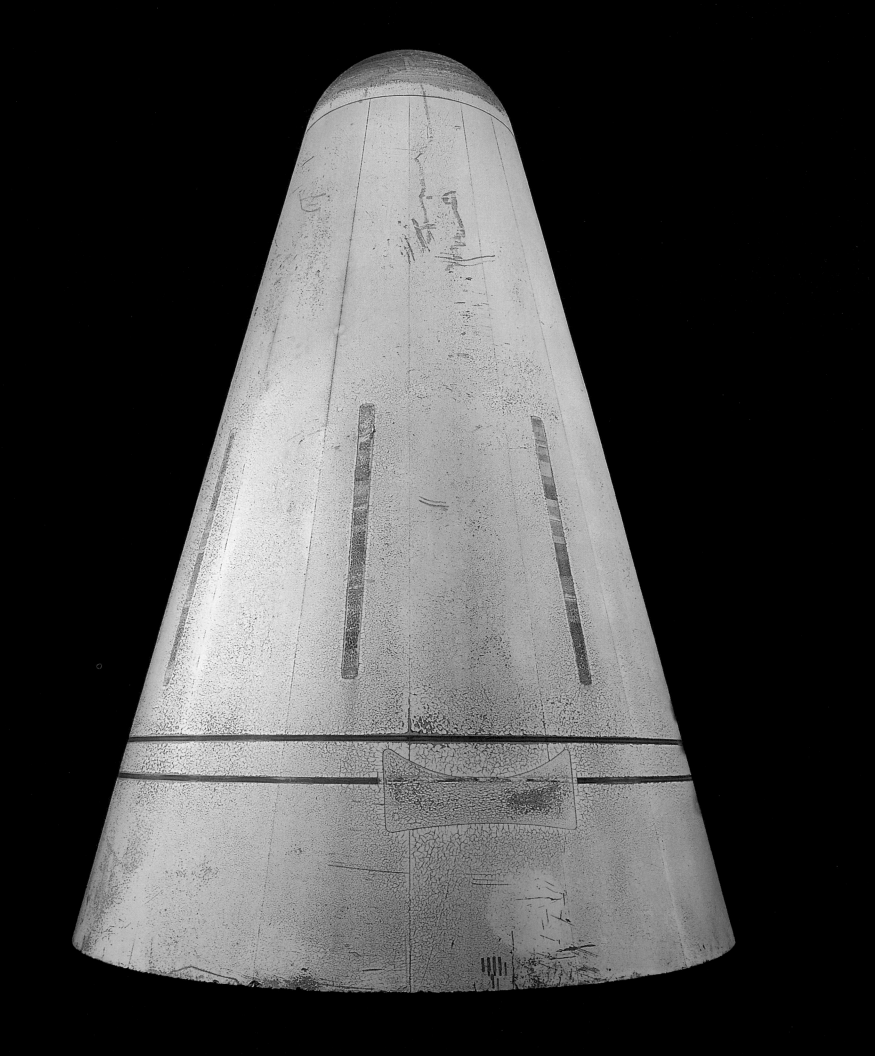

MK 6 Reentry Vehicle
1960s

"Our strategic arms and defense must be adequate to deter any deliberate nuclear attack on the United States or our allies—by making clear to any potential aggressor that sufficient retaliatory forces will be able to survive a first strike and penetrate his defenses in order to inflict unacceptable losses upon him."

—President John F. Kennedy, March 28, 1961

PRESIDENT Kennedy's March 28, 1961, message to Congress set forth the basis of U.S. nuclear strategy during the Cold War. The concept, often referred to as Mutual Assured Destruction, required the deployment of strategic nuclear forces that would survive an all-out Soviet attack largely unscathed and then strike back. By the early 1960s, the U.S. established a nuclear force structure to achieve this goal; it remained in place throughout the Cold War—a combination of long-range bombers, intercontinental ballistic missiles (ICBMs), and submarine-launched ballistic missiles. Reentry vehicles—the housing for a nuclear warhead carried on top of ballistic missiles—were a critical element in this system.

The MK 6 reentry vehicle was built by General Electric for the U.S. Air Force Titan II ICBM. America's first-generation ICBMs, the Air Force Atlas and Titan 1, proved unsuited to the nuclear strategy enunciated by President Kennedy. The missiles had to be fueled above ground shortly before firing, making them vulnerable to a first strike. Second-generation ICBMs, the Minuteman and Titan II, deployed beginning in the1960s, launched from blast-protected underground silos and used improved fuels—respectively, solid fuel and storable liquid fuel.

Titan IIs had a range of 15,000 kilometers (9,600 miles) and carried a single W-53 thermonuclear

warhead in its MK 6. The W-53 possessed an explosive power of approximately nine megatons—the largest warhead carried on any U.S. missile during the Cold War, 600 times more potent than the atomic bomb that destroyed Hiroshima. Although the W-53 threatened massive destruction, the Titan II was not particularly accurate. As a result, the Air Force reportedly aimed the missiles at large targets, usually urban areas. The Minutemans carried less powerful warheads but possessed a more accurate guidance system. The Air Force directed these missiles at Soviet ICBM bases and other targets that required a high degree of accuracy to destroy. More than 50 Titan IIs were deployed beginning in 1964. All were withdrawn from service during 1984–1987.

The Museum's MK 6 is an unflown test version, but no records exist on its donation. The Air Force launched such test vehicles to designated landing zones in the Atlantic or Pacific, where they were recovered for analysis. The ablative material on the outside gave protection against the high temperatures generated during the vehicle's reentry into the Earth's atmosphere. Instrumentation on the inside of the aluminum frame relayed temperature and other readings back to Earth via the antennas on the skin. These tests helped the U.S. maintain a high level of readiness, essential to deterrence in the Cold War years and in the event of a nuclear war.

ABOVE: An MK 6 sits atop a Titan II missile, 31 meters (103 feet) tall, in its silo in Kansas.

TOP: This test explosion in 1954, known as Bravo, was the largest thermonuclear detonation ever conducted by the U.S. (a thousand times more powerful than the Hiroshima bomb) and led to development of bombs used in warheads such as the MK-6.

Gemini Heat Shield
Mid-1960s

"A Real Fireball Outside."

—JOHN H. GLENN JR., FEBRUARY 20, 1962

BEFORE THE United States could put an astronaut in space, it needed to find a way to bring him (in those days, always "him") back. Descending from Earth orbit, any spacecraft encounters the upper reaches of the Earth's atmosphere at velocities on the order of 28,000 kilometers per hour (17,000 miles per hour), with the result that heating from atmospheric friction will soon destroy an unprotected vehicle. Thus the heat shield, such as the one from the two-astronaut Gemini spacecraft shown here, became an essential technology for the Space Race.

Like the heat shields of other early human spacecraft, Gemini's derived from ballistic missile warhead technology. The challenge in the 1950s had been to find a way to bring a nuclear bomb through the atmosphere at near-orbital velocities. Aerodynamicist H. Julian Allen, conducting pioneering work at the National Advisory Committee for Aeronautics, showed that a blunt body traveling through the atmosphere at hypersonic speeds created a leading shock wave that held off much—but not all—frictional heating. Super-hot ionized gases would still pass through the shock wave to the warhead or spacecraft, as would energy radiated by the wave itself. The question was how to dissipate the remaining, still very intense heat flux from those super-hot gases.

Defense Department laboratories and corporate contractors developed several technologies, but by far the most important was the "ablative" heat shield. The Army Ballistic Missile Agency in Huntsville, Alabama, was primarily responsible for funding early research into relatively lightweight combinations of fiberglass, epoxy, and plastic that erode by "ablation," i.e., melting, charring, and evaporation of the shield's surface. The evaporated material carries away much of the heat. Ablative heat shields are not reusable.

In the early 1960s, as the first humans returned from space, the very characteristics that made reentry dangerous also produced a visual spectacle. Just as when a pebble or rock from space enters the Earth's atmosphere as a meteor, the capsule's shock wave created a very bright glow, easily visible from the ground at night. Now astronauts saw this phenomenon up close—like John Glenn during his historic Mercury mission, when he commented on the "fireball" right outside his window.

Like the shield of the one-astronaut Mercury spacecraft that came before it, Gemini's was a gently curved dish shape, as this form created maximum atmospheric drag, slowing the capsule down enough to allow the deployment of a parachute. (Ablative warhead heat shields were by contrast much more conical, as high velocity increases accuracy and makes defensive measures extremely difficult.) The Gemini shield's ablative substance was a paste-like silicone elastomer that hardened after being poured into a honeycomb form. The origin of this heat shield's charring may come from a ground test with a rocket engine, but it is also possible that it flew on Gemini II, the last unpiloted test of that program, in January 1965. The circular marks show where NASA or the spacecraft contractor, McDonnell Aircraft Corporation, cut out samples for testing afterward. Since 2004, the Museum has exhibited it at the Stephen F. Udvar-Hazy Center beside a pristine Gemini shield, showing the "before" and "after" states of this crucial missile and space technology.

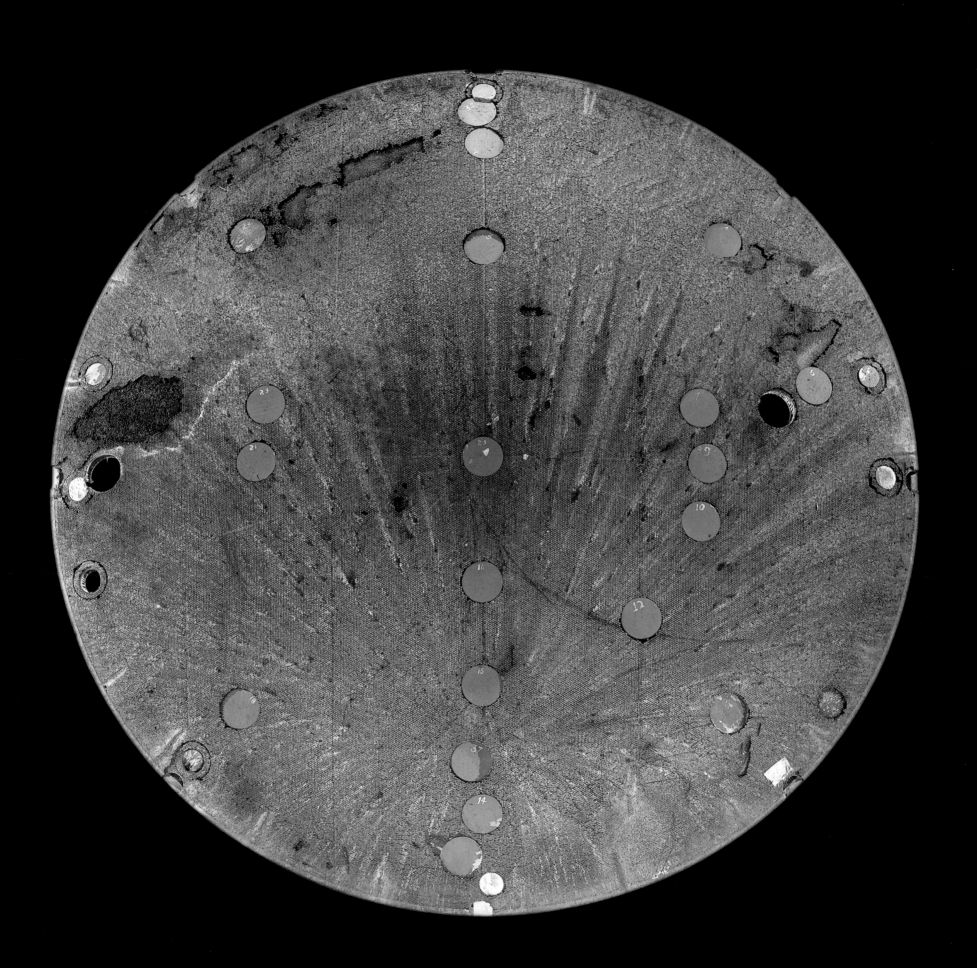

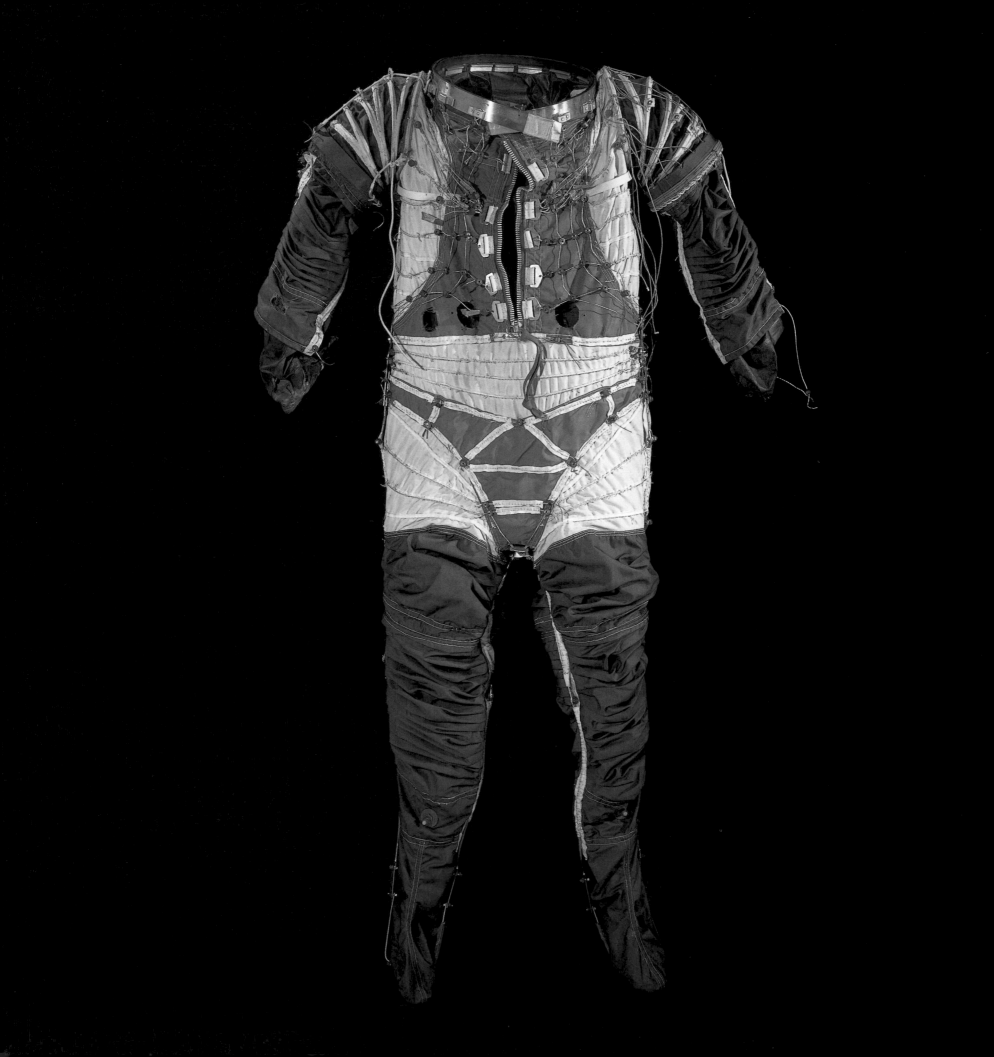

MOL (Manned Orbiting Laboratory) Spacesuit
1965

"The Nation that controlled Space would control the Ground."

—*Air Force Systems Command News Review*, OCTOBER 1965

IN THE EARLY 1960s, public attention focused on NASA's exciting Mercury 7 astronauts and plans for a historic lunar landing. However, NASA's mandate to conduct a public and peaceful program of human exploration was not the only U.S. effort to send astronauts into space. The U.S. Air Force (USAF) also looked to send personnel into space for military purposes. Their signature effort was the Manned Orbiting Laboratory (MOL), a space station. The Museum's XT-3 spacesuit, a developmental model transferred from NASA in 1980, derived from that effort.

In 1963, the Air Force began studies to develop MOL. To benefit from NASA's technology effort, the MOL program planned to use a modified Gemini spacecraft (known as a Gemini B) for transporting USAF astronauts to and from the space station. The service planned to use Titan rockets to orbit its space capsules (as did NASA in Gemini) and the 9,070-kilogram (20,000-pound) space station.

As with its NASA spacecraft cousin, the Gemini B carried a two-person crew. After reaching MOL, USAF astronauts expected to conduct missions of between 14 and 28 days. The military envisioned equipping MOL with Earth-looking cameras—the astronauts' primary objective while on board would be to conduct reconnaissance. After a mission, the crew returned to Earth, with the laboratory remaining in orbit and performing some functions automatically. In concept planning, additional crews might be sent to the station and, eventually, ground controllers would direct MOL to de-orbit and burn up in reentry.

As a human venture, MOL required spacesuits appropriate to its mission. In April 1964, Hamilton Standard, under an Air Force contract, put together a spacesuit "Tiger Team" headed by Dr. Edwin Vale, a renowned pressure suit expert. The team's goal was to design and construct full pressure suits with enhanced mobility. During the following year, they developed a series of "XT" suits—the XT-1 in September 1964, and the XT-2 shortly thereafter. The Museum's suit is the XT-3, completed in March 1965.

The XT-3 solved a crucial problem in spacesuit design. Suit designers had to balance two conflicting factors: to provide a pressure envelope to protect the astronaut from the vacuum of space and provide the astronaut with mobility while wearing the suit. Pressurization made a spacesuit "balloon," causing it to become stiff and rigid and thereby limiting mobility. One design strategy was to control this "ballooning" with a restraint system that allowed the astronaut to move and bend. This suit featured a new restraint system with greatly improved mobility features, including the ability of the wearer to rotate his torso.

Hamilton Standard incorporated this torso mechanism into the XT-4. The company then used this updated prototype as their entry into the competition for the contract to manufacture the MOL suit. The XT-4 did not win the competition, however—and MOL itself did not survive long. In 1969, the Department of Defense cancelled the program, largely because satellites already were fulfilling MOL's reconnaissance mission with great success. Sending humans into space—with all the complexities associated with such an effort, including developing sophisticated spacesuits— proved more expensive and risky than using robotic spacecraft.

Note that this suit is missing its original neck-ring. During this period, developers often cannibalized suits for parts to use in new designs. Prior to donation to the Museum, a neck-ring of less precise manufacture was attached.

Gemini IV Umbilical
1965

"This is the saddest moment of my life."

—EDWARD WHITE, JUNE 3, 1965

ABOVE: After leaving his Gemini IV capsule, astronaut Ed White walks in space, June 1965.

WHEN ED WHITE became the first American, and second human, to "walk in space," he was connected to his Gemini IV spacecraft with this 7-meter (23-foot) gold-covered tether "umbilical." Just as for a baby in a mother's womb, White's umbilical was his lifeline, carrying breathing oxygen from the capsule to his spacesuit. In the event of an emergency during his spacewalk, he had an oxygen chest pack to give him enough time to get back in and hook himself directly to the spacecraft. Wrapped inside the umbilical, along with the oxygen hoses, was a flat nylon tether to ensure that he did not float away from Gemini IV, plus communications and electrical lines. White so thoroughly enjoyed his 23 minutes at the end of his golden umbilical that when Houston ordered him to get back in, he sighed, "This is the saddest moment of my life."

Some kind of "extravehicular activity" (EVA) had long been in the planning for Gemini IV, the second piloted flight of the two-astronaut Gemini series. Initially, the mission called for White to just stand in an open hatch. But after Soviet cosmonaut Alexei Leonov exited his Voskhod 2 spacecraft on March 18, 1965, NASA examined the possibility of letting White float outside. The successful test flight of Gemini III on March 23 cleared the way for the next crew, Jim McDivitt and Ed White, to do their full four-day mission, including an EVA. The crew in particular pushed hard for a "spacewalk," as the EVA equipment was ready. That included not only the tether umbilical, but also the emergency chest pack, modifications to the Gemini spacesuit, and a handheld maneuvering gun. Squeezing a trigger activated cold jets of oxygen, giving White the thrust to push himself around.

The plan was for command pilot McDivitt to maneuver Gemini IV close to their spent booster so that White could test his ability to move in space with the handheld maneuvering unit. But with very little rendezvous training, McDivitt found it impossible to catch the tumbling rocket on their second orbit. Preparing to exit also took more time than expected, so the astronauts asked Mission Control to put off the EVA until the next circuit. Coming around the Earth again, they bled off Gemini IV's atmosphere as they passed over Australia and the South Pacific; White got out as they sped by Hawaii. Mission Control timed his walk to take place primarily over the continental United States, when he would be in direct communication with Houston for many minutes.

Free of the extremely tight confines of the cabin, White immensely enjoyed maneuvering around, in front of, and behind Gemini IV, while witnessing stunning vistas of oceans, clouds, and land below and a jet black sky above. The ease of his EVA was very encouraging, but also misleading: he experienced the advantages of zero gravity and few of the disadvantages. When Gemini astronauts next ventured outside in 1966 to do more serious experiments, they found it exhausting to accomplish any physical task without proper footholds or handholds. There had been a hint of that at the end of White's walk, when he had a very difficult time getting the hatch to lock shut. McDivitt finally held his legs down while White, with his great physical strength, cranked the latch closed. It had been a frightening moment. That exertion left him completely overheated, but in hindsight the moment of danger paled in comparison to his transcendent experience outside.

NASA transferred Gemini IV, this umbilical, and other mission materiel to the Museum in 1967.

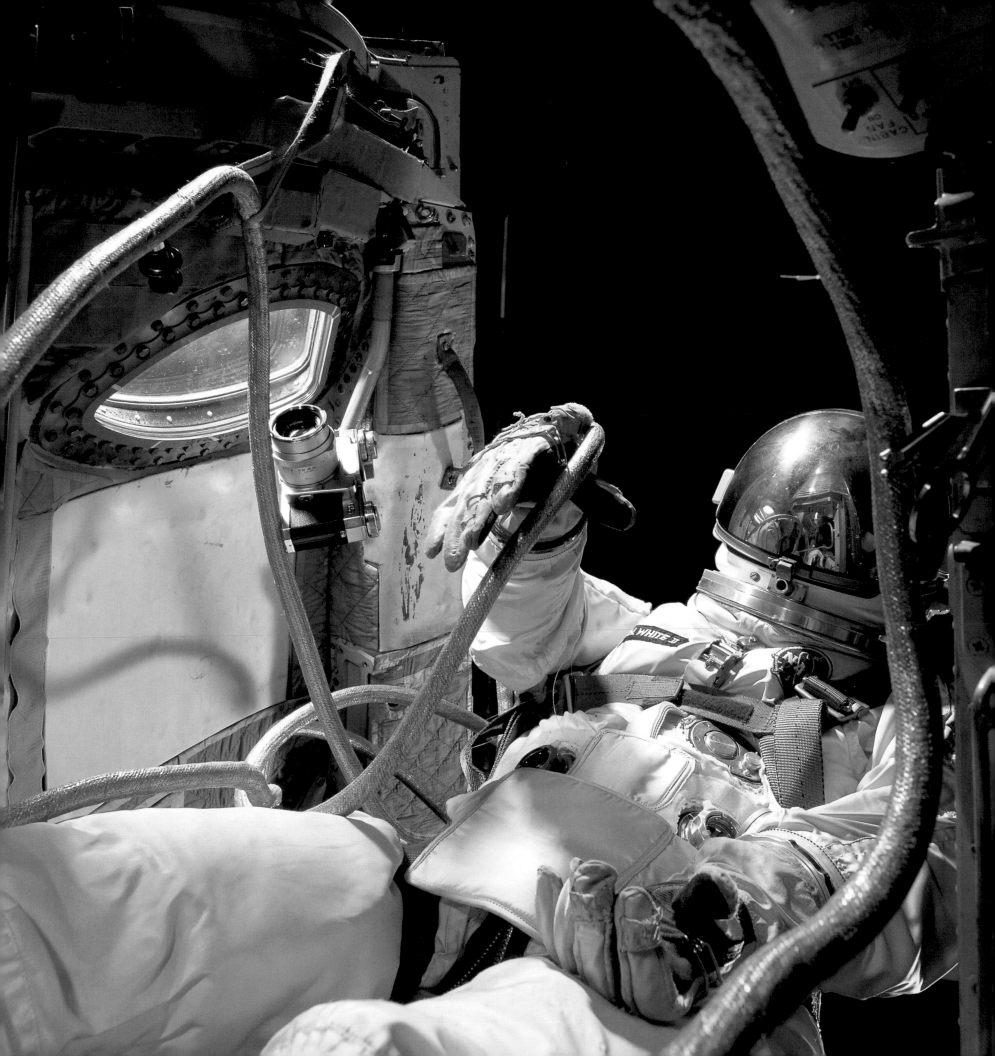

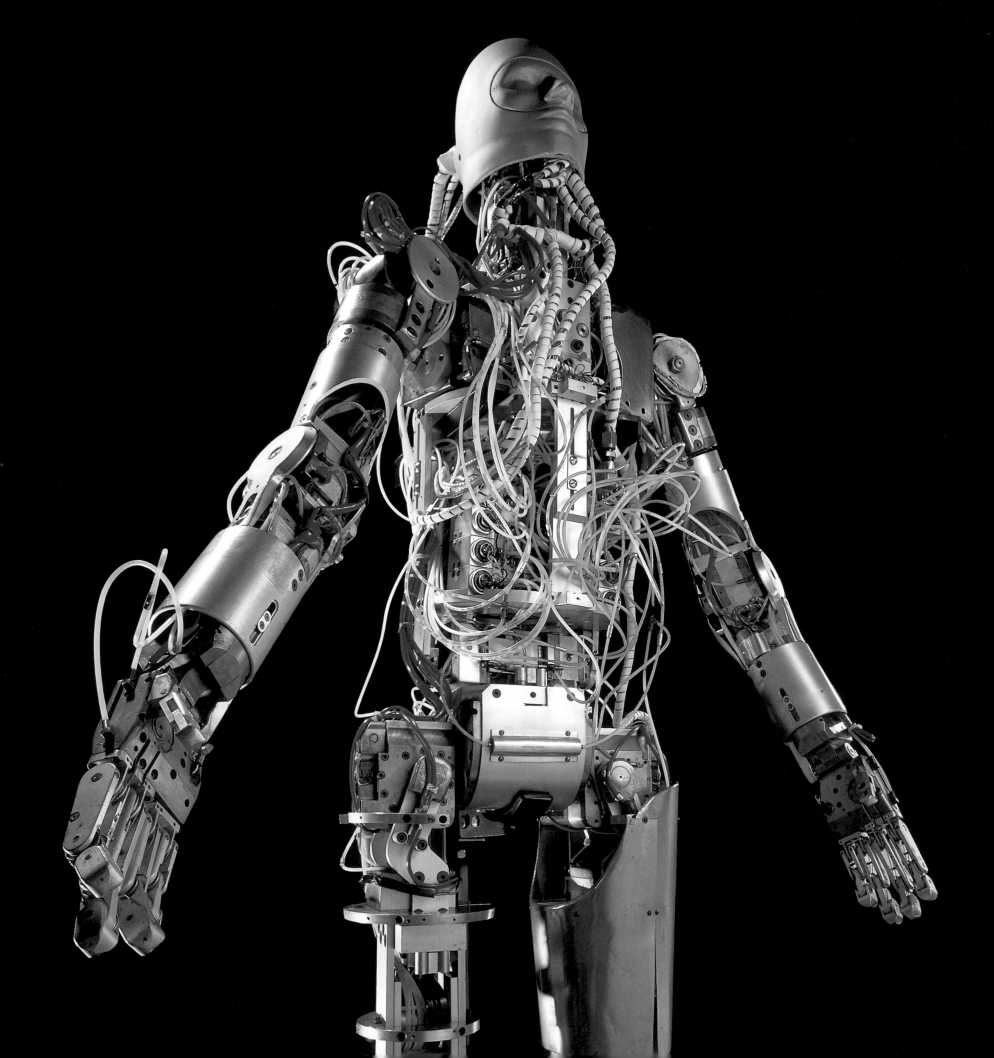

Spacesuit Android
1960s

"Add a brain and you've got the first real robot!"

—*Popular Science Magazine*, 1963

IN SCIENCE fiction, space travel and robots go hand in hand. A robot—whether as a companion, butler, or evil agent—has been a character in most science fiction movies featuring the human exploration of space. If one defines a "robot" as any machine that is capable of complex, automatic operation, then the science fiction writers are correct. Such devices have been necessary for the operation, guidance, and control of spacecraft from the beginnings of space exploration. But if one adds to that definition the requirement that such a machine also have the *physical* characteristics of the human body, the use of robots to aid the actual, as opposed to fictional, exploration of space is reduced to a very few examples. This is one of them.

A more precise, if prosaic, name for this device is "articulated dummy"—for it did not possess any independent intelligence. Nonetheless, it played a critical role in the space program. Engineers at the Illinois Institute of Technology built it in the 1960s to support the development of spacesuits for NASA (the Institute donated it to the Museum in 1980). Hydraulic and electrical actuators enabled the android to replicate the action of many of the joint motions of the human body. Sensors placed throughout the dummy measured forces that a suit might exert on a human being. The dummy enabled designers to measure how much force a human needed to move an arm or leg, or turn his head, when wearing a suit in space. By using this dummy, instead of a human being, NASA was able

to conduct tests that might otherwise be painful, tedious, or even dangerous for a human being.

But in the excitement of the race to the Moon, a machine that looked and moved like a human was not just a machine. It caused a stir—even if hydraulic fluid (which tended to leak all over the floor during tests) and actuators animated its motions rather than blood and a beating heart. Newspaper reporters and other observers gave it fanciful, if inaccurate, names, like "Art the Android," while its proud builders showed off their creation by having it do dance steps popular with teenagers of the day. It may therefore have been the first mechanical man to do an imitation of Elvis Presley.

Although a long way from the robots of science fiction, this "articulated dummy" leads us to think about real, computer-controlled robots, and how they may be used in the space program. Given the difficulties of keeping humans alive and healthy over long missions, it may be inevitable that robots will precede humans to the outer planets. Perhaps only robots, and not human beings, will experience the exploration of other solar systems. Given advances in materials and computer technologies, these devices might display intelligence and dexterity that would astonish the engineers who built this dummy. But regardless of the intelligence of these deep-space probes, unless they *look* like a human being (an unlikely but not impossible requirement), they will not evoke the emotions that this spacesuit test dummy evokes when the casual Museum visitor encounters "him."

Star Trek Starship *Enterprise* Model
1966–1969

"Space, the final frontier. These are the voyages of the Starship Enterprise. *Its five-year mission: to explore strange new worlds, to seek out new life and new civilizations, to boldly go where no man has gone before."*

—WILLIAM SHATNER, AS CAPTAIN KIRK, *Star Trek*, 1966–1969

THIS 3.4-meter (11-foot) model of the fictional Starship *Enterprise* was used to film the weekly hour-long *Star Trek* television show, which aired on NBC with mixed success from 1966 to 1969 (and later became a television classic in syndication). Originally joining the Museum's collections as an example of a fictional spaceship, the model, donated by Paramount Pictures, illustrates *Star Trek*'s impact on American culture—and on the actual space program.

Series creator Gene Roddenberry wanted a completely new design for *Star Trek*'s principal ship. It should not remind viewers of *Flash Gordon* or *Buck Rogers* rocket ships. It should not feature obvious chemical propulsion. It should not bear similarity to any design on the drawing boards at aerospace firms in 1965 and thus risk looking dated within a few years. Rather, Roddenberry wanted a design that immediately conveyed futuristic power and speed. At first glance, it should appear capable of traveling between solar systems—a *star*ship, not just a spaceship.

Designer Matt Jeffries sketched plan after plan, only to have Roddenberry reject them. Finally, Jeffries suggested a saucer-shaped main section attached to a cigar-shaped module with two thinner propulsion systems thrusting out behind. When Roddenberry came in to inspect the design, Jeffries' mock-up showed the saucer and propulsion "nacelles" on the bottom, with the ship's tubular body riding above them. Turning the prototype upside-down yielded the *Enterprise*'s now-famous configuration.

Because the model was built for television, the ship's details were designed for a single purpose: to serve filming. The original gray paint appeared whiter on screen. Bright lights obliterated some of the model's features (such as windows), making the ship seem less detailed on screen. Wiring protruding from the left side never appeared because the ship was almost always shot from its starboard side. And, of course, as a television prop, the model's makers did not envision it as a museum piece. As a result, a series of four restorations have adapted the model for 360-degree viewing while remaining faithful to the original television filming decisions. (*Star Trek*'s detail-savvy fans required no less.)

As much as particular characters or episodes, the Starship *Enterprise* became central to the franchise. Fans tracked the creation and destruction of seven different *Enterprise*s through five television shows, an animated series, and 10 films. The original identification numbers NCC-1701, created from the tail numbers on Jeffries' own airplane, were extrapolated to vessels A through E and the prototype NX-01 in the prequel series, *Star Trek: Enterprise* (2001–2005).

Even as *Star Trek* developed new fictional starships, it honored the original seagoing vessels named "Enterprise," and helped to bestow the name on a new (and very real) spacecraft prototype. More than 25 years before *Star Trek: Enterprise* decorated Captain Archer's quarters with drawings of historic *Enterprise*s (including the 1799 schooner and the 1936 aircraft carrier), *Trek* fans campaigned to give a NASA spaceship that historic name. *Trek* fans organized a letter-writing campaign that convinced the space agency to name its first Space Shuttle orbiter "Enterprise." The Shuttle *Enterprise*, which now rests in the Steven F. Udvar-Hazy Center's McDonnell Space Hangar, owes its name to *Star Trek* fans—and to the 3.4-meter (11-foot) starship model displayed at the National Mall site.

Impacts and Influences

Imagining Space Exploration as a New Frontier

Star Trek, 1966 and Beyond

EVEN IF you have never seen a single episode of the six different *Star Trek* television series, watched any of the 10 feature films, participated in any of the fan conventions, or purchased any of the merchandise, you probably still recognize *Star Trek.* William Shatner's Captain Kirk and Leonard Nimoy's Mr. Spock are icons of American culture; "Live long and prosper" and "Resistance is futile" have entered the popular lexicon. You may not speak fluent Klingon (one of *Trek's* invented languages), but you know that some diehard fans do. As one of the most successful television franchises in history, *Star Trek's* impact on American culture has been both wide and deep. Born in the late 1960s and reinvented into the twenty-first century, *Star Trek* pushed the boundaries of network television—and influenced actual space exploration.

Star Trek creator Gene Roddenberry conceived the science fiction show as "*Wagon Train* to the stars," chronicling the weekly adventures of the crew of the fictional Starship *Enterprise.* Launched in 1966 only to be cancelled in its third season, the show went off the air in 1969. Through reruns, however, *Star Trek* found a loyal and vocal following. In the 1970s, fans spurred the creation of an animated series and a feature film starring the original cast members. They even persuaded NASA to name the first Space Shuttle for the fictional *Enterprise.* Those successes allowed Roddenberry to oversee a new generation of *Star Trek* shows and films in the 1980s and 1990s.

Star Trek's long-term influence lies not only in the staying power of the franchise, which lasted almost 40 years (the last *Star Trek* series, the prequel *Enterprise,* ended in 2005), but also in its massive and coherent fan culture. Beginning in the 1970s, enthusiastic "Trekkers" found like-minded fans at conventions held across the country. Later, the Internet allowed virtual fan communities to reach around the world at any time of day. Either way, *Trek* fans earned a well-deserved reputation as detail oriented. Just as *Star Trek's* writers and producers ensured that all of the franchise's disparate pieces fit together smoothly, so also fans have worked to make their handmade costumes, fictional fanzines, or websites consistent with the franchise's appearance and logic.

Many fans admired Roddenberry's vision of racial harmony and social equality, a vision that pushed the boundaries of network television in the late 1960s. The *Enterprise* boasted a racially integrated, mixed-sex, international crew at the height of the civil rights movement, the women's movement, and the Cold War. The fictional crew's racial composition made such an impact that when African American actress Nichelle Nichols was considering leaving the show, no less than Dr. Martin Luther King Jr. himself asked her to continue, lest the model of successful racial integration be lost. The 1968 episode "Plato's Stepchildren" (original airdate November 22, 1968) included the first interracial kiss on network television.

Star Trek's illustration of diversity in space even became an example for social change in the U.S. space program. In 1977, when NASA wanted to ensure that its astronaut corps—the public face of the space agency—reflected America's diversity, they hired Nichols to conduct a six-month public relations campaign. Drawing on her role as *Star Trek's* Lieutenant Uhura, she criss-crossed the nation, encouraging women and people of color to apply to become mission specialists for the new Shuttle program. Roddenberry's fictional vision of integrated spaceship crews—begun in 1983—has become a consistent reality at NASA.

"Leave bigotry in your quarters; there's no room for it on the bridge."

—WILLIAM SHATNER, AS CAPTAIN KIRK, IN *Star Trek,* "BALANCE OF TERROR," ORIGINAL AIRDATE DECEMBER 15, 1966

Norman Rockwell, *Astronauts Grissom and Young Suiting Up*
1965

" And one there was among us, ever moved / Among us in white armor, Galahad.
'God make thee good as thou art beautifull' / Said Arthur, when he dubb'd him knight …"

—ALFRED LORD TENNYSON, *Idylls of the King*, 1859

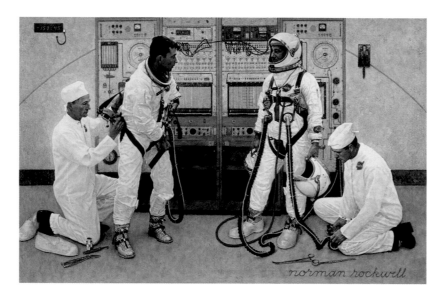

BY THE spring of 1965, the Space Race was well underway. The latest in a long string of Soviet space triumphs came on March 18, when Voskhod 2 carried the first two-person crew, cosmonauts Pavel Belyayev and Alexei Leonov, into space. On their second orbit, Leonov exited the capsule on the world's first spacewalk. The Soviets quickly trumpeted their achievement but failed to note that Leonov partially deflated his spacesuit in order to struggle back into the craft, or that the two cosmonauts barely survived a nearly disastrous reentry and were forced to spend their first night back on Earth surrounded by wolves until rescued the next morning.

In America, the astronauts were the heroes of the hour. With the Project Mercury spaceflights complete, next up were the two-person Gemini missions, with astronauts Virgil I. "Gus" Grissom and John Young leading the way.

Norman Rockwell became a familiar figure at NASA's John F. Kennedy Space Center. Anxious to find a way around rival *Life* magazine's exclusive access to personal stories of and by the astronauts, the editors of *Look* had commissioned America's best known and most-beloved artist to visit Cape Canaveral and paint

what interested him. NASA officials, who had their own art program underway, were only too happy to issue press credentials to Rockwell and provide him with the widest possible access to the facility.

Since 1916, when he painted his first *Saturday Evening Post* cover, Norman Rockwell had been, in his own words, "showing the America I knew and observed to others who might not have noticed." For more than half a century he had portrayed the everyday activities of ordinary Americans with humor, warmth, and unbounded affection. Americans returned the favor. Folks who could not recognize the work of any other artist knew a Rockwell when they saw one, and they approved.

Rockwell finished the first of his *Look* paintings in the spring of 1965. Having initially titled it *The Longest Step*, the artist changed his mind in favor of a far more descriptive title, *Astronauts Grissom and Young Suiting Up*. It was to be expected that Rockwell would be much more impressed by the people of NASA than by the giant rockets, and no one impressed him more than the astronauts of Gemini III. The sight of Grissom (right) and Young (left) donning their white spacesuits prior to the flight, assisted by technicians Joe Schmitt (left) and Alan Rochford (right), reminded the artist of "Sir Galahad girding for battle." Rather than "a background of tapestries and maidens for this classic portrait of a knight," an anonymous *Look* staff writer noted, "Rockwell found a flashing arena of electronic consoles and out-of-the-world instruments." Nevertheless, the artist "sensed that romance was still there, in the individual courage of the spacemen, and painted the first Gemini team as 'idealists in the old romantic sense, dedicated and devoted to their mission.'"

The magazine published the painting in its April 20, 1965, issue. Several years later, with the approach of the first lunar landing, Rockwell produced four more major canvases to adorn the pages of *Look*. The artist presented all of the paintings to the Museum in 1974, where they continue to remind visitors of the excitement and the deeper meaning that Americans found in the drive to send human beings to the Moon.

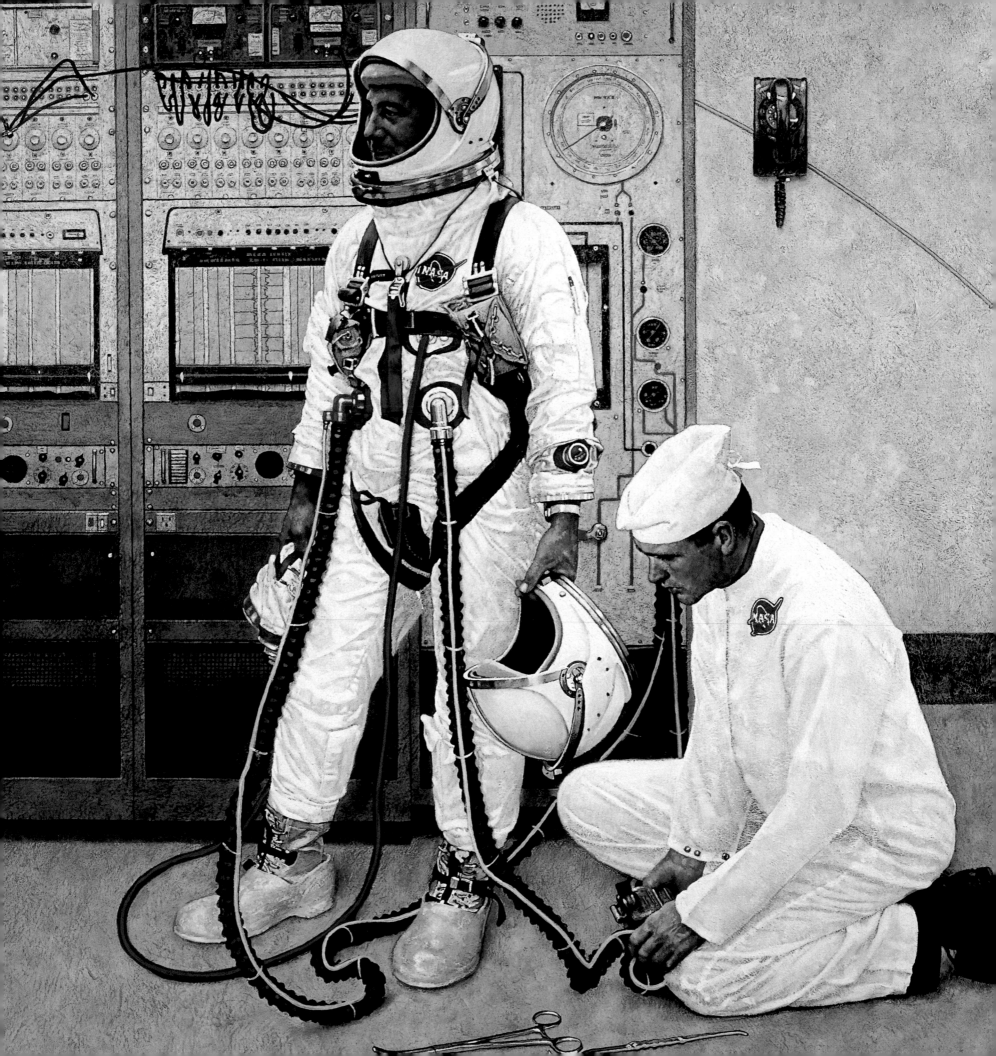

Gemini VI-A Jingle Bells and Harmonica
1965

"A little levity is appropriate in a dangerous trade."

—WALTER M. SCHIRRA JR.

WHEN "JINGLE BELLS" rang out from the Gemini VI-A capsule in December 1965, this bell set and miniature harmonica became the first musical instruments played in space. Launched 10 days before Christmas, Gemini VI-A hastily replaced a failed Gemini VI mission and headed for a rendezvous with Gemini VII, already in orbit. On December 15, Gemini VI-A astronauts Walter M. "Wally" Schirra Jr. and Thomas P. Stafford successfully piloted their spacecraft to within inches of Gemini VII, crewed by Frank Borman and Jim Lovell.

The holiday celebration came after the mission's work had been completed. According to the official NASA report: "Several hours before retrofire [Gemini VI-A's reentry maneuver], during a pass over the States, Schirra made the following report to Mission Control: 'Houston, Gemini VII, this is Gemini VI. We have an object, looks like a satellite, going from north to south, up in a polar orbit....I see a command module and eight smaller modules in front. Stand by, it looks like he is trying to signal us.' This transmission was immediately followed by 'Jingle Bells,' played on the harmonica and bells. Thus, the spirit of the season was brought into the mission."

The mission's musical interlude illustrated the human side of the dedicated astronauts. Each had two children (each with an eight-year-old daughter) and was well aware of Santa's pending visit. Schirra also had a reputation for "gotchas." As Stafford explained, "Wally came up with the idea. He could play the harmonica, and we practiced two or three times before we took off, but of course we didn't tell the guys on the ground....We never considered singing, since I couldn't carry a tune in a bushel basket."

To prepare for their historic performance, the astronauts modified the tiny instruments. Schirra attached a loop of dental floss and a small piece of Velcro to the 4-centimeter (1.5-inch)-long Little Lady Hohner harmonica, a four-hole model capable of eight notes. The bells came from Frances Slaughter, a Cape Canaveral flight crew operations officer, who had tied them onto the astronauts' shoes as a joke during training.

The whimsical additions rounded out a highly successful mission. In fact, Schirra called Gemini VI-A "the ultimate experience of my 21-year flying career." The flight not only included the first rendezvous of U.S. spacecraft but also the first computer-controlled reentry, an important milestone for the Apollo lunar landing flight plan. Gemini VI-A landed so close to its target range that television cameras aboard the aircraft carrier *Wasp* broadcast the splashdown live via satellite (another first).

In the letter giving the harmonica to the Smithsonian, Schirra, the consummate test pilot, assured the Institution that he was donating his instrument in perfect working order: "I have retested the harmonica and it performs quite well." These instruments remind us of how these cramped spacecraft served as the astronauts' homes and workplaces. And, just as many people personalize their offices or cubicles, these men brought small items aboard that allowed them to, as Schirra explained at the Gemini VIA–VII news conference, "relieve the tension." Furthermore, he added, describing their space view of the North Pole, "I think the children of this country are happier for the fact that we might have seen something there."

Gemini V Mission Patch
1965

"8 Days or Bust"

—ORIGINAL WORDING ON THE GEMINI V MISSION PATCH

WITHIN NASA and in the public mind, astronauts walked a fine line. They were ultimate team players *and* quintessential self-reliant, quick-thinking American individualists—a mix author Tom Wolfe celebrated as "the right stuff." But this heroic image often was at odds with NASA's responsibilities as a taxpayer-funded institution. After the selection of the Mercury 7 in April 1959, a NASA-sanctioned contract between the astronauts and *Life* magazine immediately brought this issue into sharp focus. As the space program unfolded, this tension expressed itself in numerous, small ways, including the creation of the first space mission patch—this one for Gemini V.

Mission patches now are commonplace for space missions, part of crew culture and valued as collectibles by the public. But the mission patch only became established with Gemini V, flown in August 1965 with astronauts Gordon Cooper and Charles "Pete" Conrad. None of the prior Mercury or Gemini missions sported a patch.

The Gemini V patch arose out of a tradition established with the first Mercury spaceflight, Alan Shepard's historic 1961 mission. Astronauts sought, and NASA agreed, to let each astronaut name their spacecraft—thus *Freedom 7*, *Liberty Bell 7*, *Friendship 7*, and on, through Gemini III, *Molly Brown*. But the last raised concern within NASA management. Gus Grissom and John Young flew the mission, and Grissom's choice of name was a joking reference to his 1961 flight *Liberty Bell 7*, which sank to the bottom of the ocean on his return from space. The popular musical *The Unsinkable Molly Brown* provided Grissom with his spacecraft name and a way of making light of the incident. But with billions of public dollars going to spaceflight, NASA saw little room for such personal humor—and for its tradition of giving astronauts free rein to name

spacecraft. On the next flight, Gemini IV, famous for the first American spacewalk, the spacecraft bore no name.

The astronauts still wanted a way to let the individual stand out from the bureaucracy. For Gemini V, Cooper and Conrad hit on the idea of creating a mission patch—a common practice in the military and from their days as test pilots. For Cooper, the patch helped to create a sense of community for everyone involved in a mission (from ground crew to astronauts) and gave the astronauts an avenue for expression: they would create their own mission patches.

Cooper later recalled, "My father-in-law had whittled a model of a Conestoga wagon…. We thought a covered wagon might be a good way to symbolize the pioneering nature of our flight. Since our mission was designed to last eight days, the longest ever attempted … we came up with the slogan '8 Days or Bust,' which we overlaid on a Conestoga wagon. We gave the design to a local patch company, and they produced hundreds of them." At a traditional preflight dinner, two days before launch, Cooper shared the patch with NASA administrator James Webb, who was less than pleased. Webb ordered "8 Days or Bust" removed from the patch—concerned that if mission failed at any point the patch could prove a public relations embarrassment. But Webb relented on Cooper's main point: he ceded to the astronauts the right to create what now were officially called "Cooper patches" but required a set review procedure before each flight. On Gemini V, Cooper and Conrad wore the Conestoga patch, minus the catch phrase—a footnote in the marriage between a bureaucracy and its public heroes.

Mance Clayton, a private collector, donated this and other patches to the Museum in 1982.

ABOVE: Astronaut Pete Conrad rides in the recovery helicopter after his Gemini V flight. Note the patch on his upper right shoulder: the words "8 Days or Bust" are absent.

Gemini VII Spacecraft
1965

"We'd like to announce our engagement."

—JAMES LOVELL JR., DECEMBER 18, 1965

ASTRONAUT Jim Lovell playfully made this announcement after he and Frank Borman reached the aircraft carrier USS *Wasp* at the end of their grueling 14-day mission in Gemini VII. Imagine sharing two weeks in the front seats of an old Volkswagen Beetle—that was the cramped experience they endured. In so doing, they set a new world's record for time in space and accomplished their fundamental mission: to prove that astronauts could survive microgravity for the longest period expected for an Apollo lunar expedition.

Along the way, their craft acted as rendezvous target for Gemini VI-A, flown by Wally Schirra and Tom Stafford. When Schirra and Stafford's robotic rendezvous-and-docking target vehicle failed to reach orbit in October 1965, NASA cancelled their launch and rescheduled them after Gemini VII. NASA used this turn of events to include a new objective for that mission: the first orbital rendezvous in history. The two craft could not dock, but on December 15, as each spacecraft orbited the Earth at over 28,000 kilometers per hour (17,000 miles per hour), Gemini VI-A accomplished the difficult part of a rendezvous—maneuvering to meet another object in space. The four astronauts waved through the windows, joked, and talked on the radio. It was a moment of uplift for the long-suffering Gemini VII crew, and an absolutely critical step on the road to the Moon.

Transferred from NASA to the Museum in 1967, the spacecraft, as displayed here, with the two hatches removed, makes Gemini VII's cockpit appear roomier than it actually was. The astronauts sat in the two large ejection seats that provided their only means of escape in the event of a low-altitude launch emergency. Once the hatches were closed on the pad, Borman and Lovell's only outside view was through the small forward-facing window in front of each astronaut's face. Inside their claustrophobic capsule,

they worked, ate, slept, and went to the bathroom—with no privacy.

In anticipation of the long mission, NASA designed spacesuits with special zippering, making it less complicated but still challenging for the astronauts to put them on and take them off. This innovation proved helpful—to a point—when the cockpit overheated. Under NASA policy, only one astronaut at a time could remove his suit. Lovell removed his, but Borman chose to endure several days of discomfort before NASA allowed both to keep their suits off at the same time. The fear was that if Gemini VII accidentally depressurized, both would be killed. The chances of that happening seemed very unlikely, however, and Borman successfully argued that he was too uncomfortable and unable to sleep properly. So they rode around the world in their underwear.

One problem Borman and Lovell had to solve in training was how to dispose of their garbage, mostly dehydrated food containers. The previous record-breaking mission, Gemini V, had already shown that to be a problem. Eventually they decided to stuff the trash from the first seven days behind the Borman's left-hand seat, and that from the second week behind Lovell's.

The two astronauts also served as laboratory rats for a number of medical experiments measuring the effects of weightlessness and radiation on the human body. Food intake and waste had to be carefully measured, and Frank Borman wore a "beanie cap" with electroencephalogram sensors taped to his skull to record his brain waves. Unfortunately, the contacts came off after a couple of days and Borman was never able to stick them back on. Despite such occasional setbacks, the Gemini VII crew demonstrated that astronauts could indeed endure a two-week flight, which was long enough to carry out the Apollo mission.

ABOVE: Astronauts Walter Schirra and Thomas Stafford in Gemini VI-A get a good view of Gemini VII through their hatch window as they conduct rendezvous and station keeping maneuvers in Earth orbit on December 15, 1965.

Gemini VIII Guidance Computer
Mid-1960s

"We'd just crossed a major milestone in getting to the moon. We'd changed a spacecraft's orbit, and we'd shown that the ship's own computer could measure velocity and orbital changes."

—CHRIS KRAFT, *Flight: My Life in Mission Control*, 2001

To many, "space age" meant advanced technology. But early spacecraft were striking combinations of the "latest and best" and "simple and just good enough" technologies. Mercury space capsules, for example, possessed only a rudimentary control system, and their human passengers could do little to alter their trajectory once in orbit. In Project Gemini, the follow-on to Mercury, designers expressly sought to give astronauts control of the spacecraft—to allow the two-person crew to "fly" the craft in the manner of a jet-propelled airplane. As high-performance aircraft pilots, astronauts expected to exert such control over their craft.

This design change coincided with one of Gemini's principal objectives: to change purposefully a spacecraft's orbit, and to rendezvous and dock with another orbiting vehicle. This capability was crucial to the Apollo plan for a lunar landing and return. After astronauts completed a lunar mission and ascended from the Moon's surface, their lunar landing craft had to dock with a command module in lunar orbit—a technical feat that had to be performed flawlessly a quarter-million miles from Earth. The change-of-orbit and rendezvous procedure is counterintuitive: if one increases a spacecraft's forward velocity while in orbit, the craft will rise to a higher orbit and paradoxically slow down. A successful rendezvous thus requires complex calculations, thorough training by the crew and ground controllers, and a precise determination of each craft's position and velocity.

To assist in executing such a complex maneuver, the Gemini capsule carried an onboard digital computer: the first for a spacecraft carrying humans. Built by IBM, the 26-kilogram (58-pound), solid-state device performed the calculations necessary to command the Gemini's rockets to alter an orbit's altitude and orbital plane. By modern standards it was a simple device. But in the mid-1960s, nearly all digital computers occupied large, air-conditioned rooms and were programmed by punched cards. The Gemini guidance computer was one of the first to package a powerful digital computer into a small space.

This computer flew on the Gemini VIII mission, one of the more dramatic in the program, and was transferred from NASA to the Museum in 1968. Neil Armstrong and Dave Scott (both of whom later walked on the Moon in separate Apollo missions), on March 16, 1966, piloted the mission, rendezvoused, and docked with an unpiloted Agena spacecraft—the first successful docking in space. That mission unexpectedly turned out to be one of the most dramatic in the U.S. space program. Shortly after docking, the two astronauts found their combined Gemini-Agena spacecraft tumbling out of control. The problem, only discovered later, was a control rocket stuck in the firing position. The crew nearly lost consciousness, but Armstrong brought the Gemini capsule under control, and the crew returned to Earth safely. Armstrong's cool handling of the crisis was a factor in his selection as commander of the Apollo 11 mission, the first to land on the Moon, three years later.

Toni Foster's Bracelet
1959–1967

"In a very real sense, it will not be one man going to the moon—it will be an entire nation. For all of us must work to put him there."

—President John F. Kennedy to a Special Joint Session of Congress, May 25, 1961

WHENEVER Robert L. "Bob" Foster, an engineer who worked for McDonnell Aviation (the contractor that built NASA's Mercury and Gemini space capsules), completed a project, he bought his wife Toni a special charm for her charm bracelet. These space-themed gifts celebrated the end of significant professional projects—and offered a present to make up for being away so often as each project neared completion. The charms also recognized how Toni Foster contributed to Bob's career by taking care of their children and home, allowing him to spend time away from home as chief engineer on Project Mercury (1959–1965) and operations manager on Project Gemini (1965–1967). Although the bracelet is now missing its clasp and is too short (approximately 13 centimeters— 5 inches—long) to be worn, the charms hanging from the delicate gold links tell an important story about the people who made human spaceflight happen in the early 1960s.

Bracelets with individually acquired charms were in fashion in the 1950s and 1960s. Women used the gold or silver open-link bracelets to accumulate individual attachments one by one, sometimes as gifts and often to mark special events. The six 14K gold charms that Toni Foster collected included, in order: a Redstone rocket charm, two engraved jeweler's charms, a Mercury capsule charm (with retropack), a Gemini capsule charm (distinguished by dual hatches), and an Atlas rocket charm. The two engraved charms were presented to Mrs. Foster by classes at the Florida schools where she taught. The octagonal jeweler's charm reads "Thank You Mrs. Foster," with "'62 " on the reverse. The heart-shaped charm reads "Cape View 5th Grade Class" with "1967" on the reverse.

As much as the tiny golden space capsules and rockets given by Bob Foster remind us of the engineers' wives and families, the presents from Toni Foster's classes illustrate the larger communities that grew up around space work. Between 1950 and 1960, Brevard County, the Florida county that included NASA's installations near Cape Canaveral and Cocoa Beach, grew faster than any other county in the United States, exploding from 23,653 people to 111,435 in only 10 years (an astonishing 371 percent growth rate). The children of space workers crowded the local schools. By 1964, the 13 schools that had existed in Brevard County in 1950 then numbered 46, and its 117 classrooms became 1,473. Community organizations flourished. When technical workers relocated to serve America's space needs, their entire families came along.

As a Museum artifact, Toni Foster's bracelet becomes much more than a private gift from a husband to a wife. Having this unique piece of jewelry in the collection illustrates the vital role played by the wives of the engineers who worked long hours in order to meet the deadlines required for space projects. It symbolizes the entire families who uprooted their lives, moved to new places, and sacrificed time with loved ones in order to participate in the national effort to put astronauts into orbit. It reminds us that entire communities made America's space program happen.

Robert and Antoinette Foster's daughter, Sarah, donated this bracelet to the Museum in 2005.

CORONA KH-4B Camera and Discoverer XIII Film Return Capsule
1967–1972

"We've spent between $35 and $40 billion on space … but if nothing else had come from that program except the knowledge that we get from our satellite photography, it would be worth 10 times to us what the whole program has cost. Because tonight I know how many missiles the enemy has."

—PRESIDENT LYNDON B. JOHNSON, 1967

ABOVE: This photograph taken by a Corona satellite in 1970 shows the area around the Kremlin in Moscow, revealing trucks, cars, and a line of people waiting to enter Lenin's Tomb in Red Square.

TOP: An Air Force C-119 aircraft captures a film return capsule, after reentry from Earth orbit, using a "hook" to capture the capsule's parachute. Notice the capsule (the small shape) below the parachute.

PRESIDENT Johnson made the above comments (probably without the knowledge or blessing of security officials) at a governors' conference in Tennessee, shedding light on one of America's most tightly guarded secrets—the CORONA photoreconnaissance satellite program. Begun in 1958, this joint Air Force-Central Intelligence Agency project aimed to photograph the USSR comprehensively and constantly—providing the United States with wide-ranging intelligence on its Cold War adversary. As Johnson's quotation suggests, CORONA symbolized the unique importance of this task.

CORONA superseded the U-2 aircraft flights, which only provided intelligence sporadically and were vulnerable to attack. But CORONA, orbiting in space, had the same goal: to look behind the USSR's "Iron Curtain" and provide essential information on the location and numbers of Soviet military forces, especially its nuclear-armed long-range bombers (and, later, intercontinental ballistic missiles). As the program could not be completely hidden from the public, authorities portrayed it as an effort to research environmental conditions in space and gave it the cover name of Discoverer.

CORONA employed satellites with cameras and film, launched into near-polar orbits to provide frequent coverage of the USSR. The exposed film wound onto reels in a special reentry capsule, built by General Electric, which separated from the spacecraft at about 160 kilometers (100 miles) altitude and then at 1,800 meters (60,000 feet) jettisoned its heat shield and deployed a parachute. Air Force planes flying over the Pacific near Hawaii snagged the parachute and capsule, returning the film for processing and analysis. If they landed in the ocean, the capsules stayed afloat for up to 48 hours before sinking.

Discoverer XIII (without cameras or film) was the program's first success. Launched on August 10, 1960, the satellite reached proper orbit and the capsule was retrieved from the Pacific a little over 24 hours later— the first human-made object to be recovered from Earth orbit. Air Force officials widely publicized the event, including bringing the capsule to the White House and giving President Eisenhower a U.S. flag that had flown in it. The Museum acquired it in late 1960.

Discoverer XIV, with cameras and film, launched one week later, and its capsule was recovered on August 19. It provided more imagery of the Soviet Union than all 24 U-2 overflights from 1956 to 1960. A total of 145 CORONA satellites were launched before the program ended in May 1972, returning more than 800,000 usable photographs of the Sino-Soviet Bloc and other regions of the world.

Early CORONA satellites carried a single panoramic camera that detected objects 11 meters (35 feet) on a side. The Museum's KH-4B, built by Itek and reconstructed from spare parts, represents the last and most advanced camera system, flown from 1967 to 1972. It had two panoramic cameras with a ground resolution of 1.8 meters (6 feet) and produced stereo photography, permitting analysts to view objects in three dimensions.

The CORONA program—its existence, technologies, methods, and photography—was one of the most secret U.S. projects during the Cold War and only was officially declassified in 1995, at which time the National Reconnaissance Office transferred this camera to the Museum. The program's value was enormous, providing U.S. policymakers with critical information on the USSR unavailable from any other source. The United States has continued operating photoreconnaissance satellites since CORONA ended, but they remain classified and little is known of them.

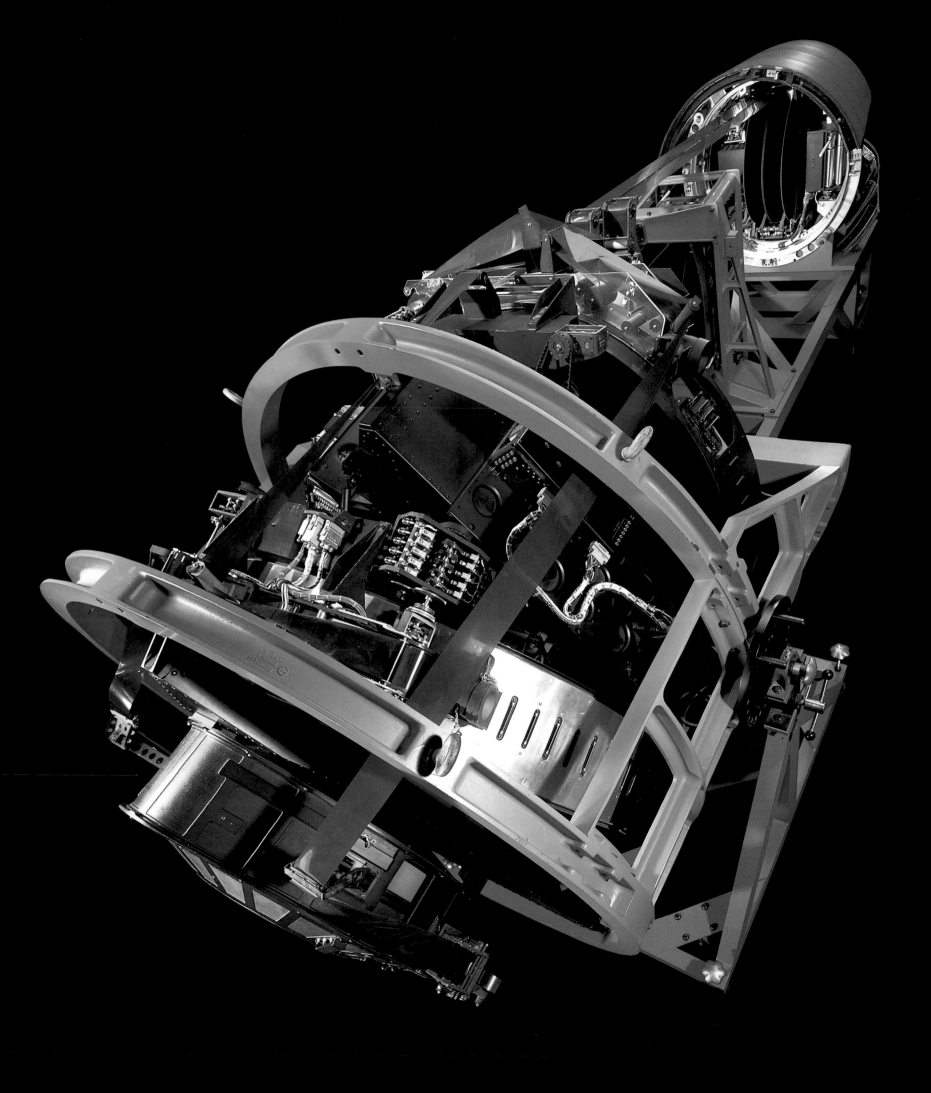

Impacts and Influences

Imagining Space Exploration as a Quest for Human Origins

2001: A Space Odyssey, 1968

"**Y**OU ARE free to speculate, as you wish," director Stanley Kubrick once remarked, "about the philosophical and allegorical meaning of *2001*." Cinema and science fiction fans have been doing precisely that since the evening of April 2, 1968, when the epic film premiered at the Uptown Theatre in Washington, D.C. Kubrick came to the project following a string of very strong and controversial films, including *Spartacus* (1960), *Lolita* (1962), and the antiwar classic, *Paths of Glory* (1957). He was determined that his next project would be "*the* science fiction film, the one against which all others would be judged." With that goal in mind, the director began to devour science fiction, quickly focusing on Arthur C. Clarke, the English master of the genre, as the ideal partner with whom to develop a story.

The two collaborators crafted a classic tale involving a journey that illuminated the human condition. While the story begins and ends with the intervention of an alien intelligence, the real theme is the nature of human evolution. Kubrick and Clarke opened their film with scenes in which hominids discover their first tool, a weapon that opens the door to the future of the species. In the core of the film, astronaut Dave Bowman thwarts the efforts of *his* tool, the HAL 9000 computer, to do him in. Our species will ultimately have to move beyond the tools that enabled us to evolve, Kubrick and Clarke seem to be suggesting, or be ruled by them.

Kubrick's masterpiece was a complete departure from the Hollywood "space operas" that had proceeded and would follow it. Frederick I. Ordway, a member of Wernher von Braun's rocket team, and NASA artist Harry Lange were brought aboard as technical and visual consultants. It was their job to ensure that the details of the film, from the orbiting space station to the food dispensers, videophones, and zero-gravity toilets, reflected the latest thinking of experts in the field and provided a touch of humor and a sense of reality to the finished product.

A thread of dark satire runs through the film and underscores the basic theme. The slow pace, spare dialogue, and understated acting combine to suggest that life in space will be a fairly banal and boring business—until something goes wrong. The astronauts, bland and emotionless fellows, stand in sharp contrast to the charismatic HAL, the most colorful and fascinating character in the film. The Heuristically programmed ALgorithmic computer (HAL) became something of a cultural icon and a focal point for discussions of artificial intelligence. When HAL's "birthday" rolled around in January 1997 (the date Clarke provided in the novel that accompanied the film), the University of Iowa celebrated with a scholarly conference on cybernetics, while *Wired* magazine marked the occasion with a special issue, and fans created a website where birthday greetings could be posted.

Though the initial reaction to *2001: A Space Odyssey* was decidedly mixed, the film quickly emerged as a genuine classic. The American Film Institute included it as number 22 on its list of the 100 most important films of the twentieth century. *Chicago Sun-Times* critic Roger Ebert, who included *2001* on his list of five films that will be familiar to audiences two centuries from now, spoke for many when he remarked that, more than any other film, "*2001* touched audiences and gave them a real feeling of what it would be like to live in space." In the end, Stanley Kubrick realized his goal. *2001: A Space Odyssey* remains the standard against which all other science fiction films are measured.

"Open the pod-bay door, please, HAL."

—ASTRONAUT DAVE BOWMAN, *2001: A Space Odyssey*

Robert McCall, *2001: A Space Odyssey*
1967

"What I try to do in my paintings is create something of the scale and immensity of space, in order to stimulate a response and enthusiasm for man's role in it.… I'm eager to transmit my astonishment to people who don't understand … because it's pleasant to consider things that are awesome, mysterious, and unbelievable."

—ROBERT McCALL, 1971

Filming on *2001: A Space Odyssey* was already well underway when director Stanley Kubrick hired a young artist, Robert McCall, to prepare some conceptual paintings of scenes from the film. A native of Columbus, Ohio, and a graduate of the Columbus School of Fine Arts, McCall came out of the Army Air Forces at the end of World War II and established himself as a successful advertising illustrator with a number of magazine covers to his credit.

McCall was invited to participate in the Air Force Art program in the mid-1950s and produced more than 40 paintings for their collection. When James Webb, administrator of NASA, created an agency art program in the early 1960s, Bob McCall was among the first artists invited to participate. Lester Cooke, curator of painting at the National Gallery of Art, who helped shape the NASA program, noted that McCall had "the quality and scope of imagination to travel in space, and carry us, the spectators, along with him." Without artists like McCall, Cooke explained, events in space that ordinary citizens could not see or experience "would remain in the realm of words, mathematical formulae, and taped electronic signals."

Given that background, McCall was a natural choice as the artist to capture the flavor and excitement of *2001: A Space Odyssey*. Kubrick arranged for him to spend three months at Metro-Goldwyn-Mayer's Borehamwood Studio, in the London suburbs, toward the end of the film's production. While not involved in the design aspects of the project, McCall produced four paintings that were used as posters in the publicity campaign that launched the film. Metro-Goldwyn-Mayer, Inc., donated all of the canvases, including this image of the space shuttle craft and Earth-orbiting space station, to the Museum in 1969, where they continue to remind visitors of the power of the film that inspired them.

Lunar Orbiter
1966–1967

"The pictures returned by the Lunar Orbiter series not only paved the way for the Apollo missions
but also gave us images of the Moon that are still used extensively by scientists today."

—PAUL D. SPUDIS, LUNAR SCIENTIST, 1996

THE LUNAR Orbiter, a robotic spacecraft equipped with a powerful camera, mapped the lunar surface prior to the first Apollo missions, paving the way for human landings. Five of these spacecraft flew to the Moon, entered orbit, and scouted the terrain to identify potential landing sites. Two prior lunar mapping missions—Ranger and Surveyor—helped in this task but the Lunar Orbiter provided the most useful and detailed data. This data allowed NASA mission planners to select a safe and scientifically interesting landing site for each Apollo mission.

Approved in 1960, the Lunar Orbiter program originally had no connection to Apollo. But after President Kennedy's 1961 call for a Moon landing, NASA redirected the mapping effort to fill in a crucial, missing piece in the Apollo plan: reliable, thorough information on lunar geography. From 1966 to 1967, on five occasions, an Atlas-Agena rocket carried a Lunar Orbiter to the Moon, each a successful mission. The returns immediately highlighted the orbiters' value—for scientists and especially for Apollo planners. In total, the orbiters mapped roughly 99 percent of the lunar surface. The orbiters proved so effective that at the completion of the third mission, Apollo officials announced that they possessed sufficient data to press on with an astronaut landing.

The Lunar Orbiter's bounty of images came from a unique onboard photographic system. Instead of sending television pictures back to Earth as electrical signals, the Lunar Orbiter took actual photographs, developed them on board, and then scanned them using a special photoelectric system. For this reason scientists referred to the Lunar Orbiter as the "flying drugstore." Owing to the possibility that radiation in space might fog photographic film, the system used a slow-speed film. To prevent blurring, the camera system moved slightly to compensate for the relatively long exposure times. The resulting images had exceptional quality and provided resolutions of up to one meter (3 feet) from an altitude of 56 kilometers (30 nautical miles).

The Lunar Orbiter, too, provides a footnote to perhaps the most famous image of the space age: the oft-seen "Earthrise" photograph taken by the astronauts on Apollo 8. Lunar Orbiter 1 actually took the first photograph of the Earth from the Moon—but the image was black and white, not color, as its more famous successor. Apollo 8's dramatic "blue Earth" captured the public imagination in a way the Lunar Orbiter image did not.

In addition to the broad photographic coverage of the Moon, the Lunar Orbiter provided additional information that also aided Apollo. Sensors mapped the radiation levels around the Moon, analysis of the spacecraft orbits finding evidence of gravity perturbations that suggested that the Moon was not gravitationally uniform. Instead it appeared as if buried concentrations of mass were under the mare deposits. By discovering and defining these "mascons," the Lunar Orbiter made it possible for the Apollo missions to conduct highly accurate landings and precision rendezvouses.

After depleting their film supplies, flight controllers purposely crashed all five Lunar Orbiters onto the Moon to prevent their radio transmitters from interfering with future spacecraft. The item on display, obtained by the Museum in 1970, is an engineering mock-up of the spacecraft built by the Boeing Company.

ABOVE: Lunar Orbiter 2, flown in 1966, provides a clear view (looking northward) of the Copernicus crater on the Moon. The crater measures roughly 100 kilometers (62 miles) in diameter.

F-1 Rocket Engine
1966

"We all realize there will be a man riding on this thing."

—Dr. Charles Davenport, Rocketdyne, March 1961

ABOVE: In a static test, NASA and Rocketdyne engineers fire a cluster of five F-1 engines— the same configuration used to power the first stage of a Saturn V rocket.

TOP: Technicians assemble F-1 engines at North American's Rocketdyne division in the mid-1960s.

IRONICALLY, the mighty F-1 rocket engine that lifted the Saturn V rocket to the Moon was conceived in 1955—before there was a space program. As part of a mid-1950s missile race, the United States pushed to develop a powerful rocket engine that would equal or surpass Russian technology, which appeared to be more advanced. Rocketdyne, then a division of the North American Aviation Company, made a feasibility study for the Air Force for such an engine, featuring 6.6 million newtons (1.5 million pounds) of thrust.

But other engine projects underway proved sufficient and the military never developed this Rocketdyne idea. After Sputnik in 1957 and the start of the Space Race, however, NASA (created in 1958) revived the F-1 concept. Again, there were many who perceived that an engine powerful enough to launch a massive rocket would be the key to catching and besting the Soviet Union in the competition to explore space. This general belief, rather a specific program, brought the F-1 back to life. Owing to its Air Force study, Rocketdyne had a head start in developing the engine and, with support from NASA, tested a full-sized prototype in 1961.

In these high-pressure years, the Rocketdyne F-1 had the advantages of using the basic design of other company engines, which already had established a track record of success. Engineers scaled up these designs to create an engine that weighed more than 8,000 kilograms (18,000 pounds), about nine times heavier than a large missile engine. But "scaling up" was not simple: the F-1 required radically new manufacturing and testing approaches. Its combustion chamber, for example, was formed of hundreds of cooling tubes brazed (soldered) together in a contoured bundle. As in other "regeneratively cooled" engines, the fuel flowed through the tubes, cooling the chamber prior to fuel injection. But because of the chamber's massive size, the tubes were brazed in a huge furnace rather than by conventional handheld torches. F-1s used standard propellants—kerosene and liquid oxygen. But the engine's massive 1,100-kilogram (2,500-pound) turbopump, capable of pumping 160,000 liters (42,500 gallons) of fuel per minute, also posed a huge engineering challenge for development and manufacture.

In May 1961, just a month after the first F-1 test, President Kennedy declared a national spaceflight goal: a human landing on the Moon before the end of the decade. This initiated Project Apollo. The Saturn V rocket now became the "national launch vehicle"—a behemoth to carry humans and all their equipment into space. As developed, a cluster of five F-1s powered the first stage of a Saturn V, producing a total thrust of over 33 million newtons (7.5 million pounds), enough to lift the 111-meter (363-foot)-long, 2.8-million-kilogram (6.2-million pound) rocket toward its goal—the Moon.

The first human mission to the Moon was Apollo 8, launched in December 1968 with the limited objective of reaching lunar orbit. Apollo 11's crew launched on July 16, 1969, and returned on July 24, achieving President Kennedy's goal. The F-1 performed without failure on each mission, and in all, Saturn V enabled a dozen astronauts to explore the Moon.

The Museum's F-1 engine was meant for Apollo 18 (the program ended with Apollo 17) or other later missions to the Moon. In preparation for flight, it underwent eight starting tests in October 1966, firing for a total of 501 seconds. With the termination of Apollo, NASA transferred the engine to the Museum.

Long after the Apollo program, the F-1 remains the most powerful liquid-fuel rocket engine ever built and a magnificent engineering achievement.

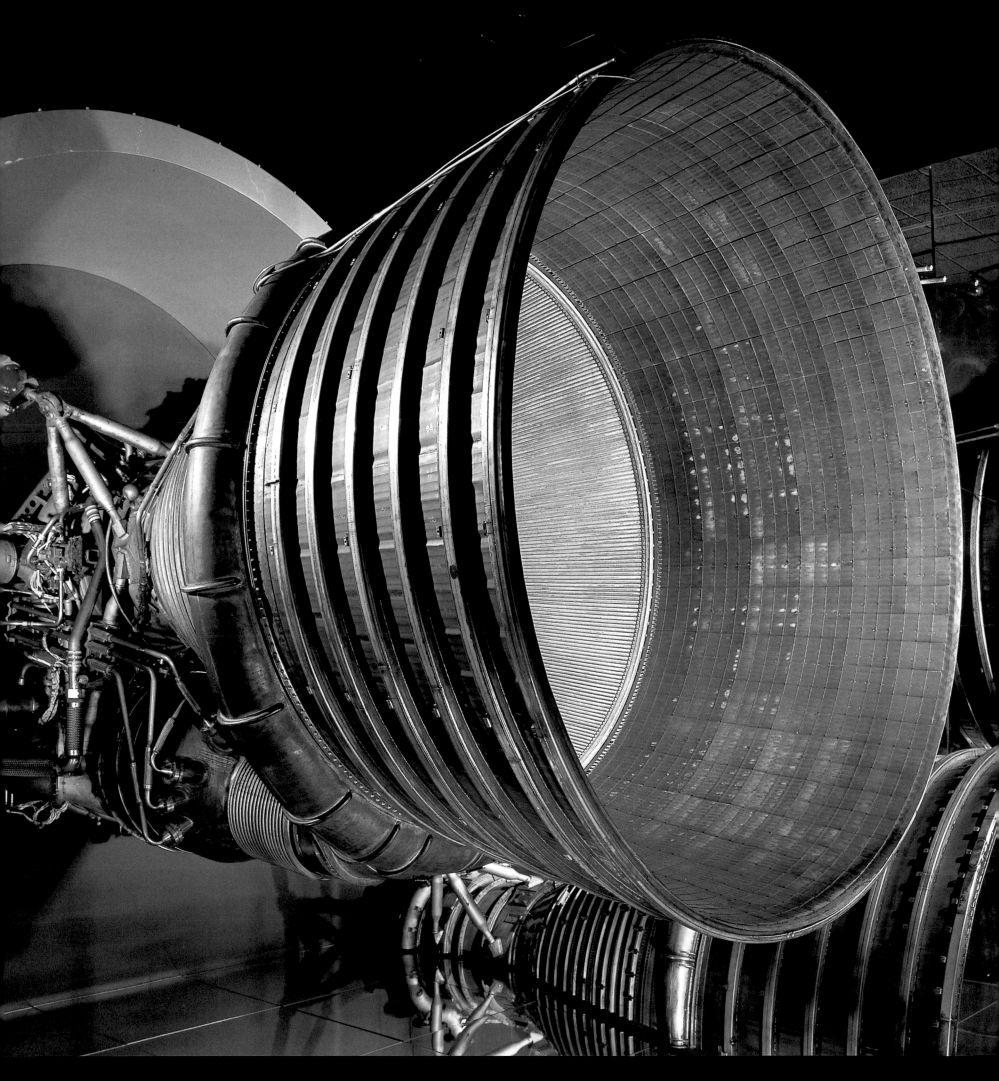

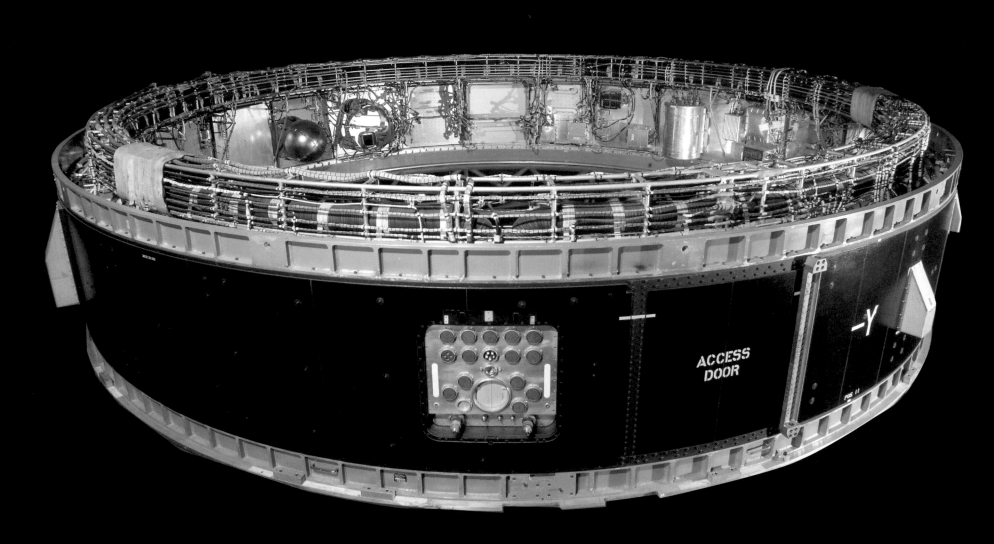

Saturn V Instrument Unit
Mid-1960s–mid-1970s

"Weighing hundreds of tons yet built with the precision of the finest watch."

—President John F. Kennedy, 1962

THE 111-meter (363-foot)-tall Saturn V rocket enabled the United States to fulfill President Kennedy's challenge to put a man on the Moon, and return him safely, by the end of the 1960s. The three-stage rocket could loft 127,000 kilograms (280,000 pounds) into Earth orbit and send 43,000 kilograms (95,000 pounds) to the Moon. Between the Saturn's third stage and its payload was the instrument unit (IU), which guided the rocket during launch and ensured that its passengers reached the Moon safely. NASA transferred this IU to the Museum in 1978.

The IU was an integral part of the rocket's structure, about one meter (3 feet) high and 7 meters (22 feet) in diameter, and made of aluminum alloy about 2.5 centimeters (one inch) thick. Within this structure were mounted a set of components that made up the Saturn's "nerve center": the guidance platform and electronics, a computer, radio and radar transmitters and receivers, and control electronics that commanded the rocket engines below. Throughout the structure was a maze of pipes that carried cooling fluids and gases. The IU was crucial to the success of the Saturn launches, guiding and controlling the vehicle until its third stage and payload were on their way to the Moon, after which the Apollo guidance computer, installed in the command module, took over.

The U.S. Army's Ballistic Missile Agency, located in Huntsville, Alabama, began work on Saturn guidance concepts even before President Kennedy's challenge. After the establishment of NASA in 1958, the center became part of NASA. It was led by Dr. Wernher von Braun, brought to the United States by the Army along with a small team of engineers from Germany after World War II. Von Braun established a close relationship with the U.S. computer company IBM, which eventually received the contract for the overall design and integration of the IU. IBM established a plant in Huntsville for the final assembly and integration; it built the unit's computer at its Owego, New York, plant. Bendix, another company with whom von Braun established close ties, built its gyroscopically stabilized platform.

Everything about the Saturn V rocket is measured in superlatives, and its guidance system was no exception. Because of its height, the Saturn rocket was subject to flexing and bending, which the IU had to monitor and compensate for, in addition to monitoring the health of the engines, their cut-off time, the vehicle's pitch, roll, and yaw, and staging. The onboard computer, with its 100 kilobytes of memory, seems primitive by today's standards. But at that time, when most computers filled rooms and were tended by a team of technicians, it was a marvel of miniaturization and ruggedness. Most of all, the computer, stable platform, and other components that made up the instrument unit did their jobs and performed reliably through all the Saturn launches from 1968 to 1975.

ABOVE: On July 16, 1969, a Saturn V rocket lifts off carrying the Apollo 11 astronauts toward their historic landing on the Moon. The dark band (lower, middle) on the rocket is the instrument unit.

Minuteman III Guidance System
1970s

"The extreme precision required of the guidance system components is evident in the fact that the miss coefficient along the range-sensitive axis is about one mile for every foot-per-second of velocity error, and about 1,000 feet for the same error on the cross-range axis."

—ROBERT C. ANDERSON, TRW, DEFENSE AND SPACE SYSTEMS GROUP, 1979

THE MINUTEMAN, a solid-fueled, intercontinental ballistic missile, had a single purpose: to deliver a nuclear warhead from bases in the north-central United States to targets mainly in the Soviet Union. Its design and deployment, beginning at the height of the Cold War in the late 1950s, illustrate, as much as any artifact from that era, how a piece of space hardware reflects not just a technical achievement but also a historical moment, its politics and concerns. The Minuteman, and its Soviet counterparts, became prominent symbols of the Cold War's greatest fear: the possibility of nuclear attack, executed swiftly, unstoppable, horrific in its destruction, changing the course of human history.

Minuteman lived up to its name—it literally could be launched on a moment's notice. One crucial part of this capability was the missile's solid-fuel rocket engines—ready to be ignited on command. Another was its guidance system, constantly running, with targets programmed into its onboard computer.

In the United States, this guidance technology became the focal point of the Cold War's most important problem, which was how to prevent, and at the same time plan for, a nuclear war with the Soviet Union. An intimate connection developed between missile accuracy (how precisely a warhead could be directed to a designated target) and nuclear strategy. Accuracy was a daunting challenge, involving many variables. An intercontinental missile had to travel several thousand miles, enduring the rigors of launch, passage through the Earth's atmosphere into space, and then plunging down toward a planned target, all in about 30 minutes. Improving accuracy took extraordinary effort.

Initially, Minuteman supported a nuclear weapons doctrine known as "Mutual Assured Destruction" (with the appropriate acronym MAD). Limitations in accuracy, for missiles as well as aircraft bombers, led to a strategy that relied on the horrible possibility of a massive exchange of weapons to deter war. As the Cold War progressed in the 1960s, an alternate doctrine emerged, partly as a result of advances in guidance technology. Better accuracy allowed an enemy's missile silos, not its cities, to serve as targets. This changed the dynamic of war planning and deterrence (but not public fears), stimulating further attempts to improve missile guidance systems and their accuracy.

The guidance system shown here, transferred from the Air Force in 1977, is from one of the later-generation missiles, a Minuteman III, first deployed in the mid-1970s, which carried three independently targeted warheads. Its guidance system delivered greater accuracy—after traveling thousands of miles the missile could land within 305 meters (1,000 feet) of its target. Most notable was its improved computer, which monitored the health and status of the missile while in its silo, and which could be reprogrammed quickly to retarget the warheads if necessary. The computer was able to perform these functions because it used the newly invented integrated circuits, or silicon chips, in its design. Minuteman guidance computers were the world's first devices to use this now-common invention in large quantities.

Missile guidance technologies, such as those developed for Minuteman, proved crucial for human space exploration. These innovations were adapted and used in launch vehicles such as the Saturn V and in the Gemini and Apollo spacecraft.

ABOVE: After launch from Vandenberg Air Force Base, California, a Minuteman III missile, in a test, releases its three warheads (each capable of hitting a separate target), which plunge toward a target zone near the Kwajalein Atoll in the Pacific Ocean.

Experimental Advanced Extravehicular Pressure Suit, EX1-A
1967–1968

"The Promise of Space: Where will it lead?"

—WILLIAM SHELTON, *Man's Conquest of Space*, 1968

FROM THE early 1960s through the early 1970s, NASA pursued two different lines of spacesuit development. "Soft" suits became the most successful approach and the type used by Apollo astronauts on the Moon. But NASA also researched "hard" suits—a type exclusively designed for extended EVAs ("Moon-walking") that usually featured a rigid exterior with highly mobile joints. As spacesuit technology still was a very new enterprise, the hard suit research served as an insurance policy in case soft suits proved inadequate for lunar exploration.

Officially known as advanced extravehicular suits (AES), hard suits maintained a constant volume, whether pressurized or unpressurized, and had important advantages: they allowed mobility, resisted damage by sharp objects, and could sustain a higher internal pressure (34 kilopascals—5 psi [pounds per square inch] compared with 25 kilopascals—3.7 psi—for soft suits). This latter feature was important: at this higher pressure an astronaut required a shorter pre-breathe time before pressurization and shorter depressurization time afterwards. (Before suit pressurization, an astronaut typically had to pre-breathe pure oxygen for about 40 minutes, which was vital in controlling the "bends," a painful phenomenon also experienced in deep-sea diving.) Hard suits, though, weighed a little more than their soft counterparts and required more physical space to store in the spacecraft—both crucial disadvantages in a spaceflight to the Moon. NASA chose soft suits for the Apollo program but continued the AES line of research into the 1980s. Some of the AES technologies developed during these years are now used in deep-sea diving suits.

NASA contracted some of the AES research to Litton Industries and the AiResearch Corporation but also conducted development at its Ames Research Center. The Museum's suit, the EX1-A, was designed and produced by AiResearch. Other types of hard suits that resulted from this effort included the RX series (Litton), the AX series (Ames), and the AES (AiResearch).

The EX1-A is a single-wall laminate (SWL) suit, developed as a prototype from 1967 to 1968. The first of only two such suits designed and built by AiResearch, it featured flexible toroidal joints in the shoulders, elbows, waist, hips, and knees. This joint system offered a full range of motion even under maximum suit pressure. In hard and soft suits, the primary cause of fatigue for an astronaut was expending energy to bend and move arms and legs. The greater the pressure in the suit, the greater the exertion required. Hard suit engineers, thus, sought design innovations that minimized this exertion. AiResearch's toroidal joint system moved without changing shape and required much less exertion by an astronaut to flex arms, waist, and legs than prior designs, thus allowing the wearer to move with greater ease.

The EX1-A suit consisted of two layers of fabric bonded together with a neoprene core and was designed to be soft in the unpressurized mode, making it more easily stowed in an Apollo spacecraft. Upon pressurization, the suit became "hard" and maintained a constant volume. The dome or hemispherical-shaped helmet provided the wearer a wide range of vision, while maintaining full pressure. Made of polycarbonate with an anodized aluminum neck-ring that fit directly into the neck-ring of the suit, the shell could be rotated if scratched, providing an unblemished field of vision.

In testing, this experimental suit primarily used Gemini gloves. NASA transferred the suit to the Museum in 1977.

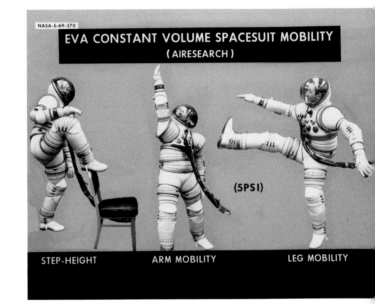

NASA-S-69-370
EVA CONSTANT VOLUME SPACESUIT MOBILITY
(AIRESEARCH)

(5PSI)

STEP-HEIGHT   ARM MOBILITY   LEG MOBILITY

ABOVE: "Hard" spacesuits excelled at protecting astronauts but could be stiff and inflexible. This graphic suggests that AiResearch succeeded in designing joints that gave astronauts a full range of flexibility.

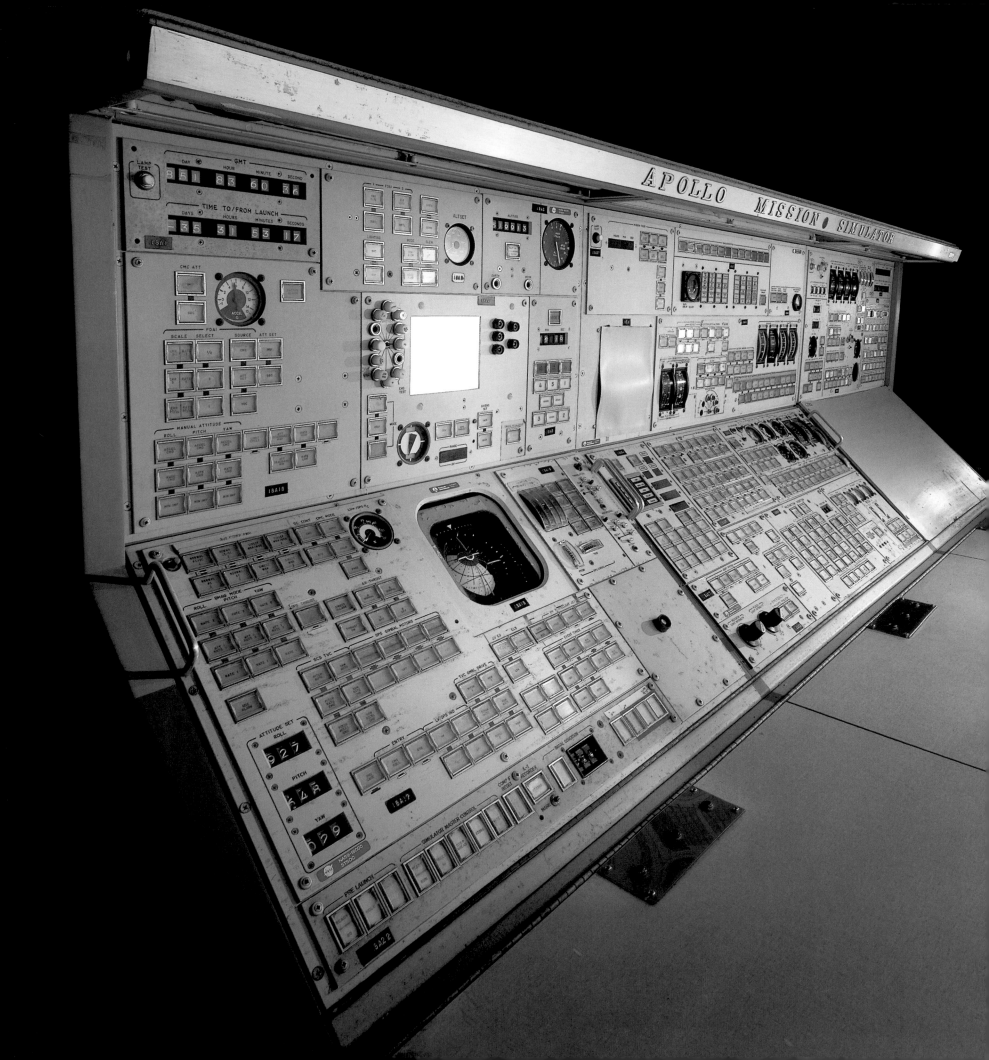

Apollo Command Module Simulator
Late 1960s

"The great train wreck."

—Astronaut John Young, circa 1967

I N A FEW short years during the 1960s, NASA and numerous aerospace companies developed the systems, large and small, required to send humans to the Moon and return them safely to Earth. The major components (in addition to the enormous Saturn V launch vehicle) were the command and service modules and the lunar module. In November 1967, the un-piloted Apollo 4 mission tested these new, complex technologies for the first time. Less than two years later—in July 1969—astronauts Armstrong, Aldrin, and Collins used these new creations to complete Apollo 11, the first historic journey to the Moon.

With these fast-paced preparations, a daunting challenge loomed: how to adequately train astronauts to live in, operate, and control strikingly complex technologies before they were fully proven in the harsh environment of space. The stakes were high. In January 1967, astronauts Grissom, White, and Chafee died during an on-the-ground launch pad test of a command module. While this tragedy was caused by faulty equipment, procedures, and design, it reemphasized the importance of astronaut training. Numerous missions dedicated to training (such as Apollo 4) were not practical, however, and, with President Kennedy's end-of-the-decade goal for a lunar landing approaching, NASA had little time.

The answer to the challenge of astronaut training was simulation.

NASA provided the astronauts and ground-based crews with a number of training devices and simulators. In combination, they were designed to replicate every conceivable contingency of a round-trip journey to the Moon. The most awe-inspiring of these, indicative of NASA's willingness to push 1960s technology to extraordinary limits, were the command module simulators (CMS).

Virtual reality headsets, plasma screens, and computer-generated images did not yet exist, so the CMS used a combination of real and imitative techniques. At the center of a CMS sat a high-fidelity replica of an actual spacecraft that included all switches, indicators, stowage compartments, seats, and windows. Surrounding the command module were boxes containing mechanical and optical gadgets designed to mimic all aspects of spaceflight. Light beams reflected off of polished metal spheres to create the illusion of stars, which were superimposed on projected TV and film images of the Earth and Moon. Loud speakers replicated sounds, and other devices introduced smells and smoke. As NASA made changes to the actual flight hardware, engineers updated the CMS.

Originally called Apollo Mission Simulators, these behemoths were assembled both at the Manned Space Center in Houston and the Kennedy Space Center in Florida. Astronaut John Young, on seeing one of these assemblages, called it a "great train wreck" in recognition of its hodgepodge look and gigantic size. Each simulator was controlled by a bank of computers, which were linked to a lunar module simulator (positioned nearby and as complex as a CMS) and to the actual launch and mission control rooms at Kennedy and Houston. Astronauts and the entire ground control team trained on these devices. The Apollo 11 crew estimated that they spent 2,000 hours in this and other simulators. The astronauts and flight controllers came to view the supervisors responsible for devising the hypothetical emergencies as diabolical geniuses. These simulations—ranging from the possible to improbable— quickly became the core of the training experience.

The Museum's CMS was the one operated at the Kennedy Space Center. The Museum stored it in pieces for years until permitting the producers of the *From the Earth to the Moon* television series to reassemble it for scenes involving astronaut training. Disassembled again, it remains in storage, pending restoration and the location of a space large enough for its display.

ABOVE: This photo reveals part of the Apollo command module (CM) simulator. At the top of the stairs is the entrance to the CM; the surrounding equipment creates "real" space experiences.

OPPOSITE: One of the consoles technicians used to program and control the simulator.

Apollo 11 Command Module *Columbia*
1969

"Spacecraft 107—alias Apollo 11 alias 'Columbia' The Best Ship to Come Down the Line. God Bless Her"

—GRAFFITI WRITTEN BY ASTRONAUT MICHAEL COLLINS
ON A PANEL INSIDE THE APOLLO 11 COMMAND MODULE, 1969

THE Apollo 11 command module *Columbia* served as the living quarters for the three-person crew during most of the first human lunar landing mission. On July 16, 1969, Neil Armstrong, Edwin "Buzz" Aldrin, and Michael Collins climbed into *Columbia* for their eight-day journey. The command module (CM) was one of three parts of the complete Apollo spacecraft. The other two were the service module (SM) and the lunar module (LM). The SM contained the main spacecraft propulsion system and consumables (oxygen, water, propellants, and hydrogen). Armstrong and Aldrin used the LM to descend to the Moon's surface. The CM was the only portion of this historic spacecraft that returned to Earth.

The cone-shaped spacecraft is divided into three compartments: forward, crew, and aft. The forward compartment is at the cone's apex, the crew compartment is in the center, and the aft compartment is in the base, or blunt end, of the craft. The forward compartment contained the parachutes and recovery equipment. The crew compartment has a volume of 5.9 cubic meters (210 cubic feet)—about the size of a standard automobile interior. It contains three couches for the crew during launch and landing. The couches are arranged so that each astronaut faces the main instrument panel. During flight, the astronauts could fold up the couches to make more room in the spacecraft. Near the feet of the couches, in the lower equipment bay, there was enough room to stand up.

Following their historic landing and exploration of the lunar surface, Neil Armstrong and Buzz Aldrin rejoined Michael Collins aboard *Columbia*. Collins, as CM pilot, fired the large engine on the service module and headed back to Earth. Several days later, on July 24, they discarded the SM and entered Earth's atmosphere.

Columbia's exterior is covered with an epoxy-resin ablative heat shield. As *Columbia* entered the atmosphere at a speed of 40,000 kilometers per hour (25,000 miles per hour), its exterior reached a temperature of 2,760° C (5,000° F). The heat shield protected the craft from burning and vaporizing. *Columbia* finished its flight with a parachute landing in the Pacific Ocean, where the USS *Hornet* retrieved it and its crew.

Following splashdown, while en route to Hawaii on the USS *Hornet*, Michael Collins crawled back into the command module (it was connected to the mobile quarantine facility by an air-tight tunnel) and wrote a short note on one of the equipment bay panels (see the image at left, top).

The Apollo command module *Columbia* has been designated a "Milestone of Flight" by the Museum and is displayed almost directly under the permanent location of the 1903 Wright Flyer, the first successful powered airplane.

ABOVE: A helicopter (not visible) hoists the Apollo 11 command module *Columbia* aboard the recovery ship USS *Hornet*. Notice the orange flotation collar (on the deck), which helped keep the module buoyant in the water.

TOP: Astronaut Michael Collins wrote this inscription inside the Apollo 11 command module *Columbia*.

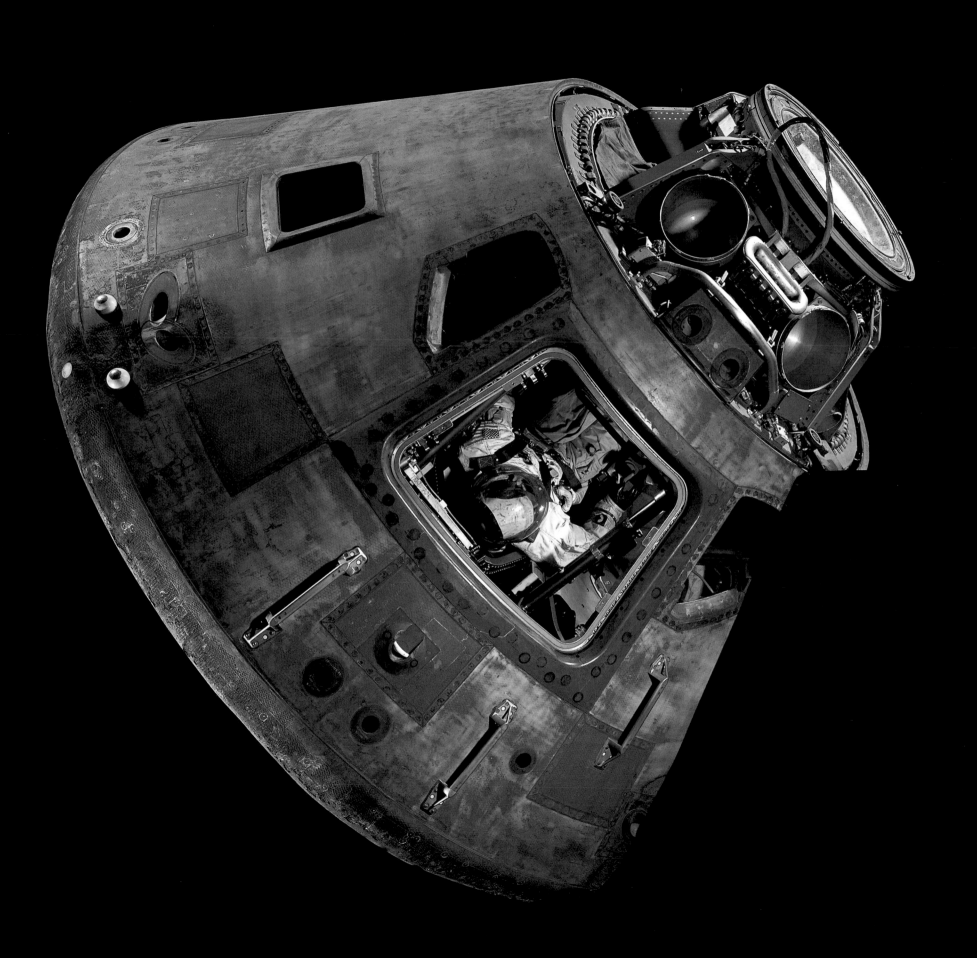

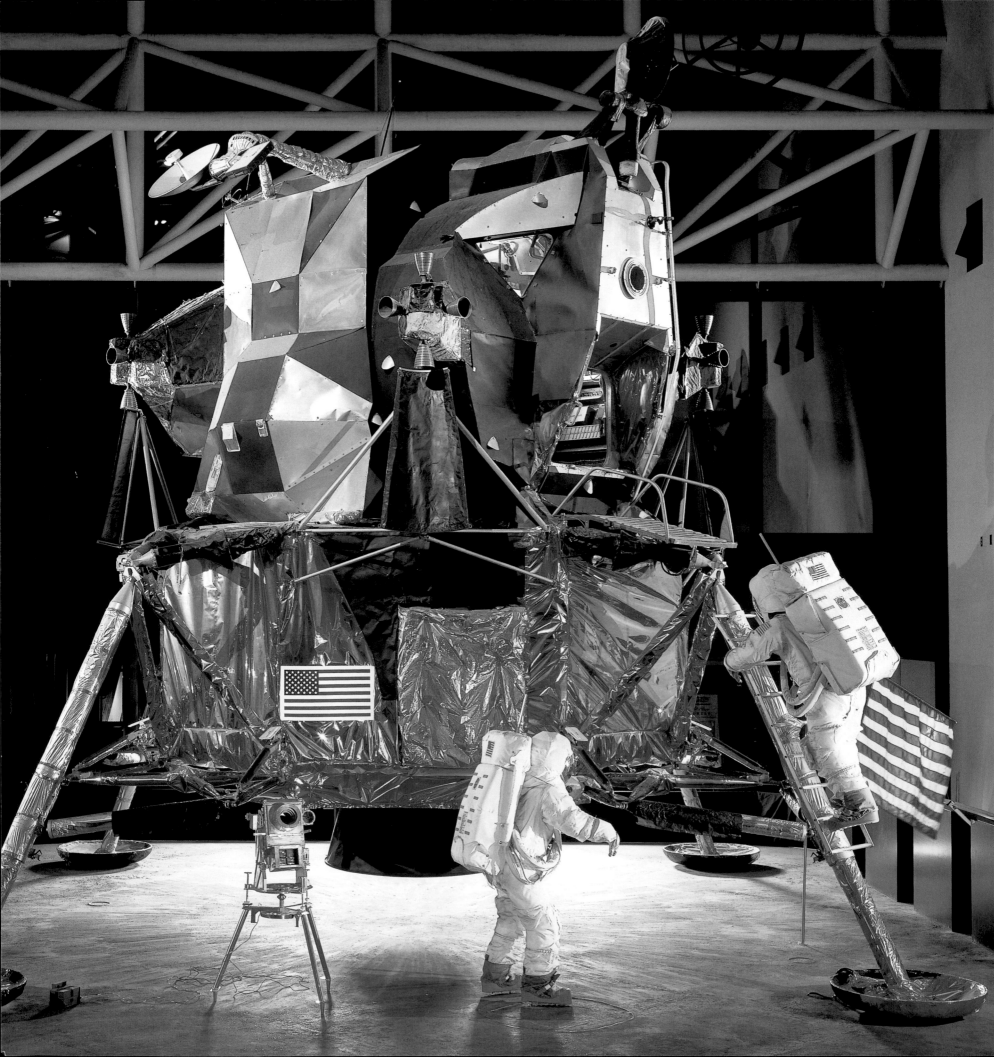

Lunar Module 2
1966

*"Additional drop tests … would enhance confidence in the strength of the LM to
withstand the impact of landing on the moon, with all subsystems functioning."*

—GEORGE MUELLER, ON THE AS-206/LM-2 HARDWARE

THE LUNAR module was the last major
component of the Apollo Moon landing
program to be designed, developed, and
tested. After President Kennedy's decision
in 1961, engineers vigorously debated the best way
to accomplish a lunar landing. Three possibilities
emerged: direct ascent, Earth orbit rendezvous, or
lunar orbit rendezvous. Each method required its
own specially designed craft for landing on the Moon.
Not until November 1962 did NASA select lunar
orbit rendezvous as the best way to achieve a landing
before the end of the decade. Built by Grumman
Aerospace Corporation under NASA contract, this
lander, LM-2, was the second unit produced.

The design and construction of a module
suited to lunar orbit rendezvous challenged the
manufacturer and caused sleepless nights for
anxious NASA managers. The structure had to
be extremely lightweight, yet strong enough to
protect the astronauts from the vacuum of space
and allow for repeated exit and entry of space-suited
astronauts to the lunar surface. The Apollo lunar
module (LM) featured two-stages: the lower descent
stage included the landing gear, the descent rocket
engine, and lunar surface experiments; the upper
ascent stage consisted of a pressurized crew
compartment, equipment areas, and an ascent
rocket engine. On descent, the entire LM carried
two astronauts from lunar orbit to the Moon's
surface; on the return, the ascent stage only
ferried them back to lunar orbit.

As the first Apollo mission to carry astronauts
neared, NASA paid increased attention to the status
and potential delivery dates of the lunar modules.
At the end of 1966, at its facility in Bethpage, Long
Island, Grumman conducted tests on the first
modules, LM-1 and LM-2 (the Museum's artifact).
NASA planned to use both for unpiloted test flights

in Earth orbit. Then, in January 1967, the tragic
Apollo 1 fire occurred. The ensuing investigation
slowed the program as NASA delayed crucial tests
and activities.

But NASA planners did not give up on the
larger goal: astronauts on the Moon before the end
of 1969. To meet that schedule, NASA dramatically
altered LM-2's role. NASA had planned to use
LM-2 in a mission to practice a crucial rendezvous
maneuver: a piloted Apollo command and service
module was to dock in Earth orbit with an unpiloted LM-2, launched separately. However, in
January of 1968, the Apollo 5 mission, with LM-1,
successfully demonstrated the flight worthiness of
the lunar module design. Soon after this success,
NASA revamped its plans and cancelled shipment
of LM-2 to Cape Canaveral—an additional practice
mission seemed unnecessary.

In April 1969, during the Apollo 6 mission, the
Saturn V revealed a disturbing tendency to vibrate
violently in an up and down direction (the so-called
pogo effect). NASA engineers immediately incorporated LM-2 into a full-scale test system to carefully
simulate the structural stability of the rocket and
spacecraft during launch. Next, engineers used
LM-2 in a drop testing program, designed to
evaluate whether the modules' electronic and
mechanical systems could withstand jarring that
might occur at lunar touchdown. The fifth and
final drop test of LM-2 occurred on May 7, 1969,
less than three months before Neil Armstrong and
Buzz Aldrin walked on the Moon.

The following year, the ascent stage of LM-2
spent several months on display at the Expo '70 in
Osaka, Japan. When it returned to the United States,
it was reunited with its descent stage and then
transferred to the Museum, where it has been
on public display since 1971.

ABOVE: Simulating a landing on the Moon,
astronaut Neil Armstrong pilots a Lunar
Landing Training Vehicle at Ellington
Air Force Base in June 1969.

Apollo Astronaut Liquid Cooling Garment
Mid-1960s

"How can man survive, much less operate, in such an unfriendly environment?"

—ROBERT SHERROD, 1975

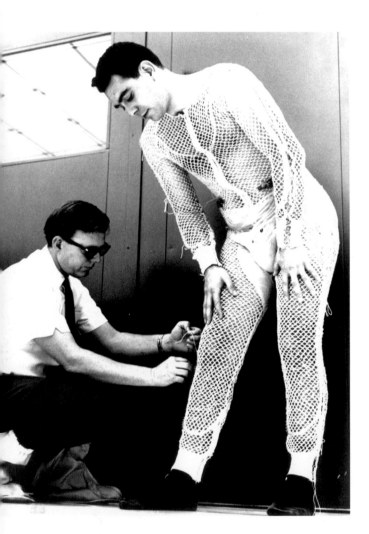

ABOVE: A technician at ILC Industries adjusts the fit of an Apollo liquid cooling garment during a test.

DURING THE early 1960s, engineers devised several techniques for insulating spacesuits from the extreme temperatures expected to be encountered in space—whether very cold or very hot. With such insulated suits, astronauts could "walk in space" or explore the surface of the Moon, but such protection presented a problem: how to dissipate the heat generated by an astronaut's body inside the suit.

During the Mercury program, astronauts maintained body temperature through an air hose attachment from their spacesuit to the spacecraft. They wore separate cotton undergarments looking very much like a pair of long johns, with "spacers" of Trilok (coiled nylon) in selected areas to keep the spacesuit from adhering to the body and blocking air circulation.

Gemini astronauts wore a plain, one-piece long john undergarment made of cotton knit. The spacesuits were ventilated by a series of air channels throughout the suit, connecting to a main inlet that attached to the spacecraft via a hose. The channels were made of stainless steel springs covered with an inner layer of lycra mesh and an outer layer of neoprene.

For the Apollo missions, the astronauts required a new system—for walking on the Moon their suits had to have completely self-contained environmental systems, independent of the spacecraft. Most importantly, the air circulation techniques used for cooling in Mercury and Gemini would not work for Apollo, so spacesuit and NASA engineers looked for an alternative method.

Coincidentally, around this time, the Royal Air Force in the United Kingdom developed a water-conditioned suit to maintain body temperatures in extreme environments—in steel furnaces, deep mines, and tanks or aircraft stationed in the Sun for long periods of time. This suit used circulating liquid to regulate internal heat levels. Unaware of this pioneering work taking place in England, Hamilton Standard adopted the same idea and began research and development into liquid-cooled suits designed specifically for wear inside a spacesuit. The initial concept was to wrap 91 meters (300 feet) of vinyl tubing against the skin of an astronaut and then overlay clothing. By late 1963, the design had evolved into a less bulky, more comfortable garment that threaded cooling tubes of polyvinyl chloride through an open mesh fabric. This design became the Apollo liquid cooling garment (LCG).

The Museum's Apollo LCG was made for but not used on an Apollo mission, and was transferred to the Museum from NASA in 1973.

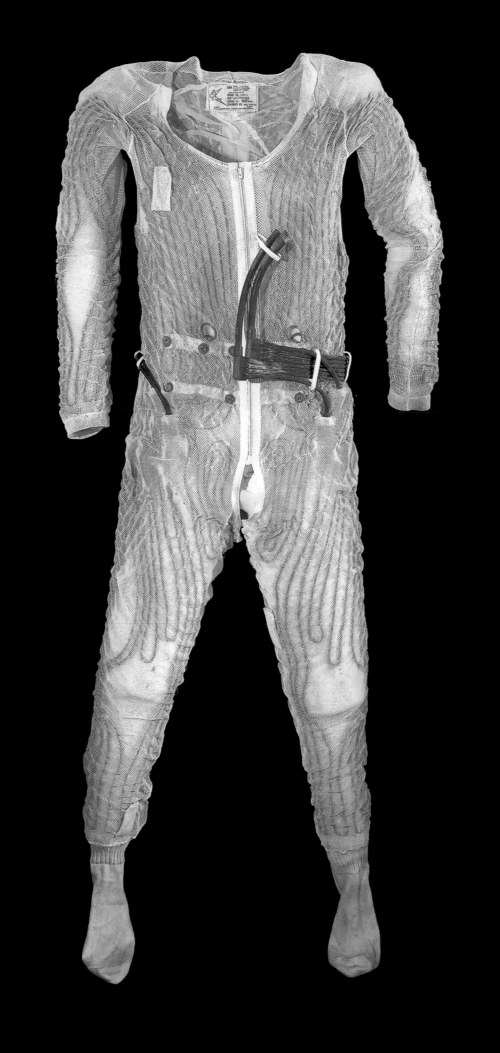

Neil Armstrong's Apollo 11 Spacesuit
1969

"Then he extended his left foot, cautiously, tentatively, as if testing water in a pool—and, in fact, testing a wholly new environment for man. That groping foot, encased in a heavy multi-layered boot (size 9 ½ B), would remain indelible in the minds of millions who watched it on TV."

—*Time*, JULY 25, 1969

ABOVE: Apollo 11 astronaut Neil Armstrong precedes fellow crew members Michael Collins and Buzz Aldrin as they head for the launch pad to begin their historic journey.

TO LAND humans on the Moon and return them safely to Earth required the invention of numerous new technologies—including spacesuit technology. In 1961, when President Kennedy announced the goal of a Moon landing no one had a clear idea of how to make a spacesuit suitable for lunar exploration. These suits had to meet many requirements. They had to protect astronauts from the deadly environment of space, allow ease of movement and flexibility, and operate reliably.

To evaluate possible designs, NASA held two competitions, one in 1962, the other in 1965, inviting companies to submit prototypes. After careful review during the 1965 competition, NASA selected a prototype called the AX5-L, submitted by ILC Industries, makers of Playtex bras and girdles. The AX5-L provided the initial design for the suits worn on the Moon. Their prototype design underwent several changes before ILC and NASA arrived at a Moon-worthy suit—called the Apollo A7-L. In 1969, astronaut Neil Armstrong wore this A7-L suit as he descended from the lunar module (LM) and took the first steps by a human on the Moon.

Changes to the prototype stemmed from two factors. An Apollo spacesuit, compared with a Mercury or Gemini suit, had to serve a much wider range of use—particularly, protecting and supporting astronauts as they explored the lunar surface. As designers and engineers thought through these requirements in more detail, they modified the suit. The tragic Apollo 1 fire in January 1967 also stimulated changes in suit design. NASA and ILC became much more conscious of the need to protect astronauts in the event of emergencies and reevaluated every component and material used in the program.

Two spacesuit models emerged from these efforts. The command module pilot (who stayed in lunar orbit during a Moon landing) wore the intravehicular (IV) suit. The mission commander and the lunar module pilot (who went to the lunar surface in the LM) wore the extravehicular (EV) model. This latter suit had additional layers of insulation and served as a self-contained life-support system—the suits did not employ an umbilical back to the LM as a lifeline.

Several principal components protected and allowed an astronaut to move about the lunar surface independently. The portable life support system (PLSS) contained the equipment necessary for communications, oxygen, and maintenance of body temperature. The complete life support system also included an emergency oxygen supply called the oxygen purge system (OPS) and a remote control unit (RCU) to monitor the PLSS/OPS functions. Combined with the spacesuit itself and other EV equipment, these technologies created a bulky apparatus that on Earth weighed approximately 90 kilograms (200 pounds), but on the Moon weighed one-sixth as much, around 15 kilograms (33 pounds). Fully dressed, an astronaut carried the PLSS on the back of his suit (with the OPS attached to the top) and RCU on the front—each was connected to the suit with buckles and small hoses. NASA tested the final design for the suits in space during the Apollo 7 mission in October 1968.

Neil Armstrong's spacesuit was transferred by NASA to the Museum in 1973. The EV helmet and gloves seen here are those he wore on his historic mission, but before he climbed into the LEM on the return journey he left his overshoes (the boots designed for additional protection, which were worn over the spacesuit boots) on the Moon.

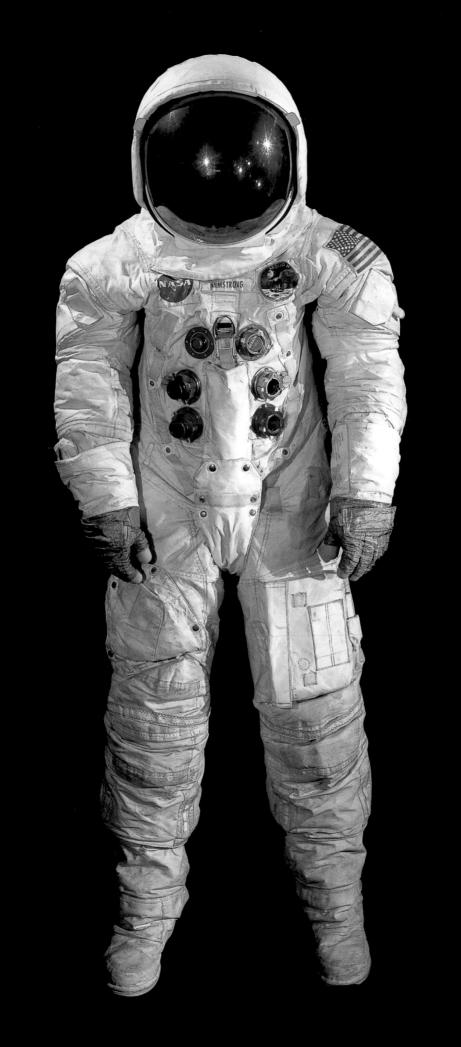

Neil Armstrong's Apollo 11 Chronograph
1969

"The wristwatches Neil Armstrong and Buzz Aldrin wore on the Moon are the everyday variety found in jewelry stores all over the world. The watches, Omega Speedmaster chronographs … remain standard production line models with no special extras, except for a strap to fit them outside the spacesuit. For practical purposes they were set on Houston time."

—*Los Angeles Times*, JULY 23, 1969

SOMETIMES a chronograph (or watch) is not just a chronograph—it also can be a symbol. Astronaut Neil Armstrong carried this Swiss-made Omega Speedmaster to the Moon on the Apollo 11 mission—but not before NASA encountered concerns about using Swiss chronographs on the nation's most epic accomplishment, landing humans on another planet.

The "watch" story began in 1961, shortly after President Kennedy's speech in which he declared that "the Nation should commit itself, before the decade is out, to landing a man on the Moon and returning him safely to the Earth." Looking ahead, a NASA official purchased five watches/chronographs from a jewelry store in Texas for the purpose of evaluating which, if any, might be used by astronauts on the Moon.

Astronauts in the Mercury and Gemini programs eventually began to test the performance of various chronographs in space. The first Mercury astronauts—Alan Shepard, Gus Grissom, and John Glenn—did not wear watches during their Mercury flights; timers on board the spacecraft sufficed for their missions. Scott Carpenter became the first astronaut to wear a watch into space, using a Breitling Navitimer made to his specifications. Wally Schirra tested an Omega Speedmaster on *Sigma 7* in October 1962, and Gordon Cooper wore both a Speedmaster chronograph and a Bulova Accutron watch on *Faith 7* in May 1963. Each used the timepieces as backups to the onboard timers.

Some, though, began to question the propriety of purchasing the Swiss-made chronographs rather than American-made Bulova watches. In the early 1960s, General Omar Bradley, a distinguished World War II leader in the Army and the first chairman of the Joint Chiefs of Staff, served as president of the Bulova Watch Company and promoted the use of the American-made Bulova watches in the space program. In 1964, the U.S. Senate held hearings to determine why NASA had chosen the Swiss Omega over the American-made Bulova. Rigorous performance tests comparing the two brands eventually trumped political concerns.

Program requirements called for a manual-winding wrist chronograph that was waterproof, shockproof, antimagnetic, and able to withstand temperatures ranging from -18° to 93°C (0 to 200°F) and accelerations of 12 g's. In 1965, after a series of rigorous tests, the Omega Speedmaster demonstrated its high level of precision and reliability under extreme conditions. NASA certified it as the chronograph for the Gemini program, and then for the Apollo and Skylab programs.

Armstrong's chronograph holds it own unique story. During the Apollo 11 mission, the onboard timer on the lunar module (LM) malfunctioned. As a backup, Armstrong left his watch in the LM as he and Buzz Aldrin explored the lunar surface. This ensured that the LM had a working mission timer in the event that Aldrin's watch was damaged while working outside the spacecraft. NASA transferred the chronograph to the Museum in 1973.

ABOVE: As part of "suiting up" in preparation for the Apollo 11 flight, astronaut Neil Armstrong wears his chronograph outside his suit, near his wrist.

Apollo 11 Lunar Sample Return Container
1969

"It took just about everything I could do to close the documented sample box. I was afraid I might have left the seal in the box. I don't think I did, because, at the time, I thought I remembered clearly taking the seal off and throwing it away; but that's what it felt like."

—Neil Armstrong during the 1969 technical debrief

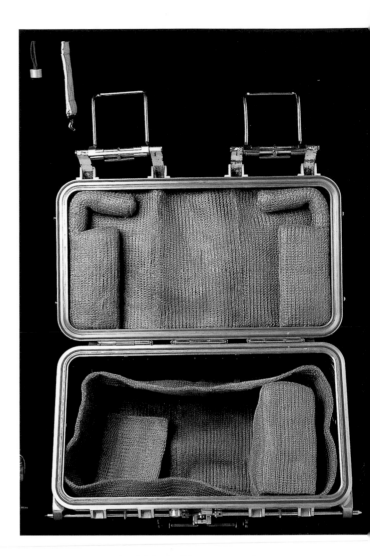

ONE OF the basic goals of the Apollo program was to obtain and study samples of soil and rock from the lunar surface. Among the most important tasks assigned to Neil Armstrong and Buzz Aldrin during their historic mission was the collection of a contingency sample. If history's first Moon-walk had to be cut short for any reason, planners wanted to be sure that the astronauts returned at least some lunar material. Later in the mission, plans called for Armstrong and Aldrin to collect more structured, "documented samples." This involved taking core samples by driving a tube into the soil and identifying and photographing specific types of surface rocks before placing them into individually labeled sample bags.

Whether "contingency" or "documented" samples, scientists at the Lunar Receiving Laboratory in Houston wanted the material brought back from the Moon isolated from Earth-born contaminants. To achieve this, Armstrong and Aldrin were provided with two specially designed aluminum "rock boxes." Manufactured by the Union Carbide Corporation's Nuclear Division in Oak Ridge, Tennessee, engineers designed the boxes to protect samples from physical damage and to prevent contamination during the return flight and recovery process. Each box lid had three separate seals: two fluorosilicone O-rings and a knife-edge seal made of soft indium metal. As the astronauts loaded a box, they had to take extreme care not to mar any of the edges with dust that could prevent an airtight seal. Teflon tape covered the indium edge and was to be removed just prior to closure. As Armstrong's remarks above suggest, in the intensity of the moment, he was unsure whether he followed procedure correctly. In fact, he had.

As the astronauts worked through their scheduled activities on the lunar surface, they took longer than anticipated to gather the contingency sample and perform other tasks. This left little time to collect the planned documented samples. As Aldrin drove the core tube to collect the sub-surface sample, Armstrong collected a diverse but not fully documented collection of rocks and placed them in the box intended for documented samples. As time ran short, Armstrong improvised further and scooped dirt into the box as a packing material to keep the rocks from shifting during transport. In all, Apollo 11 brought back about 6 kilograms (13 pounds) of lunar dirt.

Geologists later determined that this dirt was quite representative of soil from the mare regions (or "seas") of the Moon. Because the first lunar landing mission returned so much dirt, samples were later made available to a number of experimenters, including engineers trying to determine if lunar soil might someday be used to make cement or other building material for a future lunar colony.

This particular box carried Armstrong's rock and dirt samples, a total of 21.8 kilograms (47 pounds) of lunar material. Following the mission, NASA placed the box (with rocks and dirt removed) in bonded storage until transferring it to the Museum in 1970. In 1976, at the opening of the Museum's Mall building, it became one of the original items displayed in the *Apollo to the Moon* gallery.

ABOVE: A technician at the Lunar Receiving Laboratory in Houston, Texas, opens Armstrong's Sample Return Container in a specially-designed vacuum chamber (note the glove at lower right).

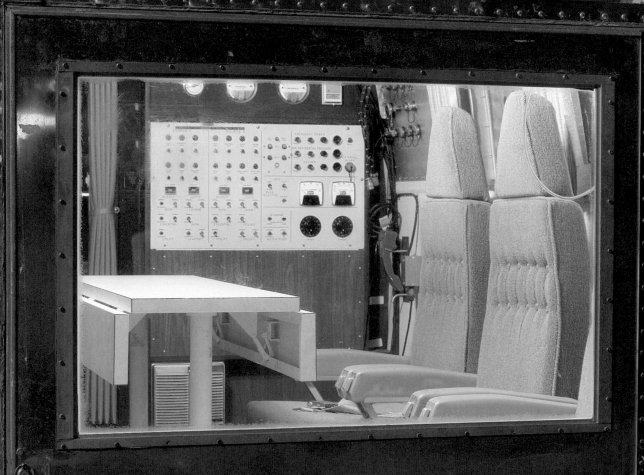

AIRSTREAM

HORNET + 3

AMERICAN STANDARD

Mobile Quarantine Facility
1969

"I did not notice any particular symptoms of claustrophobia and I am sure that the crew of Apollo 11 considered their temporary stay in the MQF as roomy when compared to the Apollo Command Module."

—JOHN HIRASAKI, 2005

S EARLY as 1960, biologists on U.S. National Academy of Science's Space Science Board (SSB) recommended the government establish policies to prevent contamination of Earth by pathogens brought back from space in lunar rocks or soil, or by returning spacecraft or astronauts. The risk was hypothetical but potentially serious. In July 1964, the board called together NASA and other government agencies to assess the so-called back-contamination problem. Concluding that "the existence of life on the Moon or planets cannot … rationally be precluded," the board recommended strongly that returning spacecraft, recovered lunar samples, and astronauts be strictly quarantined in a suitable isolation chamber for at least three weeks after their return from the Moon.

NASA and its large team of consulting scientists already had begun to design a state-of-the-art "lunar receiving laboratory" (LRL) to house and study returned rock and soil samples. Abiding by SSB guidance, NASA decided to upgrade the LRL to meet strict quarantine requirements. But this was not a complete solution. How were the astronauts and samples to be transported from splashdown in the Pacific Ocean to the LRL (at the Manned Spaceflight Center in Houston) without exposing their rescuers to possible contamination? The answer was a mobile quarantine facility (MQF), a specially modified Airstream trailer capable of supporting and isolating up to six people for several days.

After their return to Earth, the crew of Apollo 11 stayed in this MQF (serial number 3). At splashdown, the astronauts donned special biological garments before rescuers transported them and the command module to the USS *Hornet*,

where the MQF awaited. Astronauts Neil Armstrong, Buzz Aldrin, and Michael Collins entered the trailer and remained there for 65 hours. But they were not alone. Physician Dr. William Carpenter and a "recovery engineer" named John Hirasaki already were inside to monitor and assist the astronauts. Both had entered the MQF several days before. This precaution was to ensure that they did not catch and then transmit a terrestrial virus to the returning astronauts. The USS *Hornet* transported the MQF to Hawaii, then it traveled by plane to Houston, finally releasing its passengers to the more spacious confines of the upgraded lunar receiving laboratory. All told, the astronauts and their helpers remained in the MQF for 65 hours and in quarantine for 21 days.

Hirasaki performed a unique job at the start of this process. As soon as the command module landed on board the *Hornet*, personnel attached it to the MQF via a secure isolation tunnel. Hirasaki then went through the tunnel to transfer the precious cargo of lunar samples from the command module to the MQF. After this, he looked after the MQF and astronauts, which included cooking meals in a new appliance called a microwave oven.

In 2005, after the Museum put the MQF on display at its Udvar-Hazy Center, John Hirasaki provided another important service. At the request of curators, he paid his old home another visit to explain some of the MQF's features. He remembered a table cleverly stored in a special compartment under one of the sleeping bunks. In 1969, the table performed doubled duty—for dining and for use by the physician to conduct examinations. Sure enough, there it was. Now reassembled, the table can be seen on display inside the MQF.

ABOVE: President Richard Nixon, on the USS *Hornet*, greets the astronauts soon after their return from their historic Moon mission.

TOP: Exterior view of the Museum's mobile quarantine facility, used by the crew of Apollo 11 on their return to Earth.

Apollo Command Module Handbag
1969

"Okay, I can see the shape of your vehicle now, Mike."

— Edwin "Buzz" Aldrin to Michael Collins on seeing the command/service module as he
and Neil Armstrong returned to lunar orbit after the first Moon-walks, July 21, 1969

THIS UNIQUE handbag pays homage to the space vehicle that Apollo 11 command module pilot Michael Collins called "my happy home for eight days." In July 1969, while his compatriots, Neil Armstrong and Buzz Aldrin, walked on the lunar surface below, Collins orbited the Moon in the combined command module (CM) and service module (SM), together called the command/service module or CSM. After he returned to Earth, he received this purse shaped like a CSM. The unique memento celebrated the distinctive profile of the vehicle that brought America's first Moon-walkers safely back to Earth.

Inspired at least in part by the mod clothing movement that was in full swing by 1969, this streamlined fashion accessory reproduced the CSM's color and shape. The actual Apollo CSMs featured a thermal coating of silver Mylar that helped to protect them from the heat of the Sun. Photographs of the CSM taken from the lunar module (LM) show sunlight glinting off the reflective surface. On the purse, silver-colored leather with a "Morocco" grain imitated that appearance. Made in Canada by designer "Mr. Henry," the foot-long ladies handbag features gleaming silver handles and a silver clasp that completed the look.

It is perhaps not too surprising that Collins's "happy home" inspired such creativity. More than any of the previous American space vehicles, the CSM looked like the popular image of a spaceship: a classic, pointed, single-engine rocket ship. Unlike the sleek, streamlined, and aerodynamic vehicles foretold in science fiction, actual spaceships turned out to be somewhat awkward-looking. The first American astronauts flew into space in blunt-ended capsules designed by Max Faget to maximize stability during reentry. The LM, the first ship to land on another world, looked like a squat box of angular planes perched atop spindly legs. (In fact, Apollo 9's LM earned the affectionate call sign "Spider" because of its buglike appearance.) When joined to the tubular SM and its single-rocket engine, however, the conical CM became the pointed end of a classic rocket ship profile.

Although the trips made by the Apollo Program's CSMs remain the longest distances ever traveled by human beings, the CSM handbag's journey into the Smithsonian's collection was ultimately a very short one. In 1972, retired astronaut Michael Collins became the first director of the Museum and offered the bag for the National Collection. A staff member simply picked it up from his office and accessioned it.

Apollo 11 Pin
1969

"This is the greatest week in the history of the world since Creation."
—President Richard M. Nixon upon greeting the Apollo 11 crew
when they returned from the Moon, July 24, 1969

IN THE summer of 1969 the United States was in ferment. Race relations and civil rights dominated domestic affairs. The murders committed by Charles Manson and his followers seemed to signal a disturbing unraveling of the social fabric. The war in Vietnam consumed public energy and provoked deep political divisions. The Woodstock rock concert highlighted the gap in values between a youth-oriented counterculture and mainstream America. Overall American self-confidence appeared in decline, as did trust in the nation's leadership.

The overwhelming success story of the year was the Apollo 11 landing on the Moon, accomplished on July 20, 1969. Christopher Flournoy recalled that as a five-year-old he may not have understood much of what took place but nonetheless was excited. He remembered his father saying that "he was never more proud of being an American than on the day our flag flew on the Moon." One seven-year-old boy from Puerto Rico said of the first Moon landing, "I kept racing between the TV and the balcony and looking at the Moon to see if I could see them on the Moon." These experiences were typical. The flight of Apollo 11 met with an ecstatic reaction around the globe, as everyone shared in the success of astronauts Neil Armstrong, Buzz Aldrin, and Mike Collins.

President Nixon responded to this outpouring of goodwill and emotion by sending the Apollo 11 crew on a round-the-world tour to promote the accomplishment and American values. Ticker-tape parades, speaking engagements, and public relations events served to create goodwill both in the U.S. and abroad. Flying on Air Force One, the tour started in September 1969 with the crew and their wives traveling under State Department auspices to Mexico City. The crew took part in a parade and met the Mexican president and other leaders. Geneva Barnes, a NASA public affairs officer who accompanied the astronauts on this tour, recalled, "We were sort of overwhelmed and I don't really think we knew what was in store for us. There were large crowds everywhere we stopped, but Mexico City was the first where we were exposed firsthand to such large masses of people."

The tour included other parts of Latin America, the Canary Islands, Europe, Africa, Southwest and Southeast Asia, Australia, Japan, and the islands of the Pacific before returning to the U.S. on November 5, 1969. During a public ceremony at the White House, President Nixon officially welcomed them back.

The same enthusiasm expressed in the world tour generated a tidal wave of popular culture memorabilia commemorating Apollo 11. Various organizations created buttons, postcards, lithographs, coins, medallions, mission patches, ribbons, glasses, lapel pins, jewelry, and even a commemorative champagne bottle. This button is one of thousands made to celebrate astronauts Armstrong, Aldrin, and Collins, as well as the American success in reaching the Moon. It hints at Apollo's importance as a symbolic battleground in the Cold War by noting that the U.S., not the Soviet Union, was "First To Moon." Geneva Barnes may have characterized the feelings of many in the aftermath of Apollo 11: "You could not help but feel that there was something big happening and you were just glad to be a part of it." This button served as a tangible expression of being "a part of it."

Apollo 12 Maurer Data Acquisition Camera
1969

"But essentially the camera makes everyone a tourist in other people's reality, and eventually in one's own."

—SUSAN SONTAG, *On Photography*, 1974

ABOVE: Astronaut Al Bean climbs down the lunar module to the surface. The color television camera, accidentally damaged by Bean, is attached to the module below and to the right of the flag.

THE INVENTION of the camera by Louis Daguerre in 1837 inaugurated the era of the photographic image, allowing humans to visually relate to people and events they might never personally know. This sense of "being there" expanded with the advent of motion picture and television imaging. In the early 1960s, these visual technologies became integral to the NASA spaceflight program, facilitating mission documentation and scientific research, but especially allowing those back on Earth to participate in one of history's great dramas—the exploration of space by humans. This Apollo 12 data acquisition camera, transferred from NASA in 1972, highlights the role of television and film in connecting Earth-bound viewers with lunar explorers.

As the U.S. space program moved toward the Moon, NASA formulated plans to document the missions visually and transmit live images back to Earth. Still, film, and television cameras allowed the public to see the astronauts at work—and, in the case of television, to see what the astronauts *saw*, to be part of their reality. Via television from the Moon, nearly 500 million people around the world witnessed Neil Armstrong's first steps on to the lunar surface as it happened, and then continued to see much of Armstrong and Buzz Aldrin's exploration of the Sea of Tranquility.

On Apollo 12, the second lunar mission, NASA's elaborate effort to include the public in the drama of exploration met with some misfortune. The mission almost never moved past liftoff: lightning struck the giant Saturn V launch vehicle within moments of leaving the ground. Astronauts Charles "Pete" Conrad, Alan Bean, and Richard Gordon resolved the resulting electrical outages quickly and headed for the Moon. In piloting the lunar module (LM), Conrad and Bean accomplished an incredible feat: they landed within yards of the Surveyor III lunar probe (sent to the Moon in 1967), then recovered and returned to Earth one of its cameras. However, this recovery and nearly all of their other activity on the Moon never reached television viewers. Shortly after recording Conrad's first steps on the Ocean of Storms, Bean accidentally damaged the television camera, mounted on the LM. As he repositioned the camera to record other mission activities, the lens pointed directly at the Sun (or perhaps at the highly reflective surface of the LM), burning out the optics.

After this mishap, this data acquisition camera (DAC)—returned to Earth by the astronauts—provided the only means to document lunar surface activities with moving images. Typically, the DAC only documented technical mission tasks such as landing and lunar ascent. This documentation proved valuable, but as a motion picture film camera, the DAC could not provide live pictures to Earth. The public did not have the opportunity to "see" the Apollo 12 mission as it happened—only small parts of it after the three astronauts returned the film. This delay in visually experiencing the second lunar landing stood in sharp contrast to the thrilling "being there" experience of Apollo 11. As a consequence, few now remember any images from Apollo 12. The camera mishaps on the mission perhaps signaled a trend: after Apollo 11, public interest in television coverage of the lunar missions fell sharply, a decline only briefly reversed during the compelling story of Apollo 13's near disaster in space.

Television cameras on the lunar missions made possible extraordinary public participation in one of history's greatest exploration ventures—but they could not guarantee public interest.

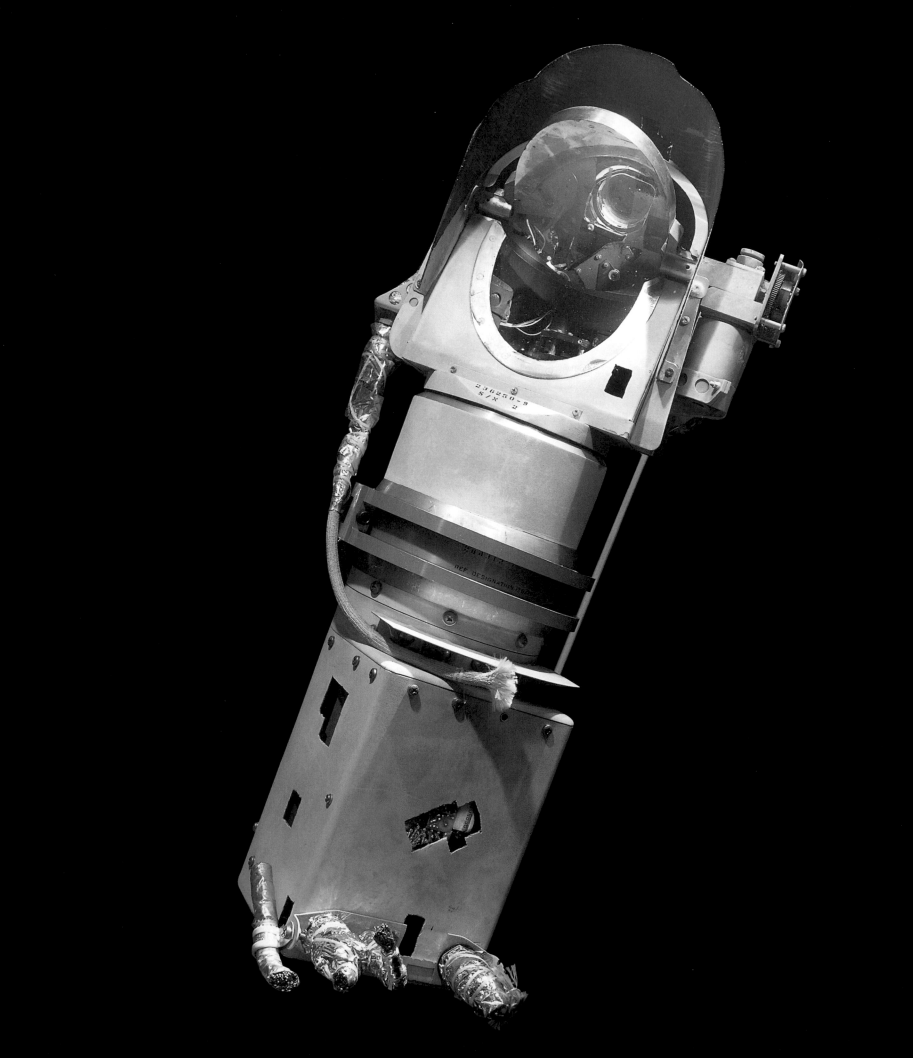

Surveyor III Television Camera
1967

"I always thought the most significant thing that we ever found on the whole goddamn Moon was that little bacteria who came back and lived and nobody ever said shit about it."

—PETE CONRAD, APOLLO 12 MISSION COMMANDER, 1991

ONE OF the few parts of a probe ever returned from space, the Surveyor III camera journeyed to the Moon on April 20, 1967, and sat exposed on the lunar surface for 31 months before the Apollo 12 astronauts retrieved it. The Surveyors conducted a series of soft touchdowns on the Moon in 1966 and 1967 to provide data about its surface and possible atmosphere in preparation for human landings. A total of five Surveyors landed on the Moon successfully, returning more than 88,000 high-resolution photographs and invaluable detailed data on the nature of the lunar surface. Bringing parts of Surveyor III back from the Moon presented a rare opportunity to analyze the long-term effects of spaceflight.

NASA planned to land Apollo 12, the second U.S. lunar mission, launched in November 1969, very close to Surveyor III. The landing required extraordinary precision, and astronaut Pete Conrad succeeded, piloting the lunar module to within approximately 160 meters (525 feet) of Surveyor III. He and fellow astronaut Alan Bean walked over and retrieved several pieces from the spacecraft, including its television camera and some associated electrical cables, the sample scoop, and two pieces of generic aluminum tubing.

In examining the camera, scientists saw evidence of micrometeoroid bombardment. But they also found the terrestrial bacteria *Streptococcus mitis*—apparently surviving for two and a half years in the vacuum of space. As only one of 33 samples from various parts of the spacecraft harbored the bacteria, the question arose as to whether it predated Apollo 12 or resulted from accidental contamination following return from the Moon.

The possibility of life surviving in space gained greater credence in the 1990s, as scientists discovered robust forms of microbial life in "extreme" places on Earth. They found microbes adapted to life in the superheated water of deep-sea vents, pools of acid, and even deep within the crust of the Earth. As scientist Leslie Orgel commented in 2000, "You could take *E. coli* and rapidly cool it to 10° K and leave it for 10 billion years and then put it back in glucose, and I suspect you would have 99 percent survival." With this understanding, finding such simple life on other bodies in our solar system now seemed at least a possibility. The idea that microbial life on Earth hitchhiked to the Moon on Surveyor III and survived dormant until returned by the Apollo 12 crew remains an intriguing possibility— one with far-reaching ramifications as we continue to explore the solar system.

There were skeptics who dismissed the presence of the Surveyor bacteria as laboratory error. A technician had inadvertently placed one of the tools used to scrape samples off the camera on a nonsterile bench and then reused it without resterilization. Dr. Leonard Jaffe, project scientist for Surveyor, wrote at the time, "It is, therefore, quite possible that the microorganisms were transferred to the camera after its return to Earth, and that they had never been to the Moon."

There still is no consensus as scientists continue to differ over how to assess the Surveyor III camera biological material. As one Cornell University scientist commented about "extremeophilic" life: "It's compelling evidence for astrobiologists that the environmental limits for living things are set pretty far apart."

NASA provided the Surveyor camera on indefinite loan to the Museum in the early 1980s. NASA regards the camera as the equivalent of a Moon rock, and U.S. regulations prevent private ownership of these historic materials.

ABOVE: Apollo 12 astronaut Charles "Pete" Conrad stands near the Surveyor III spacecraft and its camera (positioned adjacent to Conrad's right arm).

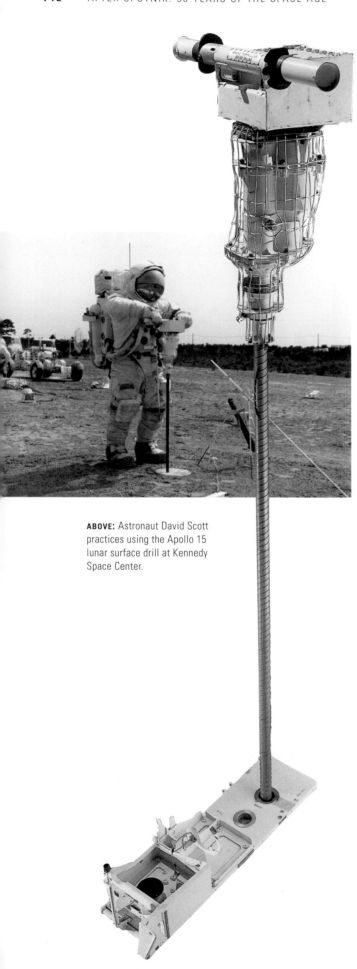

ABOVE: Astronaut David Scott practices using the Apollo 15 lunar surface drill at Kennedy Space Center.

Apollo Lunar Surface Drill
1960s

"Hey, Joe, you never did tell me that [core] was that important. Just tell me … and then I'll feel a lot better."

"It's that important, Dave.… the deepest sample out of the Moon for perhaps as long as the Moon itself has been there."

—ASTRONAUT DAVE SCOTT AND MISSION CONTROL CAPCOM JOSEPH ALLEN, APOLLO 15, 1971

A POLLO 15, 16, and 17 each carried a lunar roving vehicle (LRV). With these lightweight electric vehicles, astronauts explored a dramatically increased range of the lunar surface. The LRV also served as a simple truck, permitting the astronauts to carry and deploy new equipment. One important addition was the Apollo lunar surface drill (ALSD), which combined a cordless, battery-operated motor with specialized drill bits and modular core stems. Each core stem segment was a rigid but hollow tube measuring about 40 centimeters (16 inches) in length. Joined together and driven into the surface, they enabled astronauts to tap holes as deep as 3 meters (10 feet) into the lunar soil. This capability allowed explorers to extract core samples up to 240 centimeters (7 feet 10 inches) long and place special temperature sensors well below the surface.

The first astronauts to employ the drill were Dave Scott and Jim Irwin on Apollo 15. Before their flight, the astronauts trained extensively with the drill. Their mission plan called for them to drill two holes for the "heat flow experiment" sensors and to collect a single deep core sample. Each of the procedures required a series of steps, any one of which might pose difficulties under conditions on the Moon. The procedure for the core sample, for instance, called for astronauts to drive six separate 40-centimeter (16-inch) core stems attached one on top of the other down a total of more than 2 meters (7 feet) below the surface, extract the stems (with lunar material inside), separate the core stems, number and cap each one,

and then place the sample-containing stems in a special container for eventual return to Earth.

Once on the Moon, the first use of the ALSD proved to be a learning experience. The astronauts found drilling the two holes for the heat flow experiment far more difficult and time consuming than in the practice sessions on Earth. The lunar soil was far more compact than had been expected. As the drill bit moved deeper and deeper, the core stems did not effectively pass material upward to the surface. Eventually the drill bound, making it seem to Dave Scott, its operator, as if the bit had hit bedrock.

Similar difficulties plagued the collection of the deep core sample. Scott attached, drilled, and inserted all six core stems (just as practiced on Earth), but the material surrounding the stems became compressed around the joints. Only with Scott and Irwin working hard together could the core sample be extricated. Besides loss of precious time, the removal caused Scott an injured shoulder and a great deal of discomfort during the remainder of the mission.

Following Apollo 15, NASA engineers redesigned the core stems and other drill components. The crews on Apollo 16 and 17 had a much better experience. The core samples returned from all three missions proved to be among the most scientifically valuable of all the samples returned. Studies provided details that were critical in understanding the processes that produce soil layers on the Moon. Scott's Apollo 15 sample, for example, had penetrated over 50 distinct layers, a unique record of soil accretion at one particular site.

The Museum's ALSD was used in training and transferred from NASA in 1975.

Apollo Fuel Cell
1960s

"A flight-time capability of the spacecraft for 14 days without resupply should be possible. Considerable study of storage batteries, fuel cells, auxiliary power units, and solar batteries would be necessary."

—SPACE TASK GROUP, *Guidelines for Advanced Manned Space Vehicle Program*, JUNE 1960

DESIGNING an electrical power system for Apollo posed major challenges. Once the spacecraft and astronauts reached space and headed for the Moon, the power system was critical for mission success. It had to work, and do so under severe conditions. A round-trip mission to the Moon took at least two weeks—perhaps more if a trip encountered difficulties—requiring a system that generated electricity for the duration and in sufficient amounts. If the primary system failed, a backup (and for some components, a backup to a backup) had to kick in. But NASA did not want the system to weigh too much—in space travel, each pound mattered. To meet these criteria, NASA planners very early on focused on an experimental technology—hydrogen-oxygen fuel cells—as the answer.

A fuel cell is like a battery. It converts energy released in a chemical reaction directly to electrical power. A fuel cell continues to supply current as long as chemical reactants are available and produces several times more energy than a conventional battery per equivalent unit of weight. When oxygen and hydrogen combine, energy is released and more than half of that energy can be extracted as electricity. The only byproduct is pure water. This fuel cell feature provided an important, added benefit: the water could be used for cooling and by the astronauts for drinking and food preparation.

A year after NASA began design studies for a Moon mission, the agency awarded a contract to Pratt & Whitney Aircraft Division of United

Aircraft Corporation for the development of a fuel cell for the Apollo spacecraft. Initially, NASA asked the company to deliver a 250-watt pilot version in 12 months. By late 1962, NASA arrived at a final design. Each mission would carry three fuel cells, with each fuel cell capable of producing from 563 to 1,420 watts during normal operation. Constructed of titanium, stainless steel, and nickel, the cells sat in one of the six equipment bays of the Apollo service module, positioned next to cryogenic tanks filled with liquid oxygen and hydrogen. Two of the fuel cells carried normal electrical loads and one was available for emergency power. During a typical two-week mission, the cells produced about 220 liters (60 gallons) of potable water, much more than astronauts could possibly use.

At the end of an actual Apollo mission, astronauts jettisoned the service module (fuel cells included) prior to the command module's fiery plunge through the atmosphere. No flown fuel cells were ever recovered. Fortunately for the Museum, however, NASA did not launch all of the fully equipped service modules. In 1968, engineers conducted a "static fire" test of SM 102 at the Kennedy Space Center in Florida. They repeatedly fired the service propulsion system (SPS) engine to determine if rocket exhaust gases might damage components inside the SM. Afterward, in 1972, NASA transferred the "used" fuel cells to the Museum.

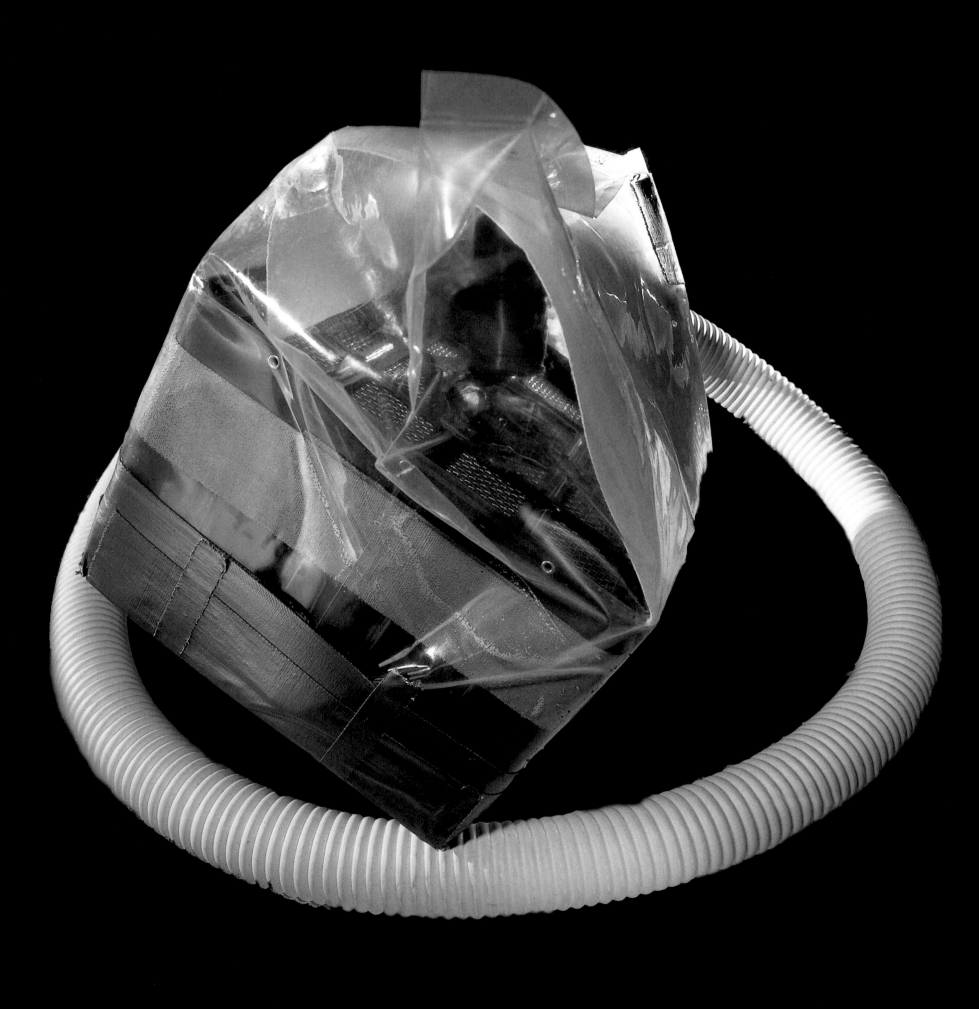

Lithium Hydroxide Canister Mock-up
1975

"Yeah, we wish we could send you a kit; it would be kinda like putting a model airplane together or something; as it turns out this contraption is going to look like a mailbox when you get it all put together."

—CAPCOM TO APOLLO 13 ASTRONAUTS DURING THE MISSION, 1970

HE APOLLO command and lunar modules had one fundamental purpose: to provide a living environment for three astronauts on the journey to and from the Moon. Inside, oxygen, water, and food sustained human life. Outside, the airless cold and radiation of space posed life-threatening dangers. A breach in an exterior wall or the failure of a life support system was potentially catastrophic.

Early on, contingency plans were made for the lunar module (LM) to serve as a "lifeboat" in case the command module (CM) systems failed during a flight and could not support the astronauts long enough for them to return to Earth. That contingency became a frightening reality as Apollo 13 headed toward the Moon. A cryogenic oxygen tank in the service module (SM) exploded, damaging the fuel cells that provided most of the CM's power. The batteries on board were intended and sufficient for use only during the relatively brief reentry period (after the SM and all its support systems were jettisoned). But reentry was four long days away. The CM systems had to be shut down until then. The astronauts—James Lovell, Jack Swigert, and Fred Haise—needed to rely on the "lifeboat" systems while they completed their trip to Moon, looped around, and headed back to Earth. To survive that long, systems in the LM intended to support two astronauts for less than three days had to support three astronauts for four days.

Despite all the contingency planning, NASA did not anticipate Apollo 13's unique chain of events, forcing engineers on Earth to improvise.

One problem that revealed itself as the astronauts settled in the LM was the gradual buildup of carbon dioxide gas. The environmental system in the LM depended on two lithium hydroxide (LiOH) filters to remove excess carbon dioxide (exhaled by humans as part of normal respiration) from the atmosphere. But with three astronauts to support instead of two, the filters did not have enough capacity to maintain safe carbon dioxide levels for the entire journey home. The CM environmental system had many unused LiOH filters that might be transferred to the LM. But there was a catch. The filters used in the CM were box-shaped and fit into box-shaped holders; those in the lunar module were cylindrical and fit into cylindrical holders. The engineers literally had to figure a way for a square peg to fit in a round hole if the astronauts were to survive.

While the ground controllers carefully monitored the CO_2 levels aboard the LM, engineers worked with astronauts at Mission Control to devise a method of modifying the square canisters, using only materials available to the crew in the spacecraft. The Mission Control team worked out a solution that cobbled together plastic bags, plastic-coated cue cards from a three-ring reference binder, hoses from the lunar spacesuits, and lots of gray duct tape, then carefully radioed instructions to the astronauts in space. After about an hour, the new device, although not very elegant, worked perfectly.

In March 1975, the Museum requested that the personnel who had been involved in the emergency effort recreate one of the jerry-rigged LiOH canisters for the collection—this prized artifact.

ABOVE: Mission Control engineers examine a prototype of the makeshift device designed to remove carbon dioxide from the atmosphere inside the Apollo 13 lunar module.

TOP: As the Apollo 13 astronauts neared Earth on their return home, they separated the command module from the service module (SM) and took this photograph of the SM, which had been severely damaged by an exploding cryogenic fuel tank.

Lunar Roving Vehicle, Qualification Test Model
1969–1971

"As there are no superhighways on the moon (yet), all vehicles must have cross-country capability. Just as on Earth, the terrain on the moon is partially smooth and flat, while other parts are rugged and mountainous."

—WERNHER VON BRAUN, *Popular Science*, 1964

ABOVE: On Apollo 17, astronaut Harrison Schmitt works near the lunar rover at the Taurus-Littrow landing site.

OPPOSITE: Detail of the control panel and steering mechanism of the Museum's lunar roving vehicle, qualification test model.

THE PRIMARY goal of the earliest Apollo Moon missions was to meet President Kennedy's challenge to land humans on the Moon and return them safely before the end of the 1960s. Science, exploration, and collection of lunar samples were important, but NASA placed emphasis on making the engineering systems perform and on the safe return of the astronauts. With the Apollo 10 lunar orbit mission and the historic Apollo 11 landing accomplished in 1969, NASA devoted increasing attention and resources to enhancing the productivity of future missions—and that meant changing the design of the lunar module. One change introduced a completely new feature to exploration: the lunar roving vehicle.

NASA designed the first batch of lunar modules to provide a very rudimentary base camp—one that supported astronauts for up to 30 hours on the surface and allowed them to explore surrounding areas by foot. But in May 1969 (just as the Apollo 10 astronauts were leaving lunar orbit) NASA approved a proposal to upgrade the lunar module and include a lightweight lunar roving vehicle (LRV), attached to the descent stage. NASA scheduled the upgraded LM for use on Apollo 15, 16, and 17. The agency released the final design specifications for the lunar rover only days before the July 16 launch of Apollo 11. In October, Boeing won the LRV contract, which called for development, testing, and delivery of flight-ready units by April 1971 (a mere 17 months in the future).

The LRV changed the character of lunar exploration. The Boeing Company developed the vehicle, with General Motors responsible for the "engine"—an electric drive system. The LRV was lightweight. Empty, it weighed only 209 kilograms (450 pounds on Earth, 75 pounds on the Moon). It folded up like a modern transformer toy to fit inside one of the storage units on the side of the lunar module. It featured a navigation system, a receptacle for tools and equipment, and a color TV camera linked to a high-gain antenna that beamed high-quality live pictures directly to Mission Control. The antenna provided a two-way connection between astronauts and Houston— allowing Mission Control to point the TV. With its ability to move at 16 kilometers per hour (10 miles per hour) or faster, the rovers greatly expanded the areas that a single crew could explore. The Apollo 15 crew, for example, logged more than three hours of total driving time, covering a distance of approximately 25 kilometers (15 miles), which was more than three times the total combined distance traversed by the astronauts on the three previous successful landing missions (Apollo 11, 12, and 14).

In addition to the flight vehicles (for Apollo 15, 16, and 17), Boeing manufactured eight nonflight units for development and testing. One, the "Qualification Test Unit," was a very close replica of the units that flew. Using special test chambers, engineers purposely subjected the qualification unit to conditions many times as severe as those expected on an actual mission. When the tests were finished, given the stresses it had been subject to, the qualification unit could not safely be used in space. In 1975, NASA transferred it to the Museum.

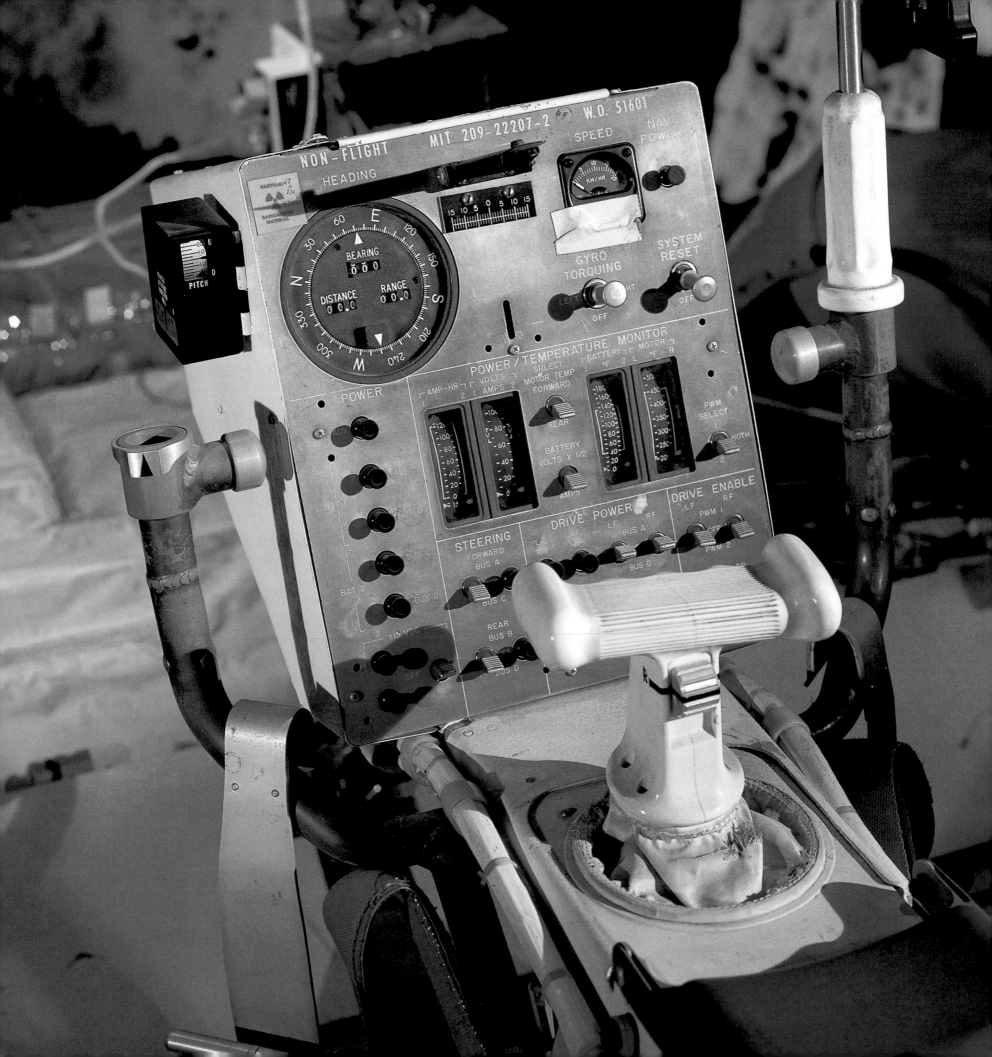

Apollo 16 Display and Keyboard (DSKY)
1972

"In 1962, computers were not considered user friendly. Heated debates arose over the nature of the computer displays. One faction, which usually included the astronauts, argued that meters and dials were necessary.... digital displays won most of the arguments because of their greater flexibility in the limited area allowed for a control panel."

—ELDON HALL, DSKY DESIGNER, 1982

W HEN President Kennedy challenged the nation to land a man on the Moon and return him safely by the end of the decade, engineers knew that the newly developing technology of digital computers would be crucial to such a mission—but they knew little beyond that. In those days, computers filled large rooms and required constant air conditioning and power, and a dedicated staff. Users fed problems into them using decks of punched cards and retrieved results up to a day later on reams of printed paper. Adapting such a device for onboard use was going to be a challenge.

By the mid-1960s, engineers had made great strides in miniaturizing circuits and in increasing their reliability. The notion of installing a general-purpose digital computer on board became feasible. But a key technical issue remained: how to design it for use by astronauts during a flight. MIT engineer Eldon Hall designed the Apollo guidance computer and recalls many conversations with Apollo astronauts. Many of the astronauts preferred round "boiler gauges," dials, and levers, like those found in the cockpits of high-performance aircraft they already knew. But Hall and his MIT colleagues argued for a general-purpose keyboard and numeric display, which allowed much more flexibility in presenting information.

The result of these discussions was the Apollo display and keyboard, known by the acronym "DSKY" and pronounced "diskey." The DSKY shown here, transferred from NASA in 1977, is one of two removed from the Apollo 16 command module, which was flown to the Moon and back in April 1972 by John Young, Thomas Mattingly, and Charles Duke. Each command module carried two identical DSKYs, each connected to the spacecraft's guidance computer; the lunar module also had a single DSKY connected to its computer. Through them the astronauts communicated with these computers to perform critical navigation, rendezvous, docking, landing, ascent, and flight management tasks throughout a mission. Measuring about 20 centimeters (8 inches) on each side, the DSKY allowed an astronaut wearing a bulky spacesuit to use the keypad.

Most of the markings on this DSKY look familiar, but it includes one set of keys never used again in the space program. That was the "verb-noun" combination to give commands. An astronaut wishing, for example, to "load the star coordinates" into the computer keyed the numeric codes for the load operation, followed by "Verb," then the code for star coordinates, followed by "Noun." In the early stages of design, MIT engineers considered this technique a temporary expedient, but the astronauts liked it and urged NASA to keep it. For noncomputer specialists, this type of operation seemed simple, intuitive, and easy to understand. This ease of use proved important; as one historian of Apollo noted, "It took about 10,500 keystrokes to complete a lunar mission; not much in the life of an airline reservations clerk but still indicative of how computer centered the crew had to be."

In the beginning, many at NASA questioned the need for the Apollo guidance computer and its DSKY. But the device ended up serving as a "fourth crew member," active at almost every phase of a flight. It highlighted the symbiosis between a human being and a digital computer, an interdependence that proved indispensable in the historic journeys to the Moon.

Apollo 16 Commander's Cuff Checklist
1972

"Neil Armstrong's first thoughts might have been, 'This is one step for a man but a giant leap for mankind,'
but I remember vividly that after climbing down the ladder and stepping on the lunar surface, mine were
'We're 20 minutes behind now and we've got to catch up.'"

—JOHN YOUNG (QUOTED BY ASTRONAUT ALAN BEAN), 1989

ASTRONAUT John Young knew that Apollo 16 was the next to the last of the missions to the Moon. As commander, he bore primary responsibility—and felt a special urgency—to accomplish the mission's many scientific and exploration goals. Once on the Moon, Young and astronaut Charles Duke faced an ambitious schedule of work. Their landing site, the Moon's Descartes highlands, was the highest of all the Apollo missions—5,475 meters (18,000 feet) above lunar "sea" level. Geologists hoped that the astronauts might find rock produced in the interior of the Moon and released through volcanism rather than samples like those collected on earlier missions that had been spread across the surface as a result of bombardment by meteors and comets. Volcanic lunar samples would reveal important, missing information about the Moon's geologic history, and potentially even contain water.

As on all of the lunar missions, the extra-vehicular activity (EVA, or "Moon-walks") for the astronauts was meticulously planned in advance. In training on Earth, the astronauts did full dress rehearsals—wearing their practice suits, they performed each step of each activity. Engineers timed each part and prepared a step-by-step agenda for each of the scheduled lunar excursions. These step-by-step procedures were then printed on small cards and bound together as a spiral notebook, attached to a Velcro strap.

During the second and third of Apollo 16's three EVAs, Young wore this "cuff checklist" near his left wrist. It details all of the maintenance, communications, and mundane tasks he was to perform, as well as the search for precious samples.

It demonstrates well the detailed planning and organization involved in the Apollo program. Of course, things didn't always go as planned. In the case of Apollo 16, the landing was delayed for several hours due to an apparent malfunction aboard the command module orbiting the Moon. When Mission Control finally authorized the lunar module landing, the total time available for the EVA was reduced, contributing to Commander Young's apprehension.

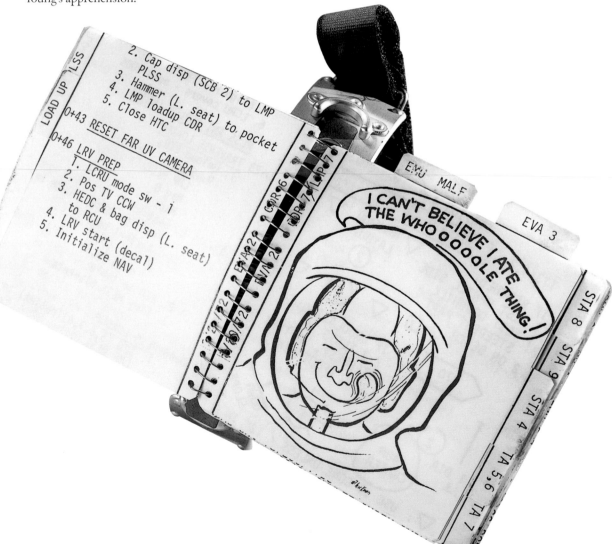

Apollo 16 Lunar Camera
1972

"Every time an Announcement of Opportunity came out, we would propose, knowing that our chances were small, but every once in a while you hit the jackpot. So that was the one time that we hit the jackpot."

—ASTRONOMER GEORGE CARRUTHERS, 1992

THIS IS how astronomer George Carruthers recalls his bid in 1969 to put a scientific experiment on board Apollo 16, an experiment that became the first astronomical observatory on another planet. Working on sounding rockets and satellites at the Naval Research Laboratory, Carruthers had developed special techniques to amplify electronically and then record photographically faint light phenomena such as ultraviolet spectra. He knew such techniques "would be applicable to a wide range of scientific objectives" in the space program as well as in the Navy. So why not try to get a ride on Apollo to the Moon?

The program had a "crash" timeline: "We submitted the proposals in 1969, the project was approved in 1970, and it flew in the spring of '72," said Carruthers. NASA teamed Carruthers with Texas astronomer Thornton Page, who had made a similar proposal. Page developed the scientific mission while Carruthers designed, built, and tested the devices. The centerpiece was an electronic telescope with an 8-centimeter (3-inch) aperture and maximum efficiency in the far ultraviolet portion of the spectrum. The Earth's upper atmosphere glows in this spectral region, as does diffuse gaseous material in the depths of space. And if the Moon possessed even a whiff of a residual atmosphere, this little device, measuring 73 centimeters (29 inches) tall, might detect it.

Carruthers's telescope imaged directly or as a spectroscope, with a simple flip of a mirror. Data was recorded on film. All the exposures were very short, from a few seconds to several minutes.

Carruthers designed the instrument for ease of use, suitable for astronauts in bulky suits. It was placed on a special tripod, much like a surveyor's telescope mounting. There was a sight on the side that the astronauts used to calibrate the directional indicators to line the instrument up with different celestial sources, such as Earth, gaseous nebulae, or bright blue hot stars.

During the mission, John Young set up the camera in the shadow of the lander, leveled it, and calibrated the directional pointers using the Earth as a test object. All he had to do next was push a button and the camera did the rest.

The astronauts brought the film cassettes back to Earth and handed them over to Carruthers. He found that his camera had faithfully recorded the ultraviolet geocorona, a vast halo of low-density hydrogen surrounding the Earth, as well as auroral activity over the planet's south magnetic pole. Other plates revealed the ultraviolet images and spectra of the diffuse gases lying between the stars, in discrete nebulae, and in the interstellar medium. The camera also looked for but did not find the remnants of a lunar atmosphere—if it existed, it was too thin to be detected even by this sensitive instrument.

The artifact in the collection is a backup to the flown camera. In 1992, Carruthers and his students restored it in preparation for display next to the Museum's lunar lander. As part of the restoration, Carruthers attached a flown film cassette at the back end of the camera. It is presently on view in the *Apollo to the Moon* gallery. He recalls that "the Apollo 16 project was probably the biggest science that I've ever been involved in … you might say that the Apollo era was the high point not only of my activities, but space science in general … it was a fast-paced program and there were no funding restrictions."

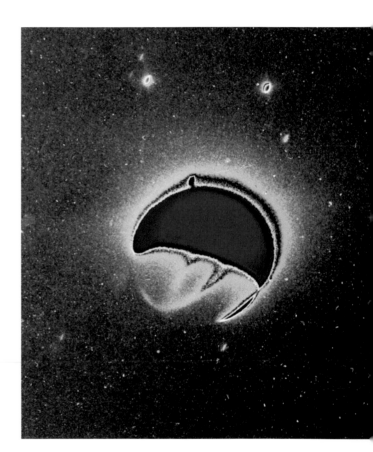

ABOVE: Astronaut John Young took this image of the Earth with the ultraviolet lunar camera, revealing the layers of the planet's geocorona.

Orbiting Astronomical Observatory III *Copernicus*, Princeton Experiment Package, Engineering Prototype
1971–1972

"… to investigate the ashes of stars that have died [and] the seeds of stars to be born."

—ASTRONOMER JOHN ROGERSON, 1972

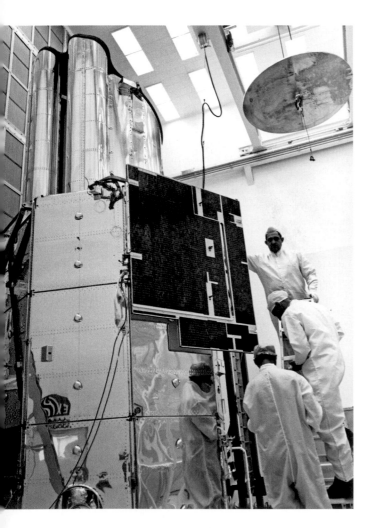

ABOVE: Technicians check out the OAO III satellite prior to launch.

AS THE fourth and last orbiting astronomical observatory launched into space in August 1972, Lyman Spitzer remembers thinking, "We'd be lucky if it lasted a year." In studying astronomy from space, Spitzer, director of the Princeton University Observatory, led a new breed of astronomers. A classical astronomer might look askance at the OAO: he or she could build a ground-based telescope of the same size for 1/500th of the cost of the $100 million NASA satellite, and then expect it to perform for at least a century. But Spitzer was not such an astronomer.

Dating back to the 1940s, Spitzer dreamed of building space telescopes. He started small. In the 1950s, his department staff built successful balloon-borne telescopes and devices for sounding rockets. As the space age started, he then campaigned for and built the largest space telescope yet thrust into orbit: its 3-meter (10-foot)-long cylindrical structure contained a reflecting telescope with a 81-meter (32-inch) mirror and a high-powered spectrometer capable of exploring the ultraviolet spectral signatures of the hottest stars and of the cool gases between the stars.

Spitzer, of course, knew that NASA missions, once in orbit, tended to outlive expectations. The Orbiting Astronomical Observatory C-1, renamed OAO III after launch and finally *Copernicus* in honor of the sixteenth-century astronomer's 500th birthday, indeed had a planned operational life of a year. But

Spitzer and NASA built it to last longer—it operated for nine years and yielded a trove of new astronomical insights. The telescope captured the ultraviolet spectra of stars, galaxies, a nova, and the interstellar medium. Serving astronomers worldwide, it provided new data on the formation and evolution of stars and greatly improved knowledge of the abundances of molecular hydrogen and deuterium in interstellar space. The latter data bore deep cosmological significance, providing evidence consistent with theories of the Big Bang.

In 1973, after NASA confirmed that the orbiting *Copernicus* worked properly, they transferred the engineering prototype of the Princeton optical telescope and spectrograph, known as the Princeton Experiment Package (PEP), to the Museum. The Princeton astronomy department designed the actual package and built it in conjunction with Perkin-Elmer, Sylvania, and the Goddard Space Flight Center. The 408-kilogram (900-pound) PEP includes guidance optics and a scanning spectrometer, the sensitivity of which reached down into the extreme ultraviolet, around 700 angstroms. *Copernicus* weighed in at 22,200 kilograms (4,900 pounds). In addition to the Princeton experiment, the telescope contained a battery of x-ray telescopes developed at University College, London. At launch, a Spitzer colleague, John Rogerson, summed up the moment: *Copernicus* was equipped "to investigate the ashes of stars that have died [and] the seeds of stars to be born."

Space In an Era of Pragmatism, 1972–1989

O N DECEMBER 14, 1972, in the early evening on the United States east coast, Apollo 17 astronauts Harrison Schmitt and Eugene Cernan prepared to leave the Moon after three days of exploration—the longest stay of the six Apollo missions to reach the Moon. Although originally several more lunar landings had been planned, this mission capped a triumphant program of exploration, begun in 1969 with Apollo 11. Schmitt climbed first into their lunar module *Challenger*. Cernan performed one last task, driving the lunar rover away from the module, then bounded back and "put a foot on the pad and grabbed the ladder." As he ascended, he offered "… as we leave the Moon at Taurus-Littrow, we leave as we came and, God willing, we shall return, with peace and hope for all mankind." In the years to follow, no human explorers returned. Astronaut Cernan became the "last man on the Moon."

Cernan's sentiment that exploration of the Moon, and of space generally, served to invigorate basic human values of fellowship and common purpose was a powerful element in Apollo. As the Moon flights unfolded, people all over the world were moved by this ideal, an antidote to a Cold War geopolitical climate in which tension and the threat of nuclear war predominated. Yet these very Cold War circumstances provided the driving rationale for Apollo, pushing the United States ahead in the Space Race of the 1960s. The triumphant national effort demonstrated, vividly, publicly, the country's success over its Soviet rivals *and* encapsulated the common human aspirations put in jeopardy by the military dimensions of the Cold War.

After the Moon landings, a key question for the future of spaceflight, especially human exploration, arose. Would such an endeavor be driven by idealism—by exploration, the search for

knowledge, and as an enterprise on behalf of humanity? Or, would it be an expression of politics, a means to practical national ends? Well before Cernan and Schmitt left the Moon, shifts in Cold War politics had begun to diminish human space exploration as a defining measure of the U.S. and USSR confrontation. In May 1972, President Richard M. Nixon and Soviet Premier Leonid Brezhnev, partially reversing Cold War tensions, reached a historic "détente" between the two superpowers. One result was an agreement to conduct a joint, cooperative mission in space—Apollo-Soyuz, which took place in 1975—a vivid contrast to years of intense competition.

This brief détente faded and the Cold War continued to dominate international politics through 1989, when the Berlin Wall crumbled, marking the end of the decades-long geopolitical conflict. Yet unlike the 1960s, *human* space exploration edged from center stage toward the wings. It was a change filled with ambiguity. The message of idealism, central to the Moon missions, still resonated with the public, but less so. After Apollo 11, the public showed a diminishing interest in the succeeding missions and television viewership plummeted (except during the near-disaster in space experienced by Apollo 13). According to opinion polls, many Americans felt that the United States need not be "No. 1" in competition with other countries, including space exploration. A 1971 *Wall Street Journal* poll noted that unlike the earlier days of the Space Race, a swath of the public felt unbothered by "whether the Russians land on Mars before we do." The prestige and pride derived from spaceflight was "part of the heritage that we should relay to our children," but jockeyed with the down-to-Earth view that "we're spending too much money flying to the moon when there are hungry people right here in our

OPPOSITE: Space Shuttle *Discovery*, mission STS-64, lifts off from Kennedy Space Center, September 1994.

own country." This shift in mood overlapped with new funding realities. The Vietnam War, then the oil "shock" of 1973, rampant inflation, and historic budget deficits in the 1980s made an idealistic vision seem a luxury. Pragmatic sensibilities reigned.

This transition, occurring over President Nixon's first term, set in motion a realignment of the various U.S. space programs that had formed in the wake of Sputnik—a realignment that largely has persisted into the twenty-first century. In this transformation, human space exploration entered into new relationships with military and intelligence space programs and commerce. Prestige and pioneering ideals now intersected, for the first time, with practical military needs and the bottom-line thinking of economics. Human space exploration moved from an "exceptional" undertaking to a "normal" one rooted in the hurly-burly of everyday politics in which it competed with powerful elements of American life, the national security establishment and the private market. In the process, two strong, opposing points of view emerged. One saw the hollowing out of the Apollo legacy as a fall from grace; the other saw the new pragmatic direction as placing space exploration on a par with many other needs of the nation. Neither view won out, but space policy tilted toward the pragmatic.

THE Space Transportation System—the Shuttle—emerged as Exhibit A of post-Apollo thinking. Approved in early 1972 by President Nixon, the Shuttle became the centerpiece of U.S. human spaceflight, from the program's inception, to first launch (*Columbia* in 1981), and still is as of the writing of this book in 2006. Perhaps the most complicated, demanding craft ever contrived, the Shuttle is, as astronaut Michael Collins noted, "unlike anything before … truly a hybrid machine, half airplane and half spacecraft." Promoters envisioned the Shuttle as analogous to a commercial airliner: the craft launched, flew into near-Earth orbit, returned, underwent maintenance, refueled, and started the cycle over—a concept that promised more and radically less expensive trips into space. In the rosiest projections, a fleet of four Shuttles might make up to 60 flights a year, about one a week. In actuality, in its best year, the fleet achieved nine flights and a total of 115 over 25 years (as of mid-2006) and reaching space still carried a hefty price tag.

The Shuttle showcased the nation's technical ingenuity, but the program's modest exploration goals emphasized the dramatic shift in perspective from the 1960s. The Shuttle represented a scaling back of a vision, first promoted in the early 1950s, that used the Western tradition of exploration as a model. In this view, humans had a destiny to move outward into space from near-Earth orbit, to the Moon, and onto Mars to scout, establish base camps, and then settle. Machines might substitute for humans and lead the way. The Viking missions to Mars in which two robotic landers touched down on the planet's surface fit with this vision. So did the two Voyager spacecraft, launched in 1977, that traveled first to the solar system's outer planets then off toward interstellar space, each carrying a "Sounds of Earth" record in the hope of encountering an intelligent species "out there." But the Apollo experience intensified the hopes for an ambitious program of human exploration and became an integral part of NASA planning for the post-Moon years. In elaborate studies, NASA planners expressed these possibilities in dry technical language, but they were animated by an almost poetic romanticism to fulfill an "extraterrestrial imperative" and prepare ourselves for the time when we "convert our earth from an all-supplying womb into a home for the long future of the human race [in space]."

Such views, though, were out of synch with the time and President Nixon's inclinations. Nixon, in approving the Shuttle, recalibrated this idealism in his official announcement. He invoked "the imperatives of universal brotherhood" and the "liberating perspectives of space" but resisted the lure of the remote frontier, seeing the future rather in "learning to think and act as guardians of one tiny blue and green island in the trackless oceans of the Universe." The new focus of human spaceflight centered on "revolutionizing transportation into near space" through cheaper costs, with the Shuttle serving as "the workhorse of our whole space effort, taking the place of all present launch vehicles except the very smallest and very largest." The Shuttle, with its strongly utilitarian objectives, thus stood in contrast to the romantic vision. Astronauts, according to critics, would not be explorers and settlers, but rather "ferry" operators. Tight budgets and stretched-out schedules for developing the new vehicle seemed only to reinforce the view that the old order had passed. The time from initial decision to first flight of the Shuttle spanned nine years, longer than it took to achieve landing humans on the Moon.

As the Shuttle began to fly in the early to mid-1980s, it embodied these contesting visions. NASA and the public tended to view the flights through the lens of the exploration ideal. The missions proved the Shuttle a remarkable machine, allowing astronaut crews to live and work in space for around two weeks, with relative ease and comfort (compared to the cramped capsules of the 1960s). Women, minorities, and foreign nationals (and even a U.S. senator and a representative) flew on missions, giving substance to a "family of man" image. Programs such as Spacelab created new bonds of international cooperation, lessening the nationalistic character of early spaceflight. These early accomplishments, too, helped spur the major civilian space policy decision of the 1980s, President Ronald W. Reagan's 1984 call to build a space station (initially named *Freedom*, now the International Space Station), an apparent "next logical step" in implementing the idealist vision of human space exploration. But its conceptualization was buffeted by the same pragmatic forces affecting the Shuttle. Some in the press quickly summarized the proposal as a "combination Holiday Inn, service station, and factory in orbit."

BUT the realities of the first years of the Shuttle program (and the newly conceived *Freedom*) clearly revealed the new framework of spaceflight. Most of the missions through the 1980s involved "trucking" either a communications satellite or a national security payload to orbit. The reason: as part of the original Space Shuttle decision NASA had to seek paying customers—private industry, other national governments, and the U.S. Air Force had the hard cash to help defray the cost of Shuttle operations. While some worried that NASA had become merely, in the words of one editorial, "a shipping agency" for industry and the military, this use of the Shuttle was one indicator of larger trends. Changes in government policy opened many public or regulated activities to private markets in areas such as telephone service and air travel. Space activities felt these political winds, too, affecting the communications satellite industry and the government-operated Landsat satellites used for monitoring the Earth. The move toward private initiatives in this period began to create the global media world that is so familiar in the early twenty-first century.

The role of the military was intimately built into the Shuttle program from its origins. NASA engineers designed the Shuttle's payload bay dimensions and the vehicle's payload lift capacity specifically to accommodate military satellites, as part of the Department of Defense's (DoD) agreement to use the vehicle. As flights began, military and intelligence payloads took up about a third of the Shuttle missions; DoD, also, had the prerogative of preempting NASA or industry payloads on non-military flights. Some of the DoD missions were secret, blacking out any press coverage, a sharp contrast to NASA's public openness. DoD spent years building its very own launch facility, called "Slick Six," at Vandenberg Air Force Base in California, to oversee military Shuttle flights from beginning to end. These multiple developments were one expression of the enhanced military role in the period, driven by the ongoing Cold War and a pragmatic national outlook. By the 1970s, the budgets for military and intelligence space activities approached and began to exceed that spent on civilian space. This shift intensified in the 1980s, with President Reagan's start of the Strategic Defense Initiative (known in the media as "Star Wars," a reference to George Lucas's popular space films) a prime example.

The history of the Shuttle, though, changed in a few horrifying moments on January 28, 1986, when the *Challenger* disintegrated 73 seconds into flight, causing the deaths of all seven crew members. The tragedy had many ramifications and drew a deep response from the public. The Shuttle era may not have carried the drama or significance of the Apollo years, but a human presence still was the emotional center and point of identification for the public in spaceflight. The accident did lead to dramatic changes in the Shuttle program. The notion of the Shuttle as a "truck" was overturned. When Shuttle flights resumed in 1988, the program began to wean itself from its commercial and military customers, focusing on science and other NASA missions. But neither this divestiture nor the outpouring of public feeling changed the basic framework of spaceflight that emerged over the 1970s and 1980s. Human spaceflight continued, including planning for space station *Freedom*, but was still torn between a romantic vision of a human destiny in the stars and political realities. Military and private market–oriented interests pursued, in different ways, their efforts to use spaceflight for earth-oriented purposes. In the post–Berlin Wall, post–Cold War era, this pragmatism forged an intimate connection between space and our everyday lives.

Space Shuttle Concept Model
1971

"There is a real requirement for an efficient Earth-to-orbit transportation system—an economical space shuttle."

—George E. Mueller, NASA's associate administrator for Manned Space Flight, 1968

THE Space Shuttle as we know it was not the only or the inevitable design. From 1967 to 1972, NASA asked several aerospace companies to explore different concepts for a reusable space transportation system. This concept model offers a glimpse into the process of realizing this novel idea.

NASA and its engineering contractors faced two main challenges: designing a reusable spacecraft that combined the features of a rocket, spacecraft, and glider, and balancing the cost of building a fleet of shuttles and operating them for many years. Spacecraft size and weight, wing shape, boosters, propellant tank location, and other important variables were among the "trade-offs" they examined as the design process evolved.

At first, NASA directed the corporate teams to study fully reusable concepts—those in which both the booster and the orbiter were capable of repeated round-trip flights. Ultimately these proved too expensive and technically challenging to pursue. Within strict cost constraints, NASA settled on a compromise design: a reusable orbiter mounted on a partially reusable booster system that could be developed with available funds.

This model of a fully reusable space shuttle concept developed by Grumman and Boeing in 1971 was called the G-3 two-stage concept. It represented the idealistic start of shuttle design studies, when NASA asked companies to scope out a "fly-back" piloted booster and orbiter combination. With internal propellant tanks and retractable jet engines for return flight, booster and orbiter looked like two huge aircraft in piggyback launch configuration. This concept had all the features NASA wanted at the outset.

As the design process progressed, certain desirable features, such as jet engines, were seen as "frills," and NASA drew back from its original goal of a fully reusable vehicle. The succession of models from 1970 to 1972 highlights how designs evolved in response to changing technical and cost constraints.

The actual operational Space Shuttle orbiter looked similar to that pictured here, but the concept booster differs from the eventual design of twin reusable solid rockets attached to a huge disposable external tank that fueled the orbiter's three main engines. This design served well for more than 25 years and, as of mid-2006, for all but two of 115 missions. As a complex, experimental machine, though, the Shuttle did not solve all the technical problems of reusable spacecraft nor reduce the cost of reaching space.

By 2005, the long commitment to a winged reusable spacecraft for missions in Earth orbit yielded to a renewed interest in missions to the Moon and to Mars. NASA decided to retire the Space Shuttles and looked back to the future for a different kind of spacecraft, a hybrid of the Apollo-era crew capsules and the Shuttle-era propulsion systems. Once again, the process of engineering design and trade-offs began, evolving toward a new look for spaceflight beyond 2010.

Models illustrate the conceptual process of engineering design. The Museum's collection includes at least 50 models of Space Shuttle designs and Shuttle-like concepts from the 1950s into the 2000s. Already the Museum has acquired early concept models of the next space exploration vehicle. Depicting different solutions to common challenges and similar requirements for spaceflight, models are symbols of the technical creativity that makes venturing into space possible.

Skylab Orbital Workshop
1973

"Here's our home in the sky."

—Jack Lousma, pilot, second Skylab mission, 1973

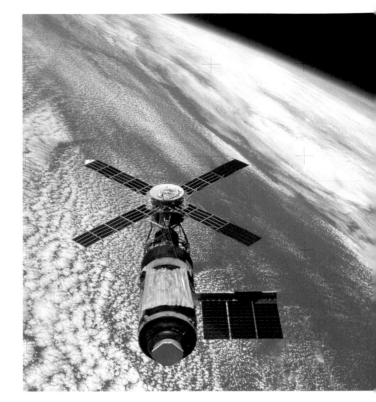

IN May 1973, just five months after the last Apollo mission to the Moon, the United States launched a space station called Skylab. NASA sought to apply the technologies and expertise of the Space Race to other uses after the first pioneering forays into space. Skylab was a trial run for long-duration human missions in space. Over a period of nine months, three successive astronaut crews experimented with living and working in Earth orbit.

To construct Skylab, engineers converted the third stage of a giant Saturn V launch vehicle into a three-story orbital workshop, complete with a galley, food lockers and freezers, sleep compartments, a toilet and shower, and more than enough workspace for three-astronaut crews. Compared to the cramped capsules in which previous astronauts flew, Skylab was huge—as roomy as a house—and the crews found it to be a comfortable temporary home. Two other modules for docking and spacewalk access plus an observatory completed the space station.

Scientists outfitted the state-of-the-art solar observatory (called the Apollo Telescope Mount) with devices to image the Sun across the electromagnetic spectrum during the 1973–1974 peak in the sunspot cycle. Other scientists devised programs of biomedical experiments to monitor crew adaptation to weightlessness and experiments to study materials in microgravity. In addition to their own ambitious research programs, Skylab astronauts conducted more than a dozen experiments developed by students.

The three astronauts on each Skylab mission were isolated, connected to civilization only by two-way radio and space-to-Earth video. Surrounded by a vast sea of space, they could see distant land but not reach it. Their prescribed routine kept them inside, with occasional spacewalks outside. Work and the constant challenges of their new environment kept any boredom at bay, and the astronauts never seemed to tire of the experience of weightlessness.

After three busy, productive missions on Skylab, the departing last crew turned out the lights and locked up tight in February 1974, hoping for an eventual return. That never happened, and the deserted Skylab tumbled back into the atmosphere and disintegrated in 1979.

Skylab served as a proving ground for equipment and techniques that might be needed for a true space station and a permanent presence in space. The missions yielded volumes of "lessons learned" about essential tasks, systems, and hardware—everything from changing film canisters and doing repairs outside the workshop to preparing meals and taking showers inside, from following checklists to dealing with stress, working hard, and relaxing in space. Planners for the Space Shuttle and later International Space Station looked to Skylab experiences for guidance about the practicalities of living and working in space.

NASA built a second Skylab space station—made up of the orbital workshop living quarters, the multiple docking adapter that served as a docking port and control center, the observatory, and the airlock module for going out on spacewalks, plus large solar arrays for electrical power. But budget constraints derailed plans for a second launch and the craft came to the Museum in 1976.

The orbital workshop stands floor-to-roof in Space Hall, where visitors walking through it can get a sense of what it was like to live as the Skylab crews did for a month or two or three.

ABOVE: Astronaut Gerald Carr demonstrates the comic possibilities of microgravity as he balances an upside down astronaut William Pogue on his finger.

TOP: After departing from the Skylab Orbital Workshop, astronauts in the Skylab 4 command and service modules fly around their home in space before returning to Earth, February 1974.

H-Alpha #1 Film Magazine
1973

"As easy as driving on the freeway."

—Astronaut Pete Conrad, 1973

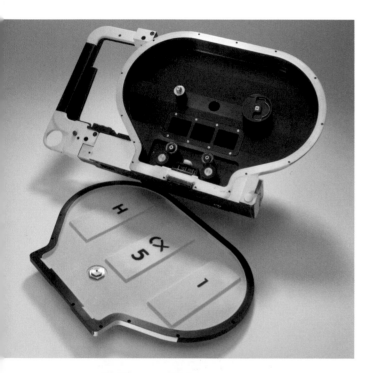

ABOVE: On Skylab 3, astronaut Owen Garriot performs an EVA to deploy a sun shield to protect the orbital workshop.

In late May 1973, when the first crew of astronauts arrived at Skylab, the first U.S. space station, they confronted a dire situation. Astronauts Charles "Pete" Conrad and Paul Weitz's first task was to conduct two emergency "extravehicular activities" (EVAs) to mitigate damage caused during launch to the station's meteoroid shield and one of its solar panels. With the repairs completed, they turned to their scheduled work—another EVA. This time they "spacewalked" to replace film magazines for many of the eight solar telescopes on the Apollo Telescope Mount (ATM), the most significant scientific apparatus on Skylab. ATM's success depended on the seemingly routine task of collecting and replacing the film magazines—which held a valuable collection of data.

ATM carried eight solar telescopes, designed to study sunspots, flares, and prominences, forms of high-energy disruptions in the solar atmosphere that can have profound effects on Earth. Examining the Sun over a spectrum of wavelengths (especially those blocked by Earth's atmosphere), the instruments included x-ray and ultraviolet cameras, spectro-heliographs, a coronagraph, and two monitoring telescopes filtered to see only the energy radiated by the first Balmer line of hydrogen, known as H-alpha, in the red region of the visible spectrum.

The film magazine featured here was one designed for the H-alpha #1 telescope monitor, developed by the Marshall Space Flight Center and operated jointly with the Harvard College Observatory. Located outside the space station, on ATM, this film magazine, among others, recorded data from each of the instruments. Skylab astronauts did EVAs to retrieve the full magazines and swap in new ones.

Astronauts Conrad and Weitz prepared carefully for the film magazine EVAs, practicing the procedure in a water tank at Marshall Space Flight

Center. In space, they suited up, left the station, and took a specially designated "EVA trail"—just as they did in practice. This consisted of a series of blue-painted single and double handrails that passed through the ATM's trusses. "Road signs" along the trail alerted them to a series of hatches and doors on various parts of the cylindrical telescope canister. They opened each door and removed the exposed film magazine. A telescoping boom attached to Skylab held a fresh magazine at the ready. Astronauts placed the new magazine in the instrument and secured the exposed magazine to the boom. Astronauts then returned to the Skylab airlock with the magazine, storing it until their return to Earth.

Conrad reported that performing the EVAs using the handrails was "as easy as driving on the freeway." The procedure was completed in "a fraction of the time planned" even with the added task of brushing clean the front end of the sensitive white light coronagraph, which had become contaminated with the pollution from the space station.

Why did Skylab use film to record valuable scientific data, rather than recording the data electronically? In the 1970s, the information density and stability of film vastly outperformed even the best electronic image detectors. The problem with photography, of course, was that it had to be retrieved physically, by astronauts. Thus a seemingly minor technical question—to use film or electronic means to capture data—profoundly affected Skylab's design and necessitated specialized astronaut training and in-space activities.

The Museum's film magazine came from inventory left over from a complete backup Skylab space station that never flew. Note the oversized handholds and other characteristics of an object that had to be manipulated by an operator in a bulky spacesuit.

Skylab Tool Kit #1
1973

"Skylab Repaired With $65.50 Cable-Cutter, Salvaging the Mission and Rest of Program"

— *Wall Street Journal*, JUNE 8, 1973

IMAGINE preparing to move into a brand new house and finding out just before the moving truck arrives that your new house already needs repairs! During the launch of the Skylab orbital workshop, the United States' first station in space, the spacecraft suffered serious external damage to its meteoroid shield and one of its solar panels, jeopardizing the entire program. This made repair of the workshop a critical task for the first crew of astronauts. NASA engineers went to work, urgently improvising tools to fix the exterior problems. As noted in the headline above, a $65.50 commercial cable-cutter became part of the repair activity.

But long before this launch-caused damage, NASA engineers had anticipated the need to make repairs to the space station that would serve as the astronauts' home. They developed tools for use inside and outside the spacecraft, trying to foresee all of the possible repair work astronauts might encounter during their weeks and months in orbit. Astronauts moved into Skylab with kits including an extensive range of tools—impressive even to the most avid of home improvement gurus. During three missions to the workshop, nine different men worked with dozens of tools to deal with urgent problems and daily maintenance tasks on America's first space station.

As the nearest hardware store was hundreds of miles straight down, Skylab was equipped with tools for every size nut, bolt, and screw holding the spacecraft together. In this set, Tool Kit #1, transferred from NASA to the Museum in 1976, astronauts found familiar household tools such as screwdrivers, clamps, wrenches, and ratchets. NASA saved time and money by buying commercially available tools designed for indoor use only. In contrast, most tools for exterior work were custom-made to withstand the extreme conditions of space and the special needs of astronauts wearing bulky spacesuit gloves. One important addition to each indoor tool, helpful as astronauts floated weightless in the spacecraft, was a small piece of Velcro. This simple but strong material kept these and hundreds of other articles taken into space from drifting

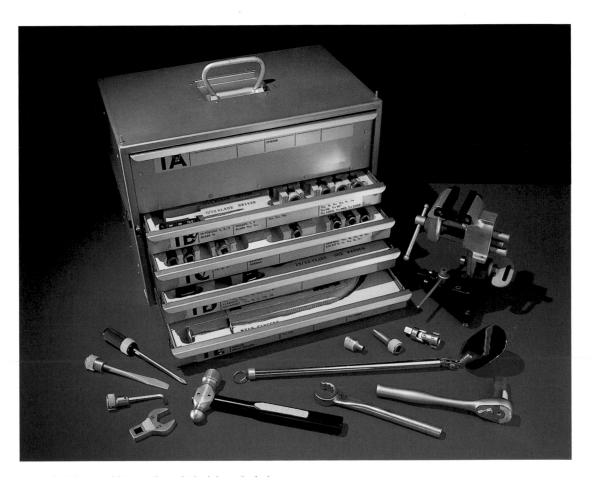

around. Other tool kits on board Skylab included materials such as twine, tape, and adhesives for making other kinds of repairs.

The Skylab mission proved the ability of humans to survive long-term space missions—a crucial part of which was being able to make major and minor repairs as needed. From daily scheduled housekeeping repairs to unexpected minor damages, Skylab tool kits proved essential resources for astronauts aboard the orbiting workshop. Engineers later standardized this equipment on the Space Shuttle. On-orbit repair became a critical skill as the United States moved out of the Apollo program and toward days-long flights on the Space Shuttle and long-duration missions to the Russian *Mir* Space Station and the International Space Station.

"Anita," the Skylab 3 Spider
1973

"Skylab astronauts mourned the death today of space spider Anita, a tiny animal which captured the interest of the orbiting pilots and earthlings with her weightless web spinning."

—*Chicago Tribune*, SEPTEMBER 17, 1973

ABOVE: Arabella, a common cross spider, sits in the web she spun in microgravity aboard Skylab 3.

ASTRONAUTS conducted experiments developed by students across the country as part of the long-duration missions to the Skylab orbital workshop in 1973. For the first (but not the last) time NASA allowed the public to participate in research in space as part of its goal to educate young people. NASA received more than 3,400 proposals from secondary school students, with just 19 qualifying for flight to the country's first space station. One, submitted by Judith Miles of Lexington, Massachusetts, and flown on Skylab 3, tested the ability of two spiders, "Anita" and "Arabella," to construct webs in the unfamiliar environment of weightlessness.

Anita and Arabella, common cross spiders of the orb weaver group, launched with the Skylab 3 crew in July 1973. Each went aloft in its own vial, along with a small supply of food and water. Astronaut Owen Garriott conducted Miss Miles's experiment with the two spiders, recording their activities with movie and still cameras in a cage specially designed for web creation. When the two spiders exhausted their food supply of houseflies, Garriott suggested sharing his filet mignon to supplement their diets. The spiders thrived, allowing the test to continue well beyond its planned three to five days.

Arabella entered the test cage first, and her initial attempts showed her disorientation during the early days of the mission; her webs were crooked and broken. After some food, water, and time for adaptation to space, she created nearly

perfect web specimens. As second in line, Anita had time to adjust to the weightless conditions before her web challenge, and she created better webs from the start. These space-made webs exhibited a curious feature: they weighed less than those created on Earth, with the threads being 20 percent finer. Without the need to support their whole weight, Arabella and Anita instinctively created webs that suited them in space.

Unfortunately, Garriott found Anita dead in her cage just a month into the mission, but he returned her to Earth with Arabella (who perished during the return flight). Examinations after flight showed that both spiders suffered from dehydration. NASA transferred Anita, preserved in this specimen bottle, to the Museum in 1974.

Astronauts—and spiders—aboard Skylab were pioneers of sorts, spending weeks in Earth orbit for the benefit of science. Following in the steps of Laika the Russian dog (launched on Sputnik 2 in 1957) and Ham the chimpanzee (launched on Mercury-Redstone 2 in 1961), the Anita and Arabella tests represented early biological research on the long-term effects of a microgravity environment on nonhumans, leading to further experimentation with animals on the Space Shuttle. Research using student-created experiments also opened space to a younger generation of Americans, and Skylab's role as an orbital laboratory gave them a unique place in which to test their ideas with the assistance of professional scientist astronauts.

Applications Technology Satellite 6 Thermal Structural Model
1974

"A satellite will bring patients to doctors."

—*The New York Times*, MAY 24, 1974

IN THE 1960s, the dramatic exploits of humans in space shared the headlines with other national undertakings—including President Lyndon Johnson's much-publicized war on poverty. Johnson saw that in a time of general prosperity and high-technology advances many U.S. citizens lacked access to jobs, education, and health care.

It was a problem that cut across the American landscape—from neighborhoods of major cities to isolated or geographically remote regions such as Appalachia and Alaska. Potent images of the poor and their communities were commonplace in newspapers and magazines such as *Life*. Government action and dollars, President Johnson felt, could and should seek to remedy these circumstances. These views found expression, too, in U.S. foreign aid to "underdeveloped" nations, but with the added intent to encourage such countries to align politically with the U.S. rather than the Soviet Union.

NASA's Applications Technology Satellite (ATS) program, initiated in the early 1960s, became a small but important symbolic part of this national and international undertaking. Communications satellites with their ability to send and receive information across vast distances seemed ideal instruments to connect "haves" with "have nots" and, thus, to offer remote communities and individuals one avenue to a better life. The ATS program sought to demonstrate the validity of this idea. Launched in 1974, ATS 6 was the last of the ATS series and the centerpiece in demonstrating that communications satellites could be powerful social tools.

ATS 6 was an imposing example of technology—it was, to date, the largest and most complex communications satellite ever built, with 19 separate communications packages. But its key feature was a 9-meter (30-foot) mesh circular parabolic antenna—a first in space communications. The reason for the antenna's size was that a larger antenna could generate more powerful signals, more carefully directed to specific regions on Earth. The more powerful the signal created by the satellite, the smaller a ground antenna needed to be to send and receive signals through the satellite. This was a crucial breakthrough for poor and remote areas—small, at around 3 meters (10 feet) in diameter, relatively portable antennas could connect communities to a larger world.

Soon after launch, ATS 6 began a broad program of education and health services to communities in Appalachia, the Rocky Mountains, and Alaska. As one example, teachers in 15 different communities in Appalachia received television instruction from the University of Kentucky on guiding their students in career options—the first time a satellite was used to provide educational services. Via ATS 6, such exchanges became commonplace across the country.

In 1975, ATS was repositioned in orbit to reach India. At the time, only four cities in all of India had ground-based broadcast television. ATS 6 provided a dramatically new capability, serving as one part of a broader program of modernization. Initially, it reached 2,400 villages in central India—all of which lacked basic telephone and television services and had high rates of illiteracy. Government programming via the satellite provided instruction on agriculture, personal hygiene, birth control, and nutrition. As planned, the ATS program ended after a year. One reporter worried that after the last television transmission "pernicious isolation could sweep in again, leaving behind yet another local legend—about the year that television dwelt in the land." Yet, in the years to follow, India's experience with ATS 6 led that country to develop its own communications satellite program, one that now reaches most of its citizens.

This artifact is a thermal structural model used in engineering tests and was transferred from NASA to the Museum in 1978.

ABOVE: Technicians from the Indian Space Research Organization install a "direct receive" antenna in the village of Kerreli. ATS 6 broadcasted instructional television to 5,000 villages in India.

CDC 3800 Computer
1960s–early 1990s

"In many cases we have operators who are younger than the computers."

—Col. Norman Michaud, USAF, 1985

THE SPECTACULAR images and stories of human exploration—highlighted by astronauts walking on the Moon—obscure other important features of the U.S. space effort, among the most significant being the role of the military and the vast ground infrastructure required to send humans and machines into space. This artifact, a CDC 3800 computer, transferred to the Museum from the U.S. Air Force in 1993, speaks to these significant but less discussed underpinnings of space exploration. Built by the Control Data Corporation (CDC) and installed at the U.S. Air Force's Satellite Control Facility at Sunnyvale, California, this mainframe computer helped control secret reconnaissance satellites from 1967 until the early 1990s.

The United States devoted large sums of money and resources to NASA's effort to put Americans into space. Over the years, though, the military (including intelligence agencies) often exceeded these highly public benchmarks, but with operations and budgets often cloaked in secrecy. The CDC 3800 was one part of that military effort. This artifact exemplifies the large infrastructure needed on the ground to operate and manage spacecraft and their missions, whether civilian or military. With one exception, the famed "Mission Control" in Houston that managed Apollo and then Shuttle flights, this element of space work has seldom been visible to the public.

The Air Force kept the CDC 3800 computer in the heart of Silicon Valley, in a facility the local press called the "Blue Cube," from its color. Little was known about the work inside this top secret facility, which housed a number of large mainframes like this one, connected to a network of smaller computers located at tracking stations around the world—including on an island in the Indian Ocean, in England, and in California and New Hampshire. The Air Force established this network around 1960 primarily to operate the top secret CORONA spy satellite program, which took reconnaissance photographs from space.

The worldwide stations tracked the orbits of the satellites and relayed that information to Sunnyvale. The CDC 3800 computed their orbits and arranged the complicated dance of commanding a CORONA satellite to orient itself properly, turn on its cameras, or eject the film (which was returned back through the Earth's atmosphere in reentry "buckets"). In those pre-Internet days, few long-distance, high-speed telecommunications lines existed to link the network; in many cases, the Air Force transferred data physically by carrying reels of magnetic tape from one computer to the other.

The CDC 3800 was a military version of a commercial computer, for which the Air Force paid about $2 million. Designed before the advent of silicon chips, it used discrete transistors and so-called magnetic core for memory—tiny doughnut-shaped pieces of magnetic material that held numeric data. The machine's core memory held up to 128,000 bytes of data. The total weight of a 3800 model was over 4,000 kilograms (4 tons). By the time it was retired, in the 1990s, such a machine might have seemed antiquated, given the dramatic advances in computer technology. But the Air Force had invested an enormous amount in developing software specific to the CDC 3800.

In the mid-1990s, the military lifted the veil from the Blue Cube but shifted many of its activities to an Air Force base in Colorado. The former location now serves as a backup site. In the early 1970s, CORONA was replaced by more advanced photo reconnaissance satellites, the most recent of which deliver their images electronically, rather than in film buckets. These "birds," and their corresponding infrastructure of ground-based stations, remain under tight secrecy—and as a large but not fully appreciated part of our spaceflight legacy.

AIL Modular Light Table 1540
Early 1970s–mid-1980s

"Mr. President, I am as sure of this as a photo interpreter can be sure of anything. And I think, sir, you might agree that we have not misled you on anything we reported to you. Yes, I am convinced they are missiles."

—ARTHUR C. LUNDAHL, DIRECTOR OF THE NATIONAL PHOTOGRAPHIC INTERPRETATION CENTER, BRIEFING PRESIDENT KENNEDY ON OCTOBER 16, 1962, CONCERNING SOVIET MISSILES IN CUBA

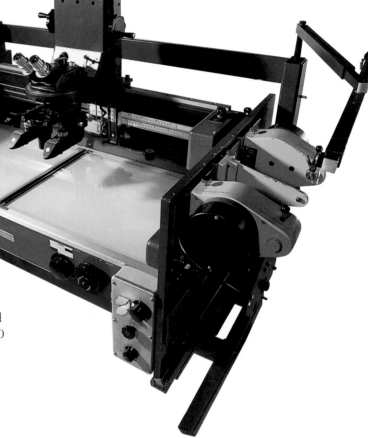

THE CUBAN Missile Crisis is an excellent illustration of the critical importance in the U.S. intelligence community of imagery analysis. U-2s and other manned aircraft took numerous photos of the island nation during the crisis in late summer and fall 1962. To untrained observers, the photos revealed little. But to analysts with lengthy experience in examining imagery of military targets in the USSR and Warsaw Pact nations, the photos clearly showed Soviet surface-to-air missiles and nuclear-armed surface-to-surface missiles deployed in Cuba.

Over several weeks, these highly trained individuals saw many objects in the photographs that led to this assessment. These included missile transport trailers, missile erector/launcher equipment, vehicles carrying fuel for the missiles, and nuclear weapons storage bunkers. Configuration of the launch sites also provided strong evidence of the missiles' presence.

The interpreters did not "read" these photographs unaided. They had a wide range of tools to assist them. The most important was the light table, which performed two critical tasks. It enabled analysts to examine film through a powerful viewer that allowed intensive study of individual frames, including the magnification of an object by many times. The table also helped interpreters to review quickly a large volume of film and to identify targets of interest in the sea of data.

Light tables used in World War II and the early decades of the Cold War were simple but effective. The Museum's MLT 1540, built by AIL Information Systems and transferred to the Museum by the National Geospatial-Intelligence Agency in 2005, entered service with the U.S. intelligence community in 1971 and represented a major improvement over existing light tables. It handled all the various types of film being generated by photoreconnaissance aircraft and satellites, either in rolls or sheets. The MLT 1540s featured motorized film drives, microscope mounts, and elevation control, making it much more automated than previous tables. Its Bausch & Lomb Zoom 240 stereomicroscope permitted analysts to view the photographic target with enhanced clarity and magnification, greatly easing their tasks.

With continual upgrades to the film drives, light sources, stereomicroscopes, and other components, the MLT 1540 continued in service far longer than its originally planned 10 years. With the introduction of digital imagery from photoreconnaissance aircraft and satellites, computers began replacing light tables as the principal imagery analysis tool in the 1980s. Nevertheless, analysts still use the MLT 1540 and similar pieces of equipment for examining historic and current nondigital imagery.

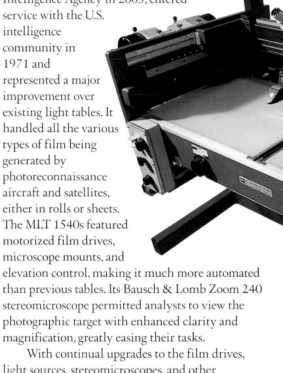

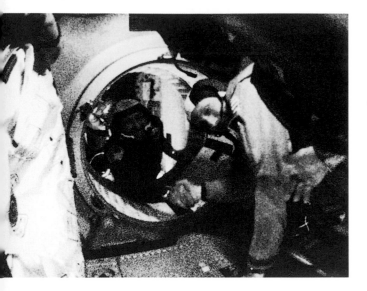

Soyuz Spacecraft
1976

"Nothing like the live Soyuz launching had ever been shown before by our television to the Soviet public—let alone the rest of the world.... I was in a state of nervous excitement ... [in] the State Department conference hall ... was the president, the secretary of state, other cabinet members, members of Congress, media, and all the rest. What if something goes wrong?"

—ANATOLY DOBRYNIN, SOVIET AMBASSADOR TO THE UNITED STATES, 1995

THE APPEARANCE of the Soyuz spacecraft belies its epic role in the history of human spaceflight. Bulbous, green, utilitarian—a contrast to the Apollo command module's clean geometric shape—the Soyuz did not fit the modern image of a spacecraft. It looked more like a bathysphere, more suited to submersion in the ocean than for launching into space. In actual operation, the Soyuz appeared even more ungainly, wrapped in a heavy, dark thermal blanket that resembled a mover's pad. Nonetheless, the Soyuz, with its distinctive look and design, has had extraordinary success—serving as the backbone of Soviet and Russian spaceflight for four decades, years longer than any other spacecraft type.

Soviet space engineer and manager Sergei Korolev originally commissioned the Soyuz as a spacecraft to send cosmonauts to the Moon. As Soviet plans for a Moon landing waned, design of Soyuz continued, but with a changed mission: to replace the Vostok spacecraft and serve as a ferry to a series of Soviet space stations. Soyuz even survived the political dissolution of the Soviet Union. As part of Russia's contribution, the spacecraft carries cosmonauts and astronauts to and from the International Space Station.

Americans learned about this long-lived spacecraft soon after it began flying. On July 18, 1975, an American Apollo spacecraft and a Soviet Soyuz joined in space for the first international human space mission, the Apollo-Soyuz Test Project (ASTP). Occurring during a period of détente in the Cold War, the Apollo-Soyuz mission culminated five years of political and technical negotiations and preparations between the two competing superpowers.

Three days before the historic docking in space, the spacecraft launched from their respective countries. Soyuz lifted off first from the Baikonur Cosmodrome in the Kazakh Republic of the Soviet Union, with cosmonauts Aleksei Leonov and Valeri Kubasov on board. Six and a half hours later, astronauts Thomas B. Stafford, Donald Slayton, and Vance Brand took off from the Kennedy Space Center in an Apollo spacecraft. Rendezvous required a specially designed docking system to link the two spacecraft. As the crew members moved from one spacecraft to the other, they conducted joint experiments, sampled each other's space food, and performed such symbolic acts as signing a joint mission certificate and exchanging gifts. To facilitate communication, each flight team's training included studying the other's language.

After the flight, 20 years passed before the next international rendezvous mission, when the American Shuttle *Atlantis* docked with the Russian space station *Mir* in June 1995.

The Apollo command and service modules at the Museum were part of a test vehicle, specifically refurbished for display. The Soyuz spacecraft originally came to the Museum as an object on loan from the Soviet government, arriving several months before the Museum opened in 1976. We know little about its history and assume that it is a high-fidelity model made from spare production components especially for the Museum's Apollo-Soyuz display. The docking adaptor is backup hardware from the Apollo-Soyuz mission.

The surviving members of the ASTP crew remain friends today. They meet periodically at the Museum to celebrate anniversaries of the mission. Trees that they planted in 1975 in commemoration of the historic flight still grow at the Museum's west end.

ABOVE: Astronaut Michael Collins, director of the National Air and Space Museum, watches as cosmonaut Valeri Kubasov and astronaut Deke Slayton plant a tree at the Museum to commemorate the Apollo-Soyuz Test Project.

TOP: Astronaut Thomas Stafford (right) and cosmonaut Alexei Leonov (left) greet each other in space with a handshake after their two-spacecraft rendezvous—the first, joint mission between the Cold War superpowers.

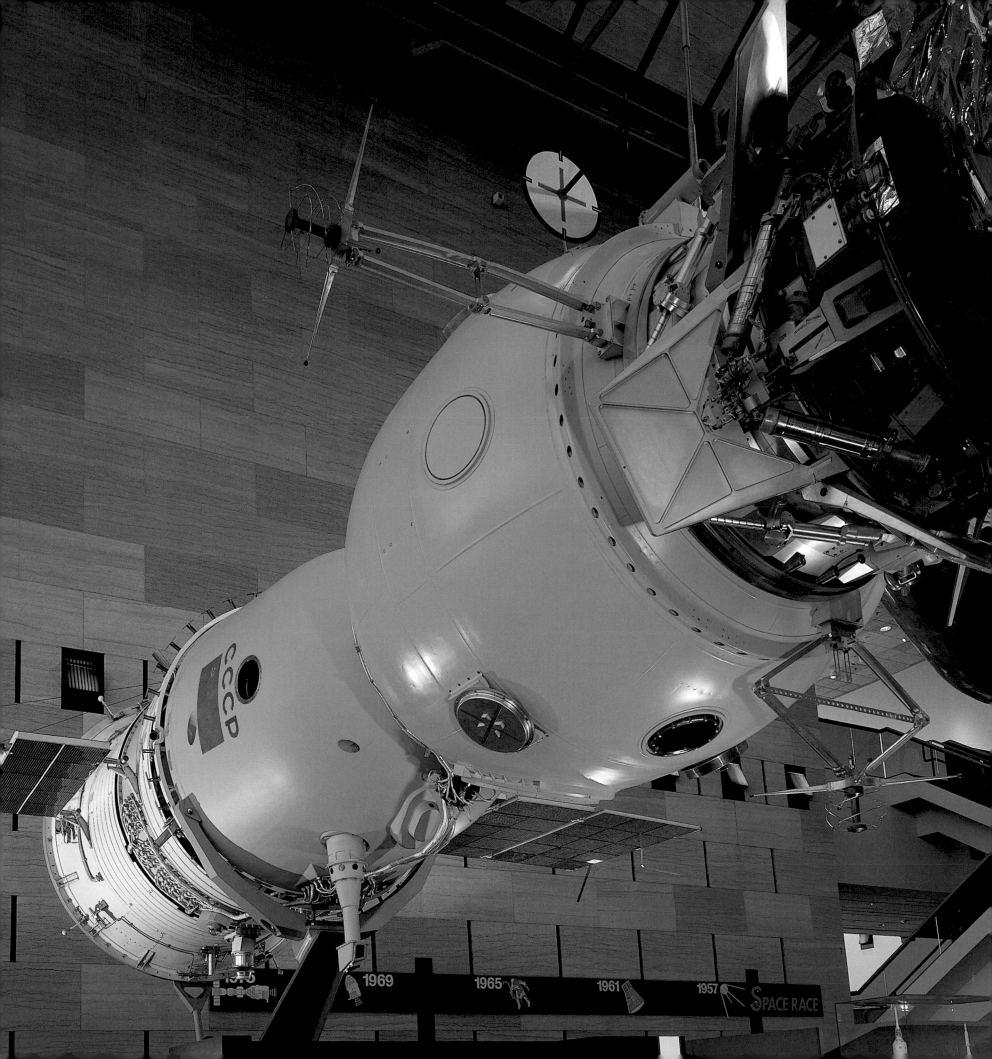

Apollo-Soyuz Cigarettes
1975

"'We supply plenty of makhorka *to all who need it,' boasts the same Mikoyan. He forgets to add that neither Europe nor America ever heard of such low-grade tobacco as* makhorka*."*

—LEON TROTSKY, *The Revolution Betrayed*, 1936

A COMMON, low-grade tobacco, *makhorka*, has had a long, prominent, and often ironic place in Russian culture, even under Soviet rule. When Soviet and American troops met at the German town of Torgau on the Elbe River in the closing days of World War II, they hugged, shared their meager rations, and exchanged cigarettes—the Soviets offering *makhorka*. Thirty years later, in the Apollo-Soyuz Test Project, when Soviet and American spacecraft docked in orbit above the Earth, they, too, embraced and shared hospitality on board their outposts in space. They did not exchange cigarettes— smoking tobacco in spacecraft is too dangerous—but cigarettes did become part of the symbolism of this historic meeting in space.

The 1975 Apollo-Soyuz stood out as a dramatic moment of cooperation between Cold War enemies, but it was just one part of a larger effort of political détente.

The United States and Soviet Union also participated in joint public and private commercial ventures on Earth to commemorate the historic mission. This package of Apollo-Soyuz cigarettes was manufactured in a partnership between the American tobacco firm Phillip Morris and the Soviet Yava cigarette factory. A NASA archivist purchased this pack while visiting the Soviet Union and donated it to the museum in 1977. At the time of the Apollo-Soyuz agreement, diplomats hoped that joint manufacturing ventures that grew of the mission could sustain détente beyond the 1975 spaceflight. Although the promise of détente soon faded, a factory in Kazakhstan still manufactures these cigarettes under license with Phillip Morris.

Apollo-Soyuz cigarettes remained popular through the collapse of the Soviet Union as a symbol of luxury and because they tasted less harsh than traditional Soviet-grade tobacco products. As more Western-grade tobacco became available in the former Soviet Union, the Apollo-Soyuz cigarettes became less popular.

Artificial Gravity Experiment
1977

"Three young rats with black felt hats,
Three young ducks with white straw flats ..."

—TRADITIONAL NURSERY RHYME

IN NOVEMBER 1957, the dog Laika became the first living being to orbit the Earth. Fifty years later, we still know relatively little about spaceflight's effects on living organisms, including humans. In an era of relatively long-duration flights on space shuttles and space stations, such knowledge is crucial, particularly with regard to resulting conditions that are difficult to reverse, such as bone deterioration through the loss of minerals. Research on human subjects is difficult and risky, so animals take their place. Researchers use monkeys, dogs, and rats to deepen knowledge of the varied effects of space because of the biological similarities that exist between these animals and humans.

Flown in 1977 on the USSR satellite *Kosmos 936* (also known as *Bion 4*), these units contained experiments to test the effects of weightlessness and artificial gravity on rats. Part of an international research project, the mission drew together scientists from France, the USSR, and the United States. Of the 30 laboratory rats on board the spacecraft, 20 experienced microgravity and 10 flew in artificial Earth-like gravity, produced through rotating mechanisms in the experimental apparatus. The apparatus consisted of four sections: two housed 10 rats each in a stationary position allowing the rats to experience the microgravity of space; two other units rotated five rats each to simulate normal Earth gravity.

After a few weeks in orbit, the satellite returned to Earth. Scientists recovered the rats, still alive, and then sacrificed them in order to take measurements of the bone densities. They sought to determine whether the two sets of space-traveling rats showed any differences in bone mass after 19 days in orbit. Scientists concluded that although there were slight differences in bone mass between the two groups, they found a greater variance between all the "space" rats and a third group of rats that had remained on Earth as a control group. These results seemed to indicate that the retention of bone mass depends on more than just the presence of artificial gravity in spaceflight. Today, scientists are attempting to design experiments that will illuminate the complex mechanisms that regulate bone growth.

The Soviet Institute for Biomedical Problems donated these two units to the Museum in 1979.

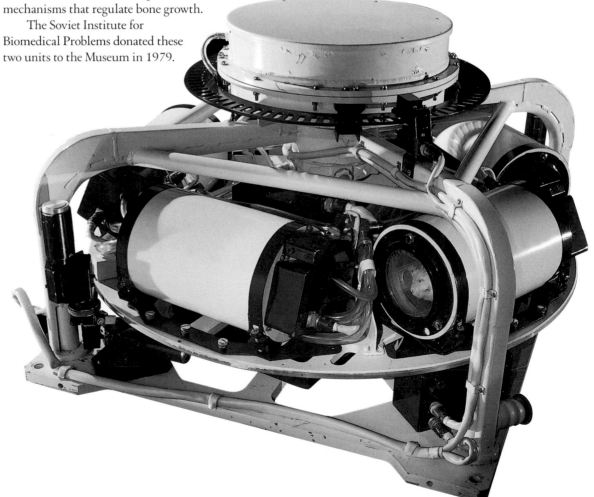

CRAY-1 Supercomputer
1977

"The performance of his machines—coupled with his well-known eccentricities—created an extraordinary mystique. Programmers wanted to attend his lecture not only to learn about supercomputing but to see Cray's face and hear his voice. They wanted to … [say] yes, they'd spoken with Seymour Cray."

—As told to author Charles J. Murray, *The Supermen*, 1997

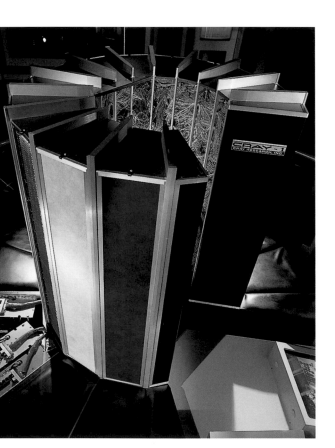

IN 1977, the National Center for Atmospheric Research (NCAR), located in Boulder, Colorado, acquired this $9 million computer, a CRAY-1. Named after its designer, Seymour Cray, the computer was believed to be about five times faster than the next fastest computer in the world, a Control Data computer, also designed by Cray. For NCAR the decision to use this expensive technology was easy: no other machine had the capability to help meteorologists solve large systems of equations necessary for understanding weather. The CRAY-1 accomplished in a few hours what an ordinary computer, even one of the better mainframes of the day, might take weeks or even years of round-the-clock number crunching to do. With a small team of engineers in a lab in Chippewa Falls, Wisconsin, Seymour Cray had found a way to design and build "supercomputers"—machines whose performance was to ordinary mainframes what a Ferrari was to a family station wagon. One of his customers called them "symphonies of tight design."

Cray computers had other applications, especially in aircraft and spacecraft design. In airplane design, the equations are just as complex as in weather modeling, and their solutions just as important. The pioneers of flight, including the Wright brothers, despaired of solving aerodynamic equations mathematically and instead took to building scale models of a wing or of a proposed airplane, placing it in a wind tunnel, and measuring the forces produced in a physical test. For decades, such methods were the bedrock of aerospace engineering. But newer aircraft and high-velocity spacecraft revealed their limits. In addition, these tunnels were expensive to operate and consumed huge amounts of electrical power. In this context, a $9 million computer seemed a bargain.

Since the late 1940s, specialty firms designed computers optimized for scientific calculations. But Cray originated and developed the "supercomputer," first at Control Data and later at his own company, Cray Research, Inc.—which donated this computer to the Museum in 1988. These computers had one design goal: to calculate arrays of numbers, called vectors, as rapidly as possible. Speed was the key to the success of CRAY-1 and other supercomputers. No one would abandon a wind tunnel for a computer that required days or weeks of round-the-clock operation to give an answer. But after the CRAY and its successors became available, the balance tipped away from physical models to models built of numbers. Wind tunnels are still used, but less and less each year, and they may follow the venerable slide rule into oblivion.

Like the world of Formula One race cars, the world of supercomputer design is not for the faint hearted. Few companies ever made money selling them. Cray Research itself eventually ran into financial trouble. One reason was that cheap personal computers kept getting faster. The CRAY-1's ability to perform about 150 million floating point operations a second is almost matched by the latest Macintosh, and its 64-bit word length is offered today by the Sony PlayStation game console. Still, the demand for the fastest possible computer remains, especially at government research and defense laboratories.

Impacts and Influences

Re-imagining the Earth and How We Live

Stewart Brand and the *Whole Earth* Community, 1968–

STEWART BRAND is a futurist and author whose creation of the *Whole Earth Catalog* led him to pioneer the Whole Earth 'Lectronic Link (the WELL), an online community that foreshadowed the Internet's capacity for connecting people to information, and to each other. A figure in the New York and San Francisco countercultures in the 1960s, Brand has also worked with scientists and researchers throughout his career, linking the ideals of communal living with the technologies of the information age.

In 1966, Brand began campaigning for NASA to make a priority of capturing an image of the whole Earth in space. He believed that such an image had the potential to change how people related to one another and to their planet. Brand made up buttons that asked, "Why haven't we seen a photograph of the Whole Earth yet?"—which he sold on college campuses and mailed to prominent scientists, futurists, and legislators. In 1972, Apollo 17 finally captured a photograph of the Earth as a complete sphere, without any solar shadow. Brand later described Whole Earth as an image that "gave the sense that Earth's an island, surrounded by a lot of inhospitable space. And it's so graphic, this little blue, white, green, and brown jewel-like icon amongst a quite featureless black vacuum."

Brand saw Whole Earth as a symbol of the importance of people finding the tools and information that they needed to live in harmony—with the environment and with each other. As a result, he took the Whole Earth image and name as the cover and title for his *Whole Earth Catalog*, created in 1968. A book that has been compared to "a space age *Walden*," Brand's buyer's guide (what he called an "evaluation and access device") reviewed a wide variety of "tools," including hand tools, philosophy books, gardening implements, and alternative energy technologies that would help each reader to "conduct his own education, find his own inspiration, shape his own environment, and share his adventure with whoever is interested." Through continual updates, including reviews contributed by readers, the *Whole Earth Catalog* energized far-flung networks of people to think differently about how they lived.

Brand also recognized that the rapid development of computers, driven in part by the space program, might enhance human interactions and improve life on Earth. By the late 1970s and early 1980s, the processing capacities and speeds demonstrated by machines like the CRAY-1 Super Computer exemplified the expanding computing power that inspired Brand. Excited about the potential that personal computing offered, Brand translated the *Whole Earth Catalog*'s "do it yourself" ethos and belief in the power of the right tools into the *Whole Earth Software Catalog* in 1984. By compiling evaluations that helped new computer users to negotiate the "daunting task" of buying computer software, Brand encouraged the use of computers by and for individuals, rather than institutions.

In 1984, Brand partnered with Larry Brilliant to found the Whole Earth 'Lectronic Link (the WELL), an electronic bulletin board system/prototype chatroom that created a virtual community among people otherwise separated by distance. Contributors could post exchanges about a variety of topics and chat using an early form of instant messaging. Intense intellectual and emotional relationships formed quickly in the new forum. As the space age matured into the information age in the 1980s, Stewart Brand helped lead the way in facilitating the adoption of computers into people's daily lives. He continues to be a leading voice in exploring the use and meaning of information technologies.

"This [Earth] is all we've got and we've got to make it work. There's no backup."

— STEWART BRAND

Close Encounters of the Third Kind Mother Ship Model
1977

"If this is just nerve gas, how come I know everything in such detail?
I've never been here before. How come I know so much?"

—RICHARD DREYFUSS, AS ROY NEARY, IN *Close Encounters of the Third Kind*, 1977

ABOVE: The Mother Ship's designers nestled a number of miniature comic details onto the model's complex exterior, including (from the top) a shark, *Star Wars*' R2-D2, a graveyard, a mailbox, a bus, and a *Star Wars* TIE fighter.

ARE WE alone in the universe? *Close Encounters of the Third Kind*, Steven Spielberg's 1977 film, offers one answer to this fundamental human question. As part of his dramatization of a close encounter of the third kind (actual alien contact), Spielberg developed several visions of possible spaceships. The most detailed and awe-inspiring was the Mother Ship—it rises over Devil's Tower at the film's climax, answering the five-note musical greeting that anchored the movie's score.

Originally, Spielberg conceived the alien Mother Ship as dark and imposing, blocking out the stars. But light rather than shadow came to define his concept. While shooting in India, he passed an oil refinery every day. Seeing the rig brightly lit against the early morning and late night sky changed his vision. He eventually ordered a model based on that image, transforming the refinery's towers and pipes into a city of light. A different experience, in the U.S., inspired the Mother Ship's base. The curved bottom portion echoed what Spielberg saw as he gazed at the San Fernando Valley reflected upside-down on the hood of his car.

Realizing Spielberg's vision fell to Gregory Jein, *Close Encounters'* chief model-maker. To get the desired effect, Jein and his team layered elements from train modeling kits onto the ship. Red and green signal lights sit among the towers and antennae. The spires along the sides are streetlights or trestles from train kits that have been bent into unfamiliar, alien shapes. It was painstaking work. Each tube (resembling old radio vacuum tubes) throughout the model was hand rolled. Thousands of small holes allowed neon light to shine through, completing the effect. To keep the film on schedule,

Spielberg and his secretaries pitched in, spending hours helping the model builders drill tiny holes all over the model.

Because the builders knew that the details would never be seen clearly in the final film, they included some inside jokes. A VW bus rests parked under an eave. A cemetery lies inside the front lip. World War II airplanes and a U.S. mailbox ride on the ship's hull. One can even find two small sharks deep inside the model, in homage to Spielberg's previous blockbuster, *Jaws* (1975). One such element does appear in the final film. Using super slow motion playback, a backlit R2-D2 robot can be spotted hanging upside down along the ship's edge (at the moment when the character Jillian first sees the Mother Ship from her rocky hiding place). Each feature demonstrates the handiwork required to build movie miniatures in the era before computer-generated imaging.

The aliens who arrive in the Mother Ship are ultimately benevolent. Through earlier UFO encounters, the visitors impart to everyday people such as Roy Neary (played by Richard Dreyfuss) detailed visions of Wyoming's Devil's Tower, the arrival point for a historic meeting between humans and aliens. Neary, like the other characters, finds himself compelled to reach this destination— even to the point of modeling the distinctive flat-topped mountain in his mashed potatoes. Riding in a vehicle of light and responding to musical greetings, the aliens provide a sharp contrast to the threatening aliens from 1950s films. Friendly and curious, their playful nature is aptly reflected in the model Jein constructed.

Columbia Pictures donated the Mother Ship to the Museum in 1979.

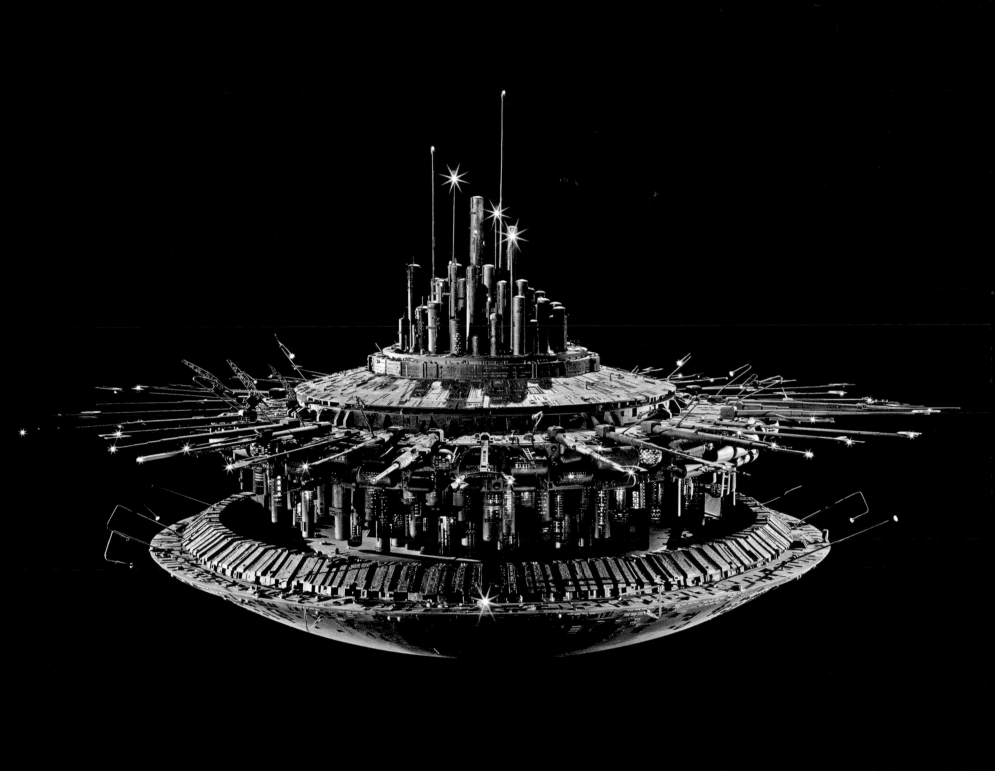

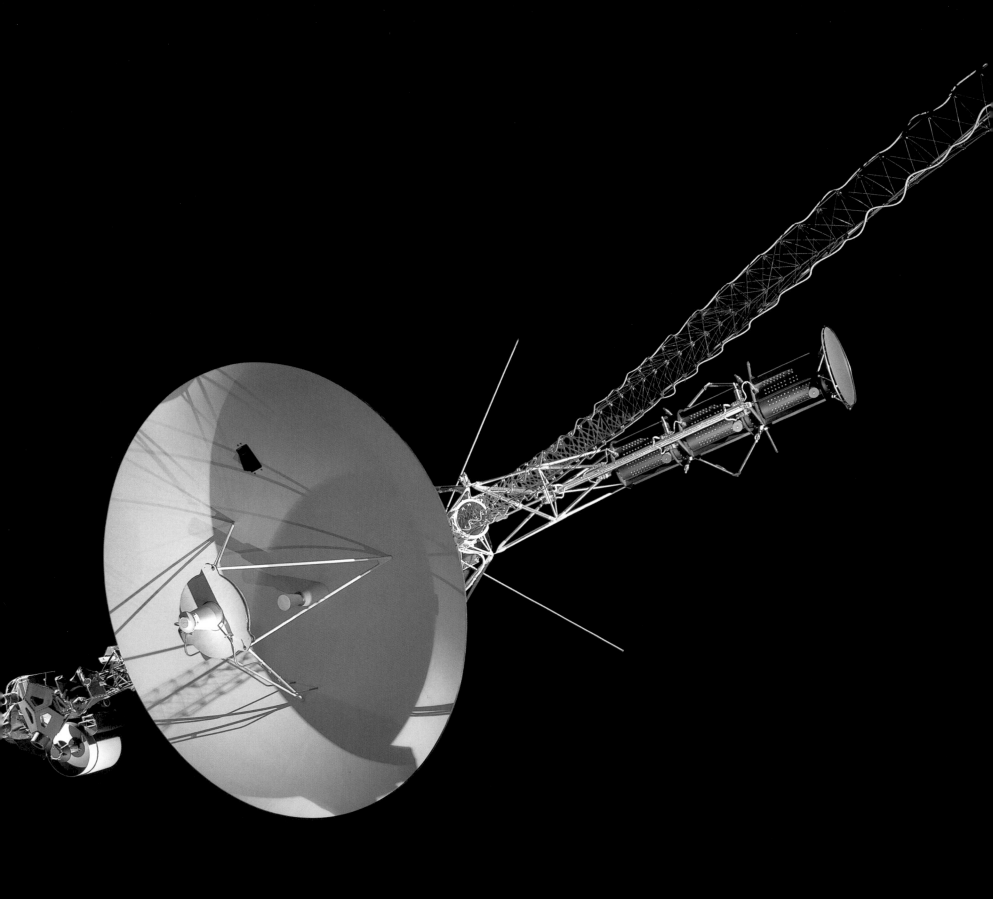

Voyager Mock-up
1977

"Voyager was the little spacecraft that could."

—Edward Stone, Voyager chief scientist, Jet Propulsion Laboratory, 1992

IN THE 1970s, NASA's leadership conceived their most daring planetary science expedition, Voyager. It took advantage of a rare astronomical event. Once every 176 years, the giant planets on the outer reaches of the solar system gather on one side of the Sun, and the next occurrence was to be in the late 1970s. The event would make it possible for a single spacecraft to make a "grand tour" of all the planets in the outer solar system (with the exception of Pluto). A technique known as "gravity assist" would aid the journey. As the spacecraft flew by, each planet would give it a slingshot boost, increasing its speed. Neptune, the outermost planet in the mission, thus could be reached in 12 rather than 30 years.

In planning for the venture, NASA benefited from its prior experience with Pioneer 10 and 11, the first probes to explore the solar system beyond Mars, launched in 1972 and 1973, respectively. Designed to last for 30 months, these craft performed for more than 20 years, returning useful data on Jupiter and Saturn and on interstellar space (our first look outside the solar system).

A four-planet tour posed a greater challenge. If undertaken by a single spacecraft, it would need to endure and operate reliably for more than a decade. Faced with this challenge, NASA scaled back the scope of the project. Rather, the new plan called for two Voyager spacecraft to conduct intensive flyby studies of Jupiter and Saturn only—in effect conducting an improved version of the flights of the two Pioneers. Launched from the Kennedy Space Center, Voyager 2 lifted off on August 20, 1977, and Voyager 1 on September 5, 1977, each taking different trajectories to the outer planets.

By December 1980, the two Voyager spacecraft had studied successfully Jupiter and Saturn and were still performing flawlessly. A grand tour now seemed possible—and irresistible—to mission scientists. Via radio communication, they reprogrammed the Voyagers to fly by the two outermost giant planets, Uranus and Neptune.

The two spacecraft returned information to Earth that revolutionized planetary science. Solving a centuries-old mystery, Voyager revealed that Jupiter's Great Red Spot was a complex storm moving in a counterclockwise direction. On Jupiter's moon Io, the spacecraft found active volcanoes—the first evidence this phenomenon existed in the solar system besides on Earth. Plumes from the volcanoes extend to more than 190 miles above the moon's surface.

The two Voyagers took well over 100,000 images of the outer planets, rings, and satellites, as well as millions of magnetic, chemical spectra, and radiation measurements. They discovered rings around Jupiter, satellites in Saturn's rings, new moons around Uranus and Neptune, and geysers on Triton. The last imaging sequence was Voyager 1's portrait of most of the solar system, showing Earth and six other planets as sparks in a dark sky lit by a single bright star, the Sun. In the first decade of the twenty-first century Voyager 1, still functioning, continued to provide important scientific data about the heliopause, the outer limits of the Sun's magnetic field and outward flow of the solar wind.

This artifact is a Development Test Model (DTM) that consists of facsimile and dummy parts manufactured by the Jet Propulsion Laboratory. It was acquired by the Museum in 1977 and placed on display in the *Exploring the Planets* gallery shortly thereafter. In 1987, JPL removed part of the artifact for use in developing the Magellan Venus spacecraft, which was similar in design, and replaced it here with a facsimile.

ABOVE: As Voyager 2 neared Neptune it caught this full view of the planet.

Voyager Record Cover
1977

"The spacecraft will be encountered and the record played only if there are advanced spacefaring civilizations in interstellar space. But … the launching of this 'bottle' into the cosmic 'ocean' says something very hopeful about life on this planet."

—CARL SAGAN, 1977

ABOVE: Technicians in an assembly "clean room" mount a *Sounds of Earth* record onto the exterior of Voyager 1.

VOYAGER 1 and 2 were the third and fourth human artifacts to leave the solar system and encounter interstellar space. Pioneers 10 and 11 had done so previously, carrying small metal plaques identifying their place of origin for the benefit of any extraterrestrials who might encounter them in the distant future. For the Voyager missions, NASA built on the Pioneer 10 and 11 experiences and decided to affix a "time capsule" to the exterior of the two spacecraft. The purpose: to communicate with life beyond Earth.

The result was the Voyager *Sounds of Earth* record, a 1970s-era analogue disk. The brainchild of Cornell University astrophysicist Carl Sagan, the record contained sounds and images selected to portray the diversity of life and culture on Earth. U.S. President Jimmy Carter recorded greetings to this life beyond, evocatively commenting, "This is a present from a small, distant world, a token of our sounds, our science, our images, our music, our thoughts, and our feelings. We are attempting to survive our time so we may live into yours."

A gold-plated aluminum cover protected the *Sounds of Earth* disk, each Voyager spacecraft carrying an identical set. The cover featured etchings providing instructions for playing the recording. Secured in plain view, on the outside of the spacecraft, the cover and record came complete with a stylus cartridge for playing the disk. Sagan believed that barring a major collision with something else, the spacecraft and record should last for as long as a billion years.

The record itself consisted of two 12-inch gold-plated copper disks, bonded back to back. The side that faced toward the spacecraft contained 122 images, as well as human greetings in 55 languages, various sounds of Earth, and music selections. The outer side consisted entirely of music. Sagan, who led the committee that decided what to include on the disk, sought to represent the broad range of the human experience. The spoken greetings included Akkadian (spoken in Sumer about 6,000 years ago) and Wu, a modern Chinese dialect, as well as many other current languages. The music included an eclectic 90-minute musical program, in which Chuck Berry's "Johnny B Goode" and Mozart's "Magic Flute" vied with a Zairian Pygmy girls' initiation song, a *shakuhachi* piece from Japan, and "Dark Was the Night," written and performed by Blind Willie Johnson.

The Voyager record captured the imagination of Americans. It represented a message in a bottle cast into the cosmic ocean, and speculation abounded about the possibilities of contact with life beyond this planet. The 1979 feature film *Star Trek: The Motion Picture* reflected the public's heightened interest in worlds beyond our own. In this film, a massive and menacing spacecraft known as *V'ger* returns to Earth. As the plot unfolds, the crew of the USS *Enterprise* defeat *V'ger*, which turns out to be a damaged Voyager spacecraft, seeking its origins, that is led back to Earth by the Voyager record. In addition, the 1984 film *Starman* featured an alien visiting Earth, invited here by the Voyager record. The potential for alien contact via the Voyager record is remote, but remains a tantalizing possibility.

The Voyager record in the Museum's collection transferred from NASA in 1978.

Impacts and Influences

Envisioning the Future as a Dark Place

Ridley Scott's *Alien* and *Blade Runner*, 1979 and 1982

A s the 1970s ended and the 1980s began, British director Ridley Scott offered movie-going audiences two disturbing visions of life in the future and in space. *Alien* (1979) presented a space thriller in which the crew of a deep space mining ship encounters a hostile alien on a deserted planet. In *Blade Runner* (1982), based on Philip K. Dick's 1968 novel *Do Androids Dream of Electric Sheep?*, a bounty hunter or "blade runner" in a post-apocalyptic Los Angeles of 2019 hunts down "replicants," genetically engineered cyborgs created as slave labor. In both films, Ridley Scott's directorial choices depicted dystopias: future worlds (or in these films, solar systems or universes) where humanity's worst characteristics shaped existence. Outer space no longer represented a haven for humanity and its best attributes, but its complete opposite, where, as *Alien's* movie posters declared, "no one can hear you scream."

In contrast to optimistic views of space exploration, such as those seen in the late 1960s television show *Star Trek*, Scott's films depict an outer space that is dehumanizing—yielding a way of life shaped by corporate and government intrigue, and "killing machine" life-forms better adapted to extreme environments. In *Alien*, the mining crew inhabits a spaceship that has been designed to create wealth—to extract ore and return it to a corporation—not to nurture humans. Indeed, humans have only been included as a business necessity. To emphasize the fragility of the humans, the planet surface shown in *Alien* is desolate, a place meant only for machines and the machine-like. Howling high winds drown out all other sounds, revealing the isolation of the individual and the planet's unsuitability for human community. Forsaken by the impersonal corporation that sent them into space, the crew fight to survive without any hope of rescue.

Likewise, Scott's depiction of post-holocaust Los Angeles in *Blade Runner* is a vision of a city where corporate capitalism dominates the urban landscape. Moreover, that single dystopian city lies within a larger dystopian solar system. Scott depicts the future Earth as embedded in a commercialized universe, where choices are driven by profit, oppression, and domination. Garish blimps adorned with lighted advertisements blare commercials for life on the outer planets, a constant reminder that Earth has become just one planet in space—and perhaps the least desirable place to live. *Blade Runner* suggests a plausible future, not far off, in which fundamental human values are in eclipse.

Both *Alien* and *Blade Runner* dramatize the struggle of individual human beings to survive in these worlds. The crew of the mining ship *Nostromo* fights for their lives against an alien that has been abetted by an artificial human-like life form and protected by orders from the mining corporation. Blade runner Rick Deckard, played by Harrison Ford, questions his own humanity—and the nature of what makes humans unique—when confronted by what he learns about the replicants. In both films, corporations, and the people who run them, seek profit and advantage above all else. Individuals are treated as expendable. Entwined in both stories is a deep suspicion of technology and power. In an age of pragmatism, Scott's films represent the darker side of science fiction, a reflection of a more cynical time.

"In space no one can hear you scream."

—Advertising tagline for *Alien* (1979)

International Ultraviolet Explorer
Mid-1970s

"The little satellite that could."

—*The New York Times,* SEPTEMBER 21, 2002

ABOVE: IUE's user-friendly operations console displays the first image of a star observed by the astronomical satellite. Note the small joystick at the lower right that allowed astronomers to easily direct IUE.

THE International Ultraviolet Explorer (IUE) was the first "video game" astronomical satellite—astronomers on Earth operated it sitting at a TV console using a keyboard and joystick. This control station had the appearance of a home video game, but the action it portrayed was real, created by a spacecraft orbiting 32,000 kilometers (22,000 miles) above the Earth. Assisted by a telescope operator sitting close by, any astronomer could direct the telescope, focus on a celestial object, and then capture its spectrum with the ultraviolet detectors on the spacecraft. The station, too, allowed the astronomer to evaluate and process the images of the resulting spectra immediately.

This operational simplicity was intended to make the IUE available to as many astronomers as possible. Until IUE (launched in 1978), astronomical satellites operated as small clubs, used mostly by the scientists who developed the principal instruments. But in the 1970s, astronomers from all over the world pushed for new procedures to allow more democratic access to satellites. The result was a broadly international venture, supported by NASA, the European Space Agency, and the British Science and Engineering Research Council.

The emphasis on utility carried over to IUE's design. The satellite combined a modest 16-inch reflecting telescope with two arrays of ultraviolet spectrographs. It remained active for more than 18 years. Thousands of astronomers used it to observe planets, stars, nebulae, and galaxies in ultraviolet wavelengths.

The Museum preserves an engineering mock-up of the IUE spacecraft and the engineering test unit of its telescope, as well as an engineering control console designed and built by the Bendix Corporation. This latter unit contains TV consoles, the façade of a PDP-11/35 computer, a keyboard, and a joystick. It is identical in appearance and function to the unit used by astronomers.

As some of IUE's stabilizing gyros started to fail, NASA felt that the satellite had reached the end of its operational life. The European Space Agency continued to support it because so many astronomers used it in research. When controllers finally turned off the satellite in 1996, it had acquired a total of more than 100,000 ultraviolet spectral images during 60,000 hours of observing. Scientists produced thousands of papers based on IUE data, and data archives will yield many more. In 1978, IUE was the only American astronomical satellite operating when scientists detected a supernova in the Large Magellanic Cloud, the brightest in almost 400 years. Joystick-using astronomers used IUE to watch this rare event, recording the supernova's most violent phases. IUE was the "little satellite that could."

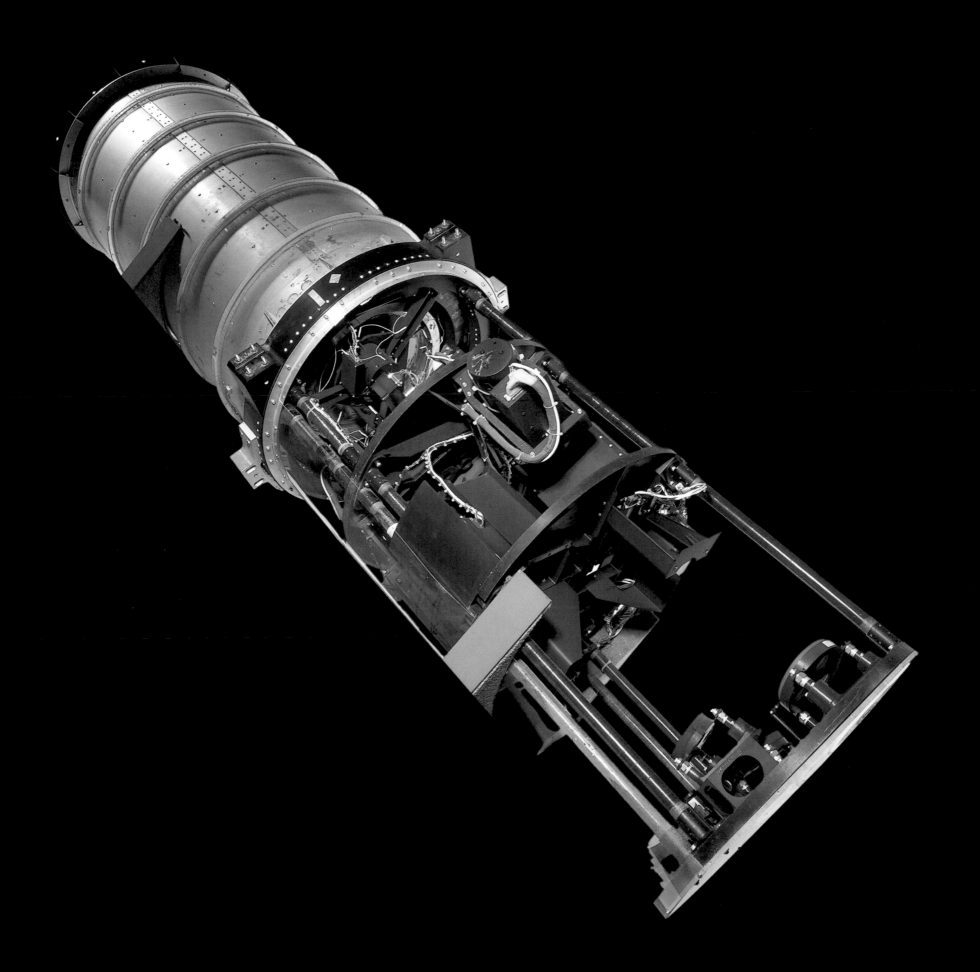

The Empire Strikes Back Figurines and Carrying Case
1980

"A long time ago in a galaxy far, far away…"

—TAGLINE FOR THE *Star Wars* FILM SERIES

I N 1977, George Lucas inaugurated his famous film series, *Star Wars*. The first installment, *A New Hope*, became one of the biggest box office hits of all time and gave rise to an innovative mass-marketing campaign for toys and other products that became an industry model. Through movie souvenirs, Lucas sought to create a strong bond between filmgoers and the *Star Wars* experience beyond the theater. This action figure set for *The Empire Strikes Back*, the sequel to *A New Hope*, highlights Lucas's formula of intermingling fun, mass marketing, and fan loyalty and commitment to *Star Wars*' fictional universe.

Beginning with the first *Star Wars* film, Lucas focused on product development, integrating it with filmmaking and promotion. His marketing strategy was simple: to give filmgoers, through toys and other merchandise, another avenue to reexperience a film. This overall strategy proved a financial success and shaped future movie-marketing across the industry. Yet in 1977, George Lucas and his production company, Lucasfilm, Ltd., initially could not persuade a single merchandiser to create the toys they envisioned. Lucas, who insisted on retaining all merchandising rights for his *Star Wars* movies, eventually signed a last-minute contract with the Kenner toy company, which managed to manufacture only a few products in time for the release of *A New Hope* on May 25, 1977. As the 1977 winter holiday season approached, merchandising efforts still lagged and Kenner was forced to sell empty action figure boxes, with a promise to send the toys to the consumer once completed. This inadvertent circumstance added a heightened sense of drama to the marketing rollout, stimulating *Star Wars* fans' desire for the movie's souvenirs. The first figurines did not hit store shelves until early 1978. As a marker of fan passion and loyalty, *Star Wars* set a new standard as parents willingly gave their children the gift of an empty box and hoped the figures might arrive later. These first figurine sets now are highly prized by collectors, especially when accompanied by the original packaging and infamous "empty box." Kenner and Lucasfilm learned from this experience and were well prepared for the merchandising boom that came with the release on May 21, 1980, of *The Empire Strikes Back*.

Perhaps one thing that made *Star Wars* toys and products so popular was the strong link between filmmaking and merchandising. The 9.5-centimeter (3.75-inch) figures, unique in size, produced by Kenner became a hallmark of the *Star Wars* toy universe, a universe that also included plush toys, fake lightsabers, children's costumes, games, and literally hundreds of other products. *Star Wars*, more than any other single film or series, inspired legions of children (and adults) to recreate their viewing experience through the purchase of products.

The billions of dollars in revenue generated by the film *A New Hope* and its souvenirs resulted in a Hollywood first in providing the financing for two additional movies, *The Empire Strikes Back* and *Return of the Jedi* (1983). In addition, it enabled expansion of the original product line, which in turn financed digitization and revision of the first three films for re-release in theaters and on DVD, and the production of an entirely new trilogy. The success of the *Star Wars* series in the late 1970s and early 1980s also inspired revival of older science fiction series such as *Buck Rogers* and *Flash Gordon*, all of which added to the public's perception of spaceflight in the first years of NASA's Space Shuttle program.

Michael O'Harro, a private collector, donated this set to the Museum in 1997.

Homing Overlay Experiment Test Vehicle
1983–1984

"The highly successful flight of the Homing Overlay Experiment (HOE) on 10 June 1984 establishes an important technological watershed for the national goal of eliminating the threat posed by nuclear-armed ballistic missiles."

—Lt. Gen. James Abrahamson, director, Strategic Defense Initiative Organization, 1984

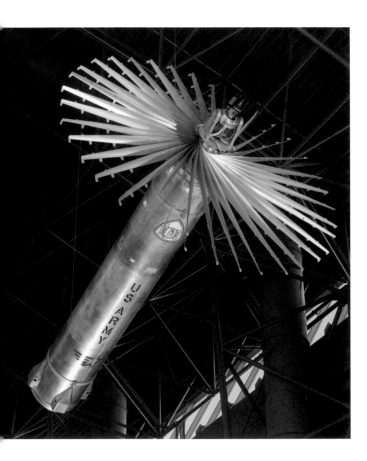

I N 1983, President Ronald Reagan announced the Strategic Defense Initiative (SDI), an ambitious, first of its kind effort to develop a comprehensive defensive shield against Soviet nuclear-armed missiles. The long-term goal was to remove the threat of nuclear attack and move the world away from the prospect of annihilation—a persistent fear throughout the Cold War. The plan (often called "Star Wars" in the media) specified a mix of ground- and space-based non-nuclear ballistic missile defenses. The Homing Overlay Experiment (HOE) was one part of this mix.

Prior to Reagan's announcement, neither the United States nor the USSR had developed or deployed extensive anti-ballistic missile (ABM) forces. Both nations recognized that an arms race in these defensive systems likely provided little or no strategic advantage, and consequently signed the ABM Treaty of 1972 limiting each to a total of 200 interceptors at two sites. Under the treaty, the U.S. built just one base in North Dakota with nuclear-armed ABMs to protect intercontinental ballistic missiles; it operated briefly, closing in 1976—the only active ABM system the nation had during the Cold War.

Among the existing ABM projects that received increased attention after President Reagan's announcement was the U.S. Army's HOE, designed to test whether an interceptor could home-in on a reentry vehicle outside the atmosphere and destroy it by physical impact rather than detonation of a warhead. Flight tests of the Lockheed-built HOE started in 1983.

During these tests, the military launched a dummy warhead from California toward a point north of Kwajalein Island, 7,240 kilometers (4,500 miles) away. Ground-based radars tracked it and passed the data on to the HOE vehicle—separately launched from Kwajalein, aiming to intercept the dummy warhead in space. After separating from the booster, the HOE's infrared tracker locked on to the dummy warhead, a small propulsion system became operational, and an onboard computer controlled the trajectory. Near the target, a metal ribbed array deployed, the purpose of which was to increase the chance of a collision.

The fourth and final test in June 1984 resulted in an interception, destroying the dummy warhead at an altitude of over 160 kilometers (100 miles). This was the world's first successful demonstration of "hit-to-kill" technology, prompting General Abrahamson, SDI director, to hail American technical ingenuity. The Museum's HOE test vehicle, transferred from the Army in 1986, was built for this series of tests but never flew.

Because the end of the Cold War greatly decreased the threat from Soviet/Russian nuclear-armed missiles, interest in the effort declined and President Reagan's goal of a defensive shield was never realized. In the 1990s, however, new ballistic missile and nuclear threats arose from North Korea and other nations. The ABM Treaty prohibited the United States from building effective defenses against these and, in 2002, President George W. Bush decided to withdraw from the treaty. Several ABM programs now are underway, using new types of hit-to-kill vehicles. As part of this policy, the U.S. deployed Ground-Based Interceptor ABMs in Alaska in 2004 and Standard Missile-3 ABMs on selected U.S. Navy ships in 2005.

Air-Launched Anti-Satellite Missile
Early 1980s

"The upside from stopping further ASAT testing and deployment far outweighs any risks … Once ASAT
systems are developed and deployed, stability is threatened. The time to curb them is now."

—SEN. JOHN CHAFFEE (R.-R.I.), OCTOBER 1985

I N THE years following Sputnik, both the Soviet Union and the United States launched a wide range of spacecraft that performed critical national security missions, among them photoreconnaissance, signals intelligence, and early warning of missile attack. Each type of satellite proved the unique value of space for spying or gathering data—these spacecraft literally were the "eyes" and "ears" of the Cold War superpowers. But this very success raised a difficult question: should spacecraft, as much as tanks or airplanes, become targets in the event of war? Senator Chaffee's comments reflected one side of a vigorous debate in the 1970s and 1980s on space, warfare, and anti-satellite weapons—a debate still not resolved.

Beginning in the 1960s, each superpower began to develop weapons capable of disabling or destroying satellites. As a very limited and interim anti-satellite (ASAT) system, the United States deployed a small number of Thor missiles with nuclear warheads in the Pacific from the 1960s to 1975. During the same period, the Soviets began tests of "killer satellites"—in trials, the "killers" apparently worked by maneuvering close to orbiting target vehicles and then detonating an explosive charge. These tests stopped, however, for almost five years in the early 1970s.

Each nation had demonstrated its ASAT capability. Yet each feared provoking a very expensive arms race in a new area of technology that ultimately provided no strategic advantage.

Despite the prospect of an arms race, the Soviets resumed testing of "killer satellites" in 1976, pushing the United States to reassess its position. In early January 1977, President Gerald Ford signed a top secret directive to acquire a non-nuclear ASAT capability that included an interceptor able to destroy up to 10 Soviet military satellites in low-Earth orbit and a system able to "electronically nullify" such vehicles at all altitudes.

The Carter administration adopted a "two-track" policy of both negotiating an ASAT ban with the USSR and acquiring an ASAT capability. One interceptor under development was the U.S. Air Force's miniature homing vehicle (MHV), which had infrared telescopes, a laser gyroscope, a computer, and small solid rocket motors. Its computer combined the telescope and gyroscope information to determine when to fire the rockets to adjust the vehicle's trajectory. The vehicle destroyed the target satellite not by detonation of a warhead but by direct collision—a concept known as "hit-to-kill." In 1978, the Air Force decided to place the MHV on a two-stage missile to be launched from F-15 fighters.

Negotiations with the Soviets on a ban were unproductive, and flight tests of the air-launched ASAT missile began in 1983. The first and only successful interception occurred on September 13, 1985, when a MHV reached and destroyed its target, the P78-1 Solwind scientific satellite orbiting 555 kilometers (345 miles) above the Earth. Congress quickly imposed restrictions on further testing and funding, putting the program on hold. It was formally cancelled in 1988. The Museum's air-launched ASAT, transferred from the U.S. Air Force in 1990, was built during the early 1980s but never flew.

Although research and development continued on other types of interceptors, the air-launched ASAT was the only one flight-tested. A ban on or limitation of ASAT weapons has remained the subject of international negotiations in the post–Cold War era, but thus far no agreements have been reached.

ABOVE: An F-15 fighter launches an ASAT during a test in September 1985. In this successful test, the ASAT went into space and destroyed a target satellite.

Enterprise, the First Space Shuttle
1976

"'Roll Out' of Space Shuttle Hailed as Start of a New Era"

—*The New York Times* HEADLINE, SEPTEMBER 18, 1976

ABOVE: *Enterprise* is released from a Boeing 747 to conduct tests of the Shuttle's landing and descent systems. These tests in 1977 prepared the way for *Columbia* to make the Shuttle's first spaceflight in 1981.

THE DEBUT in 1976 of *Enterprise*, the first Space Shuttle orbiter, signaled a revolution in space transportation. Gone were the days of small space capsules, used once and retired after parachuting into the ocean. Here was a large spacecraft with wings and wheels, designed to land on a runway like an airplane and fly again and again. NASA introduced *Enterprise* as the flagship for a fleet of reusable vehicles promising to make human spaceflight routine and economical.

After the Apollo missions to the Moon, NASA focused on developing the Shuttle as a multipurpose launch vehicle, cargo carrier, repair station, scientific platform, and crew transporter to operate in Earth orbit, arguing that a reusable, frequently launched vehicle would dramatically lower the cost of space access. The Shuttle signaled a shift from *exploration* to *exploitation* of space for practical benefits on Earth, commercial and scientific activity in orbit, and perhaps eventual passenger service.

Enterprise rolled out of the Rockwell International assembly plant in Palmdale, California, in 1976 amid anticipation about this new era in spaceflight. The next year it completed 13 successful approach and landing tests from altitudes of about 7,600 meters (25,000 feet), carried aloft on a Boeing 747 carrier aircraft at nearby NASA Dryden Flight Research Center. Flying attached to the aircraft and also released for piloted descents and landings, the orbiter passed its systems checkouts and proved its airworthiness. After these test flights, *Enterprise* moved to Marshall Space Flight Center in Alabama for vibration tests and the Kennedy Space Center in Florida for launch complex fit checks.

Upon completing these duties, NASA planned to refurbish *Enterprise* for spaceflight by installing propulsion systems, thermal protection tiles, and other features not needed for the atmospheric test

flights. As cost and schedule pressures mounted, NASA abandoned this plan, and *Enterprise* never flew in space.

Instead *Enterprise* appeared at the Paris Air Show and other sites in Europe in 1983, and at the 1984 World's Fair in New Orleans. Everywhere people thronged to see the first Space Shuttle. In 1985 its final task was fit checks at Vandenberg Air Force Base in California. NASA then transferred *Enterprise* to the Museum, where it waited in storage until going on display in the new Steven F. Udvar-Hazy Center in 2004.

In retirement, *Enterprise* continued to serve the space program. After the *Challenger* tragedy in 1986, it was briefly considered as a possible replacement orbiter. Immediately after the *Columbia* loss in 2003, NASA borrowed portions of the *Enterprise* wings and landing gear doors to help in the accident investigation. Engineering teams visited the original Shuttle many times to inspect, sample, and learn from it or to borrow components for testing.

For many, *Enterprise* heralded an optimistic era of routine spaceflight. But it also stands as a prominent symbol of wavering support for human space exploration. *Enterprise* shared a name with the legendary *Star Trek* starship but did not fulfill its own planned destiny in space, nor did the Shuttle program, with a total of five orbiters (*Columbia*, *Challenger*, *Discovery*, *Atlantis*, and *Endeavour*), succeed in making spaceflight routine or economical.

Yet when *Enterprise* first appeared in public in 1976, and when it reappeared on display almost 30 years later, it impressed—even awed—those who saw it. Its size, its wings, its very presence testify to a desire to make space travel as accessible and routine as aviation.

Columbia Space Shuttle Tiles
1981

"Every tile is different, and every tile has to be precision-machined."

—NASA's Space Shuttle program director, 1979

SHUTTLE tiles look rather simple. They are either black or white, mostly squares the size of a piece of toast, some oddly shaped and chunky, but all surprisingly lightweight. They do not appear to mark the boundary between life and death, but in fact they protect the Space Shuttle orbiter, its astronaut crew, and everything aboard from the searing heat of atmospheric reentry.

The tiles are quite complex. They are high-tech ceramics, made of silica fibers derived from sand and processed into a foam that is molded into blocks and baked in a furnace, machined into precise shapes, glazed with glass, and waterproofed. Each one is custom fit for its place on the orbiter.

About 700 white tiles (and more than 3,000 white quilted insulation blankets) cover parts of the vehicle where reentry temperatures remain below 650°C (1,200°F). More than 23,000 black tiles shield the underside expanse and certain defined areas where the heat rises up to 1,250°C (2,300°F). They fit together like pieces of a jigsaw puzzle to cover most of the Shuttle's aluminum skin, which can be easily melted.

The tiles are impervious to the blast-furnace temperatures of reentry. They dissipate heat rapidly and do not char or burn. Heat shields on all other spacecraft have been designed for one-time use—after the punishing return trip through the atmosphere, flown spacecraft with ruined heat shields are retired to museums. The Shuttle, however, is a reusable spacecraft, and heat-resistant tiles are the technical marvels that permit its repeated flights.

The Space Shuttle program itself has taken a fair amount of heat over the years. The vehicle's rows of tiles inspired some benignly critical, even humorous, depictions of the Shuttle as a flying brickyard. But more often the critique was barbed, especially during the 1979–1981 period when the first Shuttle launch was delayed repeatedly owing to difficulties bonding the tiles to the vehicle. The nation's technological reputation seemed to be riding on ceramic tiles and glue, and each setback fueled concerns about the cost and value of the Space Shuttle effort.

As the April 1981 launch of the first Shuttle mission approached, the media focused on anxiety about the tiles. If they popped off as they had during assembly, would the crew be doomed? Indeed, the orbiter *Columbia* did lose some tiles during liftoff, but the vehicle and crew completed their mission and returned safely. As technicians readied the Shuttle for its next mission, they inspected every tile and replaced any damaged during flight or landing. The two tiles pictured here are from a small batch of tiles removed during that repair and transferred to the Museum from NASA.

Over the years, the issue of the tiles faded. Regular inspections, improved manufacturing and application processes, and a successful track record kept the tiles themselves out of trouble. Yet the tiles also served to represent the myriad technical problems that plagued the complex Space Shuttle, raising concerns about NASA's management and expertise. Both the agency and the vehicle faced constant scrutiny that grew most harsh when equipment failed, as it did fatally when seven astronauts died in the 1986 *Challenger* launch tragedy and seven more perished during *Columbia*'s return from space in 2003. In both instances, flawed technology and flawed decision-making compromised safety.

The Shuttle's tiles perform well as the thin layer of protection between safety and incineration. But NASA itself bears responsibility for ensuring the safety of astronauts and Shuttles—and NASA takes the heat of public opinion and political scrutiny when failure occurs.

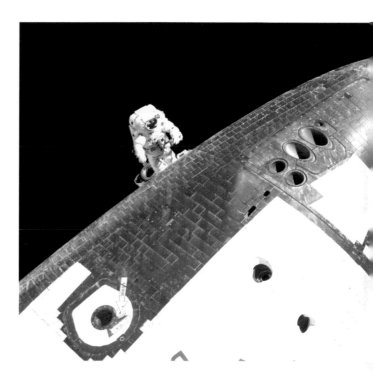

ABOVE: On STS-114 mission, August 2005, astronaut Stephen Robinson prepares to remove loose "gap fillers" from between the tiles as a safety precaution before the Shuttle returns to Earth.

The Space Shuttle Main Engine (SSME)
1980s–1990s

"The SSME is the highest performing and first reusable large rocket engine in production in the world."

—*Space*, 1989

ABOVE: During a launch, the SSME (located in the Shuttle orbiter) ignites first, followed by the "strap on" solid rocket boosters.

TOP, LEFT AND RIGHT: The Space Shuttle main engine fires up during tests at NASA's Stennis Space Center.

THE SPACE Shuttle main engine (SSME), the most advanced liquid-fuel rocket engine ever built, evolved from a long line of engines that date back to the 1940s. A cluster of three SSMEs, each producing a thrust of about 1.6 million newtons (375,000 pounds), powers the orbiter part of the Space Shuttle, or about 5 million newtons (1,125,000 pounds) total. Two huge solid rocket boosters (SRBs) strapped to the Shuttle each develop nearly 13 million newtons (2.9 million pounds) for a total liftoff thrust of more than 30 million newtons (6.9 million pounds). Both the SSMEs and SRBs are reusable, but the SRBs are ejected after burning, recovered, cleaned out, and refilled. Built into a Shuttle, an SSME is used in a succession of flights. After a certain number of uses (just like a high-performance automobile engine) it must be overhauled. Other liquid-fuel engines are used for one time only.

Another big difference from typical liquid-fuel engines is the SSME's electronic controller. Using a solid-state digital computer, the controller automatically monitors and regulates all engine functions every second and keeps a record of the engine's operational history for maintenance. This continuous control improves engine efficiency dramatically, as does the use of liquid oxygen and liquid hydrogen—the highest performing chemical propellants available—directed into the engine at extremely high pressures. In combination, these innovations make the SSME the most efficient rocket-engine technology ever developed.

Yet despite its ultra-high technology, the SSME looks and operates in its fundamental design much like its ancestor rocket engines. This evolution began with the Navaho missile project of North American Aviation in 1946. The World War II German V-2 served as the initial inspiration for North American, but in 1949 North American developed a revolutionary new engine design—cylindrically shaped, with a single flat plate injector on top and a conical nozzle. During the 1950s, they augmented this design with more powerful pumps and a bell-shaped nozzle that increased the engine's performance. This engine configuration made North American's Rocketdyne division the premier builder of rocket engines—it propelled the Thor, Atlas, and Jupiter missiles, and the massive Saturn V that took humans to the Moon.

In 1971, as the Apollo program neared completion, NASA selected Rocketdyne as the prime contractor for the engine for the next effort in human space exploration—a reusable space vehicle known as the Space Shuttle. To simplify development, Rocketdyne retained its basic engine design, refined over two decades, rather than undertake an entirely new and costly experimental approach. But to create the SSME, they added significant improvements, including giving the astronauts a kind of throttle to increase or decrease engine thrust, and, most important, to make it reusable. On April 12, 1981, NASA launched the Shuttle for the first time and the engines performed flawlessly. The U.S. entered the age of the reusable space vehicle.

The Museum's artifact is built from engine parts taken from SSMEs that flew in space on several different missions: the first four Shuttle flights, the second Hubble Space Telescope repair mission, John Glenn's Shuttle flight, and missions that launched the Magellan and Galileo space probes to the outer planets.

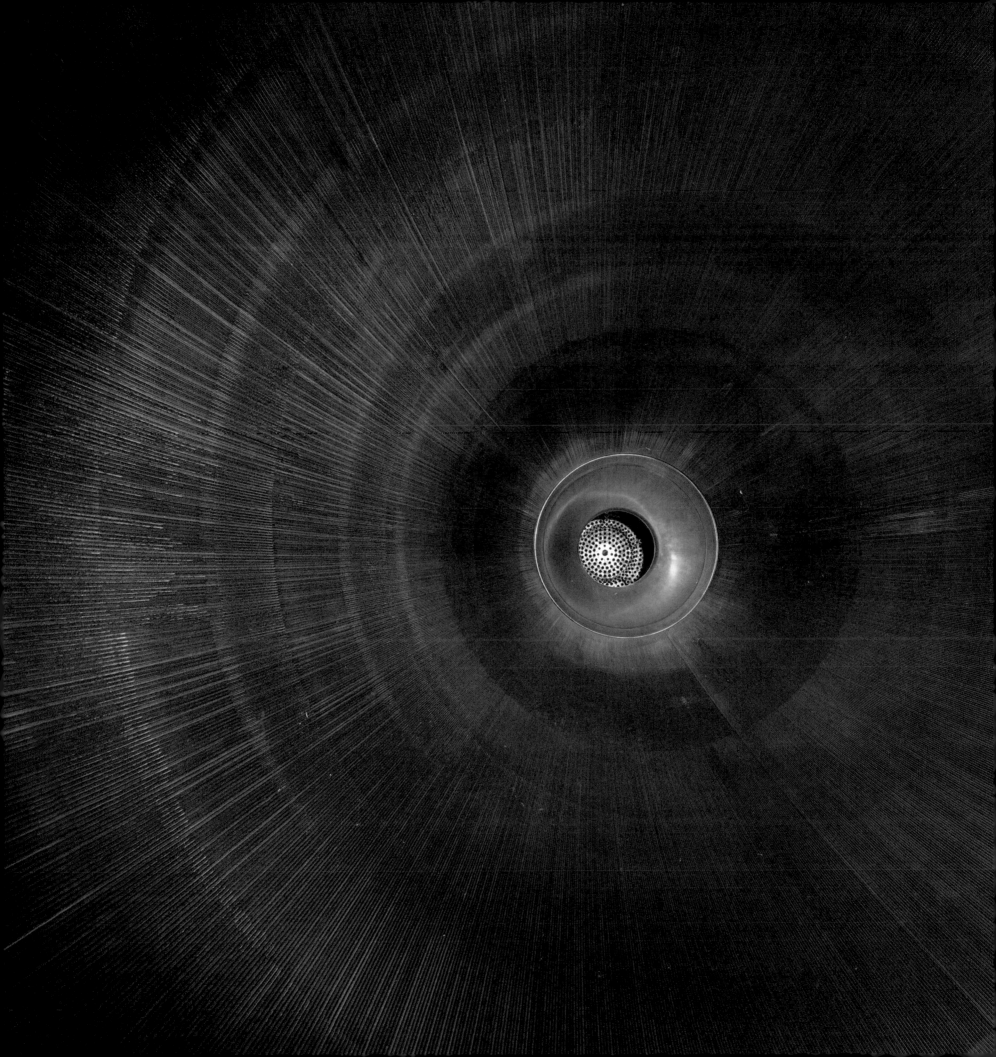

Sally Ride's Flight Suit, Space Shuttle Mission STS-7
1983

*"I didn't come into this program to be the first woman in space.
I came in to get a chance to fly in space."*

—ASTRONAUT SALLY RIDE, 1983

ABOVE: Astronaut Sally Ride, wearing her coveralls and communications headset, floats alongside the Shuttle's mid-deck airlock during her flight in June 1983.

To LOOK at these garments, you wouldn't know that they had been worn by the first American woman to go into space—or even that they belonged to a woman. The jacket, trousers, polo shirt, and even the Velcro appear decidedly unisex.

Until 1978, the U.S. astronaut corps also was unisex—all male. In the high-pressure years of the Space Race in the 1960s, NASA recruited astronauts from the ranks of the best combat and test pilots in the military services. Barred from such military programs, women could not acquire the necessary flight experience to become astronauts. Women entered space in science fiction long before they flew in U.S. spacecraft.

As the Shuttle era dawned in the 1970s, changes in American society affected the space program. Feminism, civil rights advocacy, and equal opportunity legislation pushed open doors that had been closed to women and members of racial groups. NASA and the public decided that it was important to have an astronaut corps that better represented the diverse U.S. population, and it actively sought women and racial minorities.

The Space Shuttle, roomier than the early space capsules, offered the possibility for integrating the sexes in space. Previously, the men who flew in space together did everything together—ate, slept, changed clothes, relieved themselves—within inches of each other and with no privacy whatsoever. The larger Shuttle and larger crew complement offered a modicum of privacy along with camaraderie, thus making it socially acceptable for men and women to share space in space.

The Shuttle era also called for a new type of astronaut: mission specialists. They had to be scientists or engineers and were trained to conduct experiments, spacewalk, manage the spacecraft's systems, and contribute to everything but piloting. This change opened the astronaut corps to a wider range of candidates and expanded opportunities for women and members of racial minorities.

In 1978, NASA selected its inaugural class of Shuttle astronauts, among them six women, and soon made Dr. Sally K. Ride, a physicist, the first woman assigned to a mission crew. Although they acknowledged their status as "first women" in a highly visible role, these women steadfastly considered themselves astronauts, not women astronauts, and they insisted on being treated as "one of the guys." Neither Sally Ride nor the other women wanted special accommodations, relaxed requirements, or any deference based on gender. Nor did they want different clothing or uniforms. Like the male astronauts, they saw themselves as members of a team. Sally Ride especially was reluctant to stand in the limelight apart from her crewmates or the other members of her astronaut class.

Twenty years after the Soviets first sent a woman into space, Sally Ride flew on the seventh Space Shuttle mission (STS-7) in 1983. She deployed satellites, operated the Shuttle's robotic arm, conducted several experiments, and, perhaps most important, showed that a woman in space and a mixed-sex crew had no problems.

After the mission, NASA sent Sally Ride to the Museum to deliver her flight suit for display. From time to time the Museum receives an inquiry from a visitor demanding to know why the name badge on her flight suit says *Sally*, not *Sally Ride*—suggesting that it is disrespectful not to use her full name. Actually, astronauts may choose first names, full names, initials, or nicknames for their uniforms; Sally chose *Sally*.

Guy Bluford's Flight Suit, Space Shuttle Mission STS-8
1983

*"We have selected an outstanding group of women and men who represent
the most competent, talented, and experienced people available to us today."*

—NASA ADMINISTRATOR DR. ROBERT A, FROSCH, JANUARY 1978

I F ONLY one member of the first class of Shuttle astronaut candidates had to be recognized as an example of "the most competent, talented, and experienced people available," the first person in the alphabetical list easily qualified. Among the 35 impressive women and men chosen in 1978 to launch the Shuttle era in space, Guy Bluford excelled in many ways.

Guion S. Bluford Jr. applied to the astronaut corps while serving in the United States Air Force. His military career began in the Reserve Officer's Training Corps (ROTC) while he was an aerospace engineering student at Pennsylvania State University. He became a fighter pilot and flew 144 combat missions in Vietnam, and then returned to become an Air Force instructor pilot and executive officer. He also completed a master's of science and a doctoral degree in aerospace engineering, giving him both the academic and piloting competencies required for astronauts.

Guy Bluford also stood out as one of three African American men in the first class of astronauts to include women and members of racial minorities. Others had similar credentials, but those three men had met distinct challenges growing up and advancing in a society where discrimination based on skin color was all too common. They reached the astronaut corps, then and still widely admired as a pinnacle of

achievement, through their inherent abilities and steady perseverance in the face of barriers.

In 1983, Guy Bluford wore this flight suit into history as the first African American in space. (An African Cuban, Arnaldo Tamayo-Mendez, flew on a Soviet mission in 1980.) He served as a mission specialist (a scientist-engineer astronaut) on the eighth Space Shuttle mission and on three later missions, in 1985, 1991, and 1992. On his first mission, STS-8 flown on *Challenger*, Bluford spent six days in orbit conducting science experiments and deploying a satellite with his crewmates. He carried out scientific research and engineering activities on his later missions, as well. NASA transferred his first flight suit to the Museum in 1984. When he left NASA to start a career in business, Guy Bluford had spent almost a month of his life in space.

Asked about his place in history 20 years after his first spaceflight, Guy Bluford noted that his goal was not personal fame. "I felt I had to do the best job I could for people like the Tuskegee Airmen who paved the way for me, but also to give other people the opportunity to follow in my footsteps. I wanted to set the standard … so that other people would be comfortable with African Americans flying in space and African Americans would be proud of being participants in the space program." Dr. Guy Bluford (Colonel, USAF) accomplished his mission, serving NASA and his country well.

ABOVE: Astronaut Guion Bluford, restrained by a harness and wearing a blood pressure cuff on his left arm, exercises on a treadmill during his flight on *Challenger* in August 1983.

Disposable Absorption Containment Trunk (DACT)
1980s

"How do you go to the bathroom in space?"

—A FREQUENT QUESTION

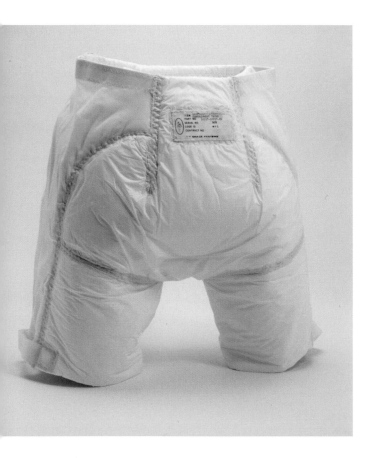

WHENEVER astronauts speak in public, to kids or adults, at home or abroad, someone always asks this question. There seems to be a universal fascination with the how and where of eliminating bodily wastes in space.

Two short answers are: "Normally" and "Very carefully!" A person can urinate and defecate without any physical problems in weightlessness, but going to the bathroom in space is complicated because the environment is different. In microgravity, fluids and solids "float" around. The Space Shuttle and International Space Station have specially designed toilets that work with air flow rather than gravity to keep wastes under control.

A related question is, "How do you go to the bathroom in space when no toilet is available?" What happens when the spacecraft is too small for a toilet, when the astronaut is walking on the Moon or doing an extravehicular activity (spacewalk) in a spacesuit, or when the astronauts are strapped into their seats awaiting launch?

When all astronauts were men, they used a rubber cuff connected to a plastic bag strapped to their body. In the world of NASA acronyms, these were UCDs or urine collection devices. A plastic fecal containment bag with an adhesive seal applied directly to the body captured solid waste. These methods seem primitive, but they were carefully engineered to minimize spills, leaks, odors, and other unpleasantness.

When women joined the Space Shuttle astronaut corps in 1978, crew equipment engineers faced a new challenge: designing a urine collection device that would work with female anatomy. The first attempt was a feminized version of the men's assembly. Tests of several different shapes led to one conclusion: unacceptable. The devices were difficult to use, messy, and uncomfortable.

Trying another approach, designers produced an undergarment that was a cross between a girdle, bicycle shorts, and a diaper. Made of stretchy fabric for comfort and a close fit, with a zipper for ease in donning and doffing, the pant had built-in absorbent padding and could be thrown away after use. Women tested it, found that it didn't leak or chafe, and judged it to be comfortable for a long countdown or a spacewalk. As it was not quite a panty or a diaper, NASA coined a new acronym, DACT or disposable absorption containment trunk. NASA transferred this DACT to the Museum in 1997.

By the time NASA invented the DACT, disposable underwear products had entered the marketplace, first as baby diapers in the 1960s and then for adult incontinence in 1980. These practical innovations caught NASA's eye, and by the late 1980s the custom-designed DACT gave way to commercial products in a variety of styles and sizes. NASA stocked them, but to avoid using trade names, created new acronyms, such as MAG for maximum absorption garment.

At the same time, marketing and education lessened the social stigma of adults wearing absorbent pants, and male astronauts decided to try them, too. Now fitted cuffs and plastic bags are relics of the past, as all astronauts choose to use comfortable disposable underwear for those times when a toilet is not available.

The DACT is thus a marker for gender-related changes in spaceflight. It was, first, an accommodation to the basic needs of women in space, a different solution to a common problem that is not simply a female version of a male device. More broadly, it signifies a trend toward a unisex solution that suits all astronauts. Women's use of disposable underwear helped make it acceptable for men to do likewise.

Onboard-Shuttle Laptop
1980s

"Imagine, though, a portable PC that's really portable. It's the size of a standard notebook ... you keep everything in it—notes from meetings and phone conferences, to-do lists, schedules, address and phone lists. You carry it around the office, you take it home, you take it on business trips."

—Bill Gates, "Information at your Fingertips," November 12, 1990

ARTIFACTS do not speak for themselves—but sometimes they come close. This laptop, a backup identical to those flown aboard early Shuttle flights, is a good example. It speaks to aspirations for the Shuttle program, fulfilled and unfilled.

As the Apollo program was underway in the late 1960s, NASA planned for a follow-on program that was not aimed at a spectacular "first" but, rather, toward achieving routine and low-cost access to space. They envisioned an ensemble consisting of a space station in low-Earth orbit, and a reusable launch vehicle to take crews to and from the station on a regular schedule. NASA gave the launch vehicle the name "Space Shuttle," suggesting the wooden bar that carries woof threads back and forth in a loom, or the short subway train that runs between Grand Central Station and Times Square in New York. Other public descriptions of the system called the orbiter a "Space Truck"; astronauts on one mission carried a sign they held up for the cameras: "We Deliver."

For a spacecraft with such quotidian aspirations, it seemed natural that as laptop computers became available as everyday tools Shuttle crews would carry them into space. When NASA conceived the Shuttle in the 1970s, neither the personal computer nor its portable cousin, the laptop, had been invented. Onboard computers used 1970s-era technology. These devices, critical in flying the Shuttle, went through an elaborate certification process, making it impractical to upgrade them frequently—unlike consumers who constantly seek the latest chips from Silicon Valley.

By the time of the first Shuttle flights in 1981, the laptop was just beginning to appear, and NASA quickly recognized it as a device to supplement—not replace—the orbiter's built-in computers. NASA chose a laptop made by the Silicon Valley firm Grid Systems, Inc., mainly because it used a memory that stored data in magnetic "bubbles" and had no moving parts. Shuttle engineers rightfully deemed laptops that used rotating disks as too

fragile to withstand the vibrations of launch. The Grid-built device also had a bright yellow screen in place of the low-contrast liquid crystal screens common at that time. NASA made a few changes to the commercial computer—adding Velcro fasteners, for example—but otherwise it remained "as built" for the consumer market.

The software was another story; it was custom-written for specific Shuttle tasks, including loading crew schedules, and backup reentry and landing routines in the event of a main computer failure.

This type of laptop flew on Shuttle flights from 1981 through 1988, after which it was replaced by a more common model with an ordinary disk memory. NASA transferred this backup device to the Museum in 1992. Its use on the Shuttle missions shows the agency's ingenuity in adapting advancing technology to its needs. The Shuttle itself never achieved NASA's hopes of "routineness," but the laptop showed that simple, everyday technologies could play a vital, useful role in space.

Manned Maneuvering Unit
1984

*"It may have been one small step for Neil [Armstrong],
but it's a heck of a big leap for me."*

—Astronaut Bruce McCandless, 1984

ABOVE: Strapped into the nitrogen-propelled MMU, astronaut Robert L. Stewart uses the unit's hand controls to pilot himself in a test of the MMU a few meters away from the cabin of the shuttle *Challenger* (not shown) on a 1984 mission.

OTHER THAN pictures of astronauts on the Moon, no photograph better captures the adventure of human spaceflight than that of astronaut Bruce McCandless as the first free-flying human satellite. In this flight, the Buck Rogers fantasy of darting through space in a jet-propelled backpack became real.

Astronauts tested handheld and backpack personal mobility devices on Gemini and Skylab missions, but joy-riding was never the intent. NASA designed the Shuttle-era manned maneuvering unit (MMU) for serious tasks in orbit. An astronaut might use it to venture away from the Shuttle to inspect or retrieve a nearby satellite, do repairs or move equipment, carry out a rescue operation, or assemble an observatory platform, large antenna, or space station.

The battery-powered manned maneuvering unit had 24 tiny nitrogen gas thrusters, each producing less than 9 newtons (2 pounds) of thrust, for forward/reverse, up/down, and left/right motion, as well as roll, pitch, and yaw. The astronaut used hand controllers on the arms of the unit to operate it and typically moved about 1.6 kilometers per hour (one mile per hour) faster than the Shuttle's 28,000 kilometers per hour (17,500 miles per hour) orbital velocity. For safety, the unit latched to the life-support pack on the spacesuit, and a lap belt kept the astronaut secure. The MMU could operate for six hours without a recharge.

Martin Marietta built three MMUs for NASA, two for flight and one for testing. The manned maneuvering unit in the Museum, transferred from NASA in 2001, is the one pictured. On February 7, 1984, on Space Shuttle mission STS 41-B, astronaut

Bruce McCandless took it on a spacewalk, or "space ride," some 90 meters (300 feet) from the Shuttle. This MMU also flew on missions STS 41-C for the Solar Max satellite retrieval and STS 51-A for retrieving the Palapa communications satellite. Astronauts Bruce McCandless, Robert Stewart, James van Hoften, and Joseph Allen logged a total of six hours 29 minutes on this unit during these three 1984 missions.

Planners viewed the manned maneuvering unit as an essential tool for ambitious work in space, but it never flew again after the 1984 missions. Tightened rules for astronaut safety and reduced emphasis on satellite servicing after the 1986 *Challenger* accident, as well as delays in space station construction, idled the manned maneuvering unit. In the 1990s, a much smaller version meant only for rescue entered service.

In its single year on duty, the manned maneuvering unit made history by enabling astronauts to move about in space without tethers. It also quickly came to symbolize something more: spacefaring as individual adventure. Although its flights were carefully choreographed, the MMU looked like freedom. It distilled the especially American urge for mobility and snazzy vehicles, the "don't fence me in" motive for travel, the appeal of pristine environments, and the allure of the extreme. In NASA's imagery, its use suggested an escape to peace and quiet, with a hint of peril. It made the space program individual—a single person, independent, at one with the technology of spaceflight. Who might not imagine, and perhaps yearn for, the thrill of a solo ride in wide open space?

Impacts and Influences

Making a Home in Space

Gerard O'Neill's Space Colony Plans, 1975–1986

"*The sure survival of all races of humanity, and of the plant and animal life forms we cherish as part of our Earthly heritage, is colonies dispersed throughout our solar system and beyond it.*"

—GERARD K. O'NEILL, 1977

I N THE 1970s, enthused by the Apollo lunar explorations, Gerard K. O'Neill, a professor of physics at Princeton University, embarked on a mission: to combine his expertise in science with his belief that humanity's future rested on establishing a home in space. His idea was utopian and unique—he envisioned bustling, complex human settlements not on other planets, but inside gigantic, manufactured habitats orbiting in space.

To begin, he undertook studies aimed at answering the question, "Is a planetary surface the right place for an expanding technological civilization?" As he calculated factors of energy, land area, size and shape, atmosphere, gravitation, and sunlight, O'Neill saw no limits to the possibilities for human colonies in space. But rather than live on a planet surface, he suggested that settlers live inside gigantic cylinders a few miles in length. His design pictured each cylinder as an independent, self-sustaining biosphere, rotating to provide artificial gravity. Earth-like, each featured a breathable atmosphere, with trees and lakes along the inner rim helping to perpetually recycle oxygen, water, waste, and other materials. Animals and plants endangered on Earth would thrive in these cosmic arks, with insect pests left behind. Solar power, directed into each colony by huge mirrors, would provide a constant source of nonpolluting energy. Millions of such colonies might be built.

O'Neill positioned each colony at a specific location between the Earth and the Moon where the gravitation fields are equalized, known as LaGrange Point 5 (L-5). More important, he saw each colony as a fresh start, a chance to create a perfect society without the problems of Earth. But even utopian colonies served practical needs. Settlers would build and operate solar-power satellites from the raw materials of asteroids and beam energy collected from the Sun to an energy-hungry Earth, providing an economic as well as social rationale for this vision.

O'Neill developed his concepts through his teaching, using questions about the future of human life as tools to excite his physics students. His work generated a following and catapulted him into the spotlight of the space community. In 1975, he and like-minded thinkers formed the L-5 Society to advance his vision. It also found an audience in NASA. O'Neill received funding from NASA's Advanced Programs Office—only $25,000—to develop his ideas more fully. Intrigued, NASA convened two study groups in 1975 to review the idea of space colonization. They concluded, "It appears that space colonization may be a paying proposition"—despite start-up costs of nearly $100 billion. But, colonization was about more than money; it offered "a way out from the sense of closure and of limits which is now oppressive to many people on Earth." The study groups recommended an international project led by the United States to establish a space colony at L-5.

O'Neill publicized these findings exhaustively, but with the political tide for an aggressive space effort at a low ebb in the latter 1970s, nothing came of it. In 1985–1986, O'Neill served on the National Commission on Space, chartered by President Reagan, that produced a study, "Pioneering the Space Frontier," which incorporated some of his work. The report, issued within days of the loss of the *Challenger* Space Shuttle, did not find an enthusiastic audience. It became O'Neill's last public effort to advance his ideas. He died in 1992, after a seven-year battle with leukemia. Some of his ideas for space colonies live on in the National Space Society, which merged with the L-5 Society in April 1987.

Space Station *Freedom* Model
1988

"America has always been greatest when we dared to be great. We can reach for greatness again.... I am directing NASA to develop a permanently manned space station.... NASA will invite other countries to participate so we can strengthen peace, build prosperity, and expand freedom for all who share our goals."

—President Ronald Reagan, State of the Union Address, 1984

Once the Space Shuttle entered service, NASA turned its sights to a program that it promoted as "the next logical step"—a space station in Earth orbit. Spaceflight visionaries had long imagined a future outpost where people lived and worked in the unique environment of space. An orbital station could serve as a laboratory, observatory, military post, industrial plant, dry dock, and launch platform. In the 1980s that outpost seemed within reach. Served by the new Shuttle transportation system, a space station would expand opportunities to use space for practical and scientific purposes. Humanity could establish a permanent presence in space.

Before and after President Reagan's 1984 approval, NASA studied various space station concepts—different sizes, shapes, missions, power capabilities, costs—and by 1988 a concept named *Freedom* emerged. As this model shows, the architecture included a cluster of living-space and laboratory modules, large solar array wings, and radiator panels, all attached to a long spine. This proposed orbital station extended the length and width of a football field.

President Reagan highlighted international participation in his mandate for a space station and, in 1988, NASA entered into agreements with the European Space Agency, the Science and Technology Agency of Japan, and the Canadian Space Agency as partners in the development and operation of *Freedom*. Sixteen nations would share responsibilities and costs for the space station—a daunting challenge in technical and managerial coordination.

Not everyone agreed, however, that *Freedom* was indeed the next logical step in space. Despite the president's direction, the project garnered little political backing and was in constant jeopardy of cancellation. Although some scientific groups supported research laboratories in space, others representing different disciplines found automated satellites more suitable and economical for their studies. Skeptics worried that a space station was not an end in itself but a disguised stepping stone to an even more expensive program—a human mission to Mars. *Freedom*'s future was precarious.

After years of sporadic progress and redesign, in 1993 President Bill Clinton directed NASA to revamp the space station program. With the collapse of the Soviet Union and the end of the Cold War, the United States brought in the Russian Federation as a partner. To curb costs and complexity, the station planners simplified the design, eliminating some components and rearranging the configuration. Down-sizing was the rule.

Amid these changes, the name *Freedom* evaporated. At its genesis, President Reagan envisioned the space station not simply as a place in space but also as a symbol of peace and freedom through international cooperation. A decade later, when it was time to build the hardware, such ideals paled in the light of practicalities.

A smaller, less capable, and less costly variant emerged as the International Space Station, and assembly began in 1998. The first occupants came aboard in 2000, and the fifth anniversary of continuous residency was reached in November 2005.

Freedom as originally envisioned did not become a home for humanity's permanent presence in space, but the namesake concept model, transferred from NASA in 2005, has a permanent home in the Museum.

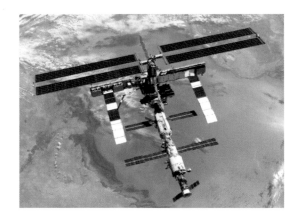

ABOVE: The Space Shuttle *Discovery* took this view of the International Space Station (the Caspian Sea is in the background) in August 2005.

Vega Model
1984

"I believe that the Soviets are more interested in international recognition that the United States sees them as equals."

—Sam Keller, NASA's deputy associate administrator, Office of Space Sciences, 1988

In 1984, the Soviet Union launched the Vega 1 and Vega 2 spacecraft, which flew by Venus, dispatched scientific instruments into its atmosphere, then traveled farther, to pass through the tail of comet Halley, transmitting data back to Earth. The name *Vega* came from a contraction of the Russian names for the planet Venus and comet Halley. The mission involved scientists and instruments from Bulgaria, Czechoslovakia, France, East and West Germany, Hungary, Poland, the United States, and the Soviet Union, and marked a new era of international cooperation for the Soviet space program. The Soviet Space Research Institute planned the mission after a series of moderately successful missions to the planet Venus. They appreciated that the widespread interest in the famous comet, ready for another pass through the solar system, might augment multinational participation in their planned return mission to Venus.

Vega's ambitious mission did attract much international interest. French scientists designed the main experiment for the atmospheric investigation of Venus. A scientific balloon that was housed in the large sphere at the top of the craft was released into the atmosphere of Venus to measure cloud movement. While the balloon relayed information on the atmosphere, a Soviet-designed lander headed to the planet surface and collected data on conditions near and in the soil. The American instrument on Vega was a dust detector, designed by University of Chicago physicist John Simpson, that measured interplanetary and cometary dust around the comet.

When the Vega craft flew within 7,724–8,690 kilometers (4,800–5,400 miles) of comet Halley's nucleus, data about the comet's structure and chemistry was collected. This mission provided valuable information to scientists of the European Space Agency who later navigated the Giotto spacecraft to within 603 kilometers (375 miles) of the comet's nucleus. In addition to Vega and Giotto, two Japanese probes, Sakigaki and Suisei, also studied comet Halley.

The spacecraft itself was not the only part of the operation that was multinational. The Soviet Union openly recruited and imported American ground-based technology for the mission. The U.S. offered the assistance of its Deep Space Tracking Network to assist in navigation. As the data streams came to Earth, American-made XT-type personal computers in Moscow did the processing that translated raw data into astounding images and navigation data.

Soviet scientists hoped that the Vegas' remarkable success would pave the way for long-term international cooperation and thereby provide support to their national spaceflight programs. Two years after the encounter with comet Halley, the Soviets launched the first of a planned series of missions to explore Mars and its moons. The initial Mars missions suffered from navigation and communications failures and never completed their objectives. The government cancelled subsequent missions owing to declining economic conditions, ending Soviet exploration of Venus.

This engineering model originally came to the Museum in 1992 on loan from the Lavochkin Scientific Production Association after almost five years of negotiations. In 1995, Texans John Buckner Hightower and Gregory Schnurr purchased the Vega from Lavochkin and then generously donated it to the Museum.

Coca-Cola and Pepsi Cans, STS 51-F
1985

"One Giant Sip for Mankind"

—Pepsi-Cola "Space Can" advertising slogan, 1985

Amid the "cola wars" between Coca-Cola and Pepsi-Cola in the 1980s, NASA took both companies into the space age with a test of the two soft drinks on board the Space Shuttle *Challenger* (STS 51-F) in 1985. Each company designed special cans and drinking attachments for use in the microgravity of Earth orbit, and astronauts tested them during their seven-day mission. The effort heightened the intense corporate rivalry of the time by giving the products a high-tech design and visibility in the new Shuttle program in hopes of attracting a young audience for their products. Pepsi's taste test concept, originated in 1975, brought their "choice of a new generation" campaign up against Coca-Cola's "America's real choice." In anticipation of the space showdown, each company sponsored local competitions and celebrities proclaiming their love of each product in television commercials. A journey into space gave these elite American corporations the chance to win the loftiest taste test of the decade.

Coca-Cola, the self-proclaimed leader of industry innovation, triggered the cola "space race" in 1984 by requesting that NASA fly an experiment with their design to test the feasibility of drinking soda in space. When Pepsi learned of the selection of the Coke trial just months before its flight, NASA agreed to delay the flight of the experiment in the spirit of fairness so Pepsi could also participate.

To make their products suitable for space travel each company needed to take into account the physics of carbonated drinks. Opening a typical can in space caused the can to spray soda everywhere as the pressurized contents vented into the microgravity of the Shuttle compartment. To correct this problem, each cola maker engineered a special can and attachment to allow for easy consumption in space. The Pepsi and Coke cans pictured here, transferred from NASA to the Museum in 1985, were one of four of each brand flown on the mission, each only partially consumed—none of the astronauts enjoyed the soda enough to finish a single can! Coke spent over a year developing their special contraption, investing more than $250,000 to prepare the cans for flight with four different levels of carbonation. The unique top of the Coca-Cola can functioned better, according to astronauts, but none of the carbonation levels available tasted as good as the product did on Earth. Pepsi spent only a few months prior to the STS 51-F mission developing their dispenser design. Unfortunately, it provided a very foamy soda via a simple shaving cream–style dispenser, an even less desirable result than the Coca-Cola version. While both products fizzled on the technology side, Pepsi managed a short marketing coup by throwing itself into a massive space-themed campaign, with commercials, advertisements, and an education alliance to aid young people interested in learning about space.

The journey of these two soft drinks into Earth orbit was a tale of two corporate giants competing over the chance to capture the imaginations of the next generation of space enthusiasts. While neither won the first test in space, Pepsi-Cola proclaimed itself the winner of the decades-long wars when its national sales surpassed that of Coca-Cola in the mid-1980s. As an event in commercial culture, the cola wars left their mark in another way: they established the first link between old-fashioned, Earth-bound marketing and life in space.

ABOVE: Astronaut Anthony England drinks from a specially designed Coke can during a flight on the Shuttle *Challenger* in 1985.

Challenger Memorial Plaque
1986

"Oh! I have slipped the surly bonds of earth / And danced the skies on laughter-silvered wings … I've trod the high untrespassed sanctity of space, / Put out my hand and touched the face of God."

—FROM "HIGH FLIGHT," JOHN GILLESPIE MAGEE JR., 1941

WITH WORDS from this poem, a somber President Ronald Reagan addressed the nation on January 28, 1986, after a launch explosion destroyed the Space Shuttle *Challenger* and claimed the lives of seven astronauts. Penned by a young pilot in World War II, the poem celebrates the sheer, even spiritual, joy of flight, and it is often quoted as a tribute to pilots.

In the aftermath of the Shuttle tragedy, the Museum itself became involved in memorializing the flight and the crew. Shortly after the event, NASA presented commemorative plaques to the families of the *Challenger* crew and to the Museum. Each plaque bears likenesses of the astronauts, as well as a small U.S. flag and mission patch that had been stowed in the official flight kit aboard the Shuttle. These items, always carried on Shuttle missions to be given away after flight, were recovered undamaged in the crew cabin debris hauled up from the ocean depths. The Museum's plaque is displayed near a large Space Shuttle model as a quiet reminder of the sudden loss of commander Dick Scobee, pilot Mike Smith, mission specialists Ron McNair, Judy Resnik, and Ellison Onizuka, payload specialist Greg Jarvis, and teacher Christa McAuliffe.

The final *Challenger* mission seemed to augur an era of space exploration in which anyone might become an astronaut. Commentators noted that the crew represented the diversity of America— men, women, an African American, an Asian American, Caucasians. The mission received heavy advance publicity because the crew included a teacher, and also an engineer, chosen from outside the astronaut corps.

Occurring in view of spectators, the media, and student audiences watching live television broadcasts, the accident was unforgettable. It shocked most observers into a realization that flying on the Space Shuttle was not as safe and routine as assumed. In the wake of the tragedy, many people questioned America's technological prowess and the space agency responsible for the Shuttle—a spacecraft regarded by the public as a signature national accomplishment.

This questioning deepened as investigators realized that the "accident" was not quite accidental. The anomaly in the solid rocket booster had been observed before and, not yet remedied, was exacerbated by the cold winter night preceding *Challenger*'s launch. Investigation of the causes of the tragedy—flawed technology and flawed decision-making—raised doubts about the human spaceflight enterprise and prompted a long delay before flights resumed in 1988.

The president's closing remarks on the loss of the *Challenger* crew blended honor with elegy in this allusion to the "High Flight" poem: "We will never forget them, nor the last time we saw them, this morning, as they prepared for the journey and waved goodbye and 'slipped the surly bonds of earth' to 'touch the face of God.'" The *Challenger* memorial plaque reminds us of the fine line, the brief moment, between triumph and tragedy, joy and sorrow, for those who venture into space.

Space Shuttle Launch-Entry Suit
Late 1980s–late 1990s

"I know spaceflight is a risk, and I know I may not come back from a flight.... And the risk of losing your life is outweighed by what space exploration is going to bring ... to the world."

—ASTRONAUT ANDY ALLEN, 2001

I N 1981, the Space Shuttle made its debut as the world's first reusable spacecraft. The first four flights tested the new vehicle's ability to launch, orbit, land, and fly again. For those flights, astronauts wore tan military high-altitude pressure suits in case they had to eject during a launch or landing emergency. As Shuttle flights moved from tests to operational missions, NASA relaxed the dress code, and Shuttle astronauts wore a simpler flight suit into space—an unpressurized light-blue zip-front coverall—with a regular helmet and oxygen supply.

The Shuttle embodied the goal—or hope—of making spaceflight routine with safe, reliable, frequent service modeled on airline travel. The blue coverall uniform helped to convey the message that riding to and from orbit was safe enough not to require a pressure suit. Astronauts no longer had to dress especially for space; they wore coveralls similar to those worn while flying their T-38 aircraft, training for missions, and making public appearances.

The belief in routine, completely safe spaceflight shattered abruptly on January 28, 1986, in a catastrophic explosion during *Challenger's* launch on the 25th Shuttle mission. All seven members of the crew perished. This loss prompted a new look at many issues, including crew safety.

During the Shuttle's development in the 1970s, NASA evaluated various schemes for ejection seats or a crew escape pod but rejected them as impractical or too expensive. After the *Challenger* accident, the agency developed two new measures: an escape system to enable the crew to bail out through the hatch under certain conditions, and a protective suit to wear during launch and entry, the most dangerous phases of spaceflight.

First worn on the 1988 return-to-flight mission, the launch-entry suit (LES) protected an astronaut if the cabin depressurized during ascent or descent or if he or she had to bail out over the ocean. The suit exerted mechanical pressure at 22 pascals (3.2 psi) on the crew member's body by inflation of an internal bladder. It was a partial-pressure suit. The helmet visor and neck dam had pressure seals but the gloves did not; they connected to the suit by a tube and needle valve to maintain air pressure on the hands. The suit included an antigravity suit (g-suit) to prevent blood from pooling in the lower body during reentry.

The fire-retardant, waterproof orange outer layer prompted the nickname "pumpkin suit." It protected the astronauts during a fire or bailout. The parachute harness backpack worn over the suit also carried survival gear including extra oxygen, an inflatable personal life raft, flares, emergency radio beacon, and other necessities.

Astronauts used the LES until the late 1990s, when NASA introduced a newer version, the full-pressure advanced crew escape suit (ACES). The Museum's launch-entry suit, transferred from NASA in 2001, was never worn in space. Suits were made in a range of sizes and this one did not fit any member of the astronaut corps. So far, astronauts have not had to depend on the LES or ACES suits during an actual flight emergency.

The last views of departing astronauts, all dressed in bright orange and happily waving goodbye as they board the bus to the launch pad, belie the nature of spaceflight. Wearing the suit is a tacit acknowledgment of the risks ahead, particularly during the critical phases of launch and entry.

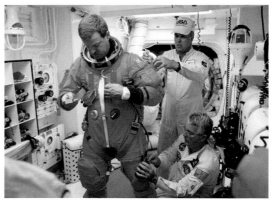

ABOVE: Prior to launch of the Space Shuttle *Atlantis* in 2000, technicians help pilot Scott Altman don his launch-entry suit.

SS-20 "Pioneer" Surface-to-Surface Missile
1987

"We are at the point of producing missiles like sausages."

—Nikita Khrushchev, 1957

ABOVE: In the Soviet Union, a giant press crushes a section of a SS-20 missile, in accordance with the INF Treaty, as official onlookers verify its destruction.

Over the course of the Cold War neither the Soviet Union nor the United States produced missiles in the numbers implied in Khrushchev's famous and ominous quote. But each side did build missiles in the thousands capable of carrying nuclear weapons. In 1987, under the terms of the Intermediate-Range Nuclear Forces (INF) Treaty, the U.S. and the USSR took one step to reduce those numbers and agreed to eliminate an entire class of missiles. This marked the first time that either country removed operational weapons from their nuclear arsenals. Such elimination stood in contrast to prior treaty efforts that merely sought to limit the growth of new weapons. The INF called for the destruction of all missile systems with a range under 5,500 kilometers (about 3,300 miles)—including the Soviet SS-20 and the American Pershing II missiles. The Museum has one of each in its collection.

The SS-20, also know as the "Pioneer" in Russian, stood almost 16.5 meters (54 feet) tall. Composed of two-stages, the solid propellant missile carried three warheads, each capable of hitting independent targets. The first and second stages of the missile, each constructed of yellow fiberglass, display numbers and Cyrillic letters printed along the circumference. Engineers used these as reference data in the manufacturing process as they covered the solid fuel in the rocket's core with fiberglass. At the top sits the reentry vehicle, enclosing the three warheads. The white fins at the base assisted in stabilizing the rocket during flight.

After the signing of the treaty, the Votkinsk Machine Building Plant, USSR, constructed this missile from factory components for exhibition at the Museum. The Museum missile never held any solid fuel—operational missiles had to be purged of their very toxic propellant before display. The Museum received the missile from the USSR Ministry of Defense through a special provision in the INF, which allowed the preservation of 15 each of the SS-20 and Pershing II missiles to commemorate the first international agreement to ban an entire class of nuclear arms. Under conditions of the treaty, the precise locations of this missile and the Pershing II in the Museum collection are listed with the U.S. Defense Department and the Russian Ministry of Defense.

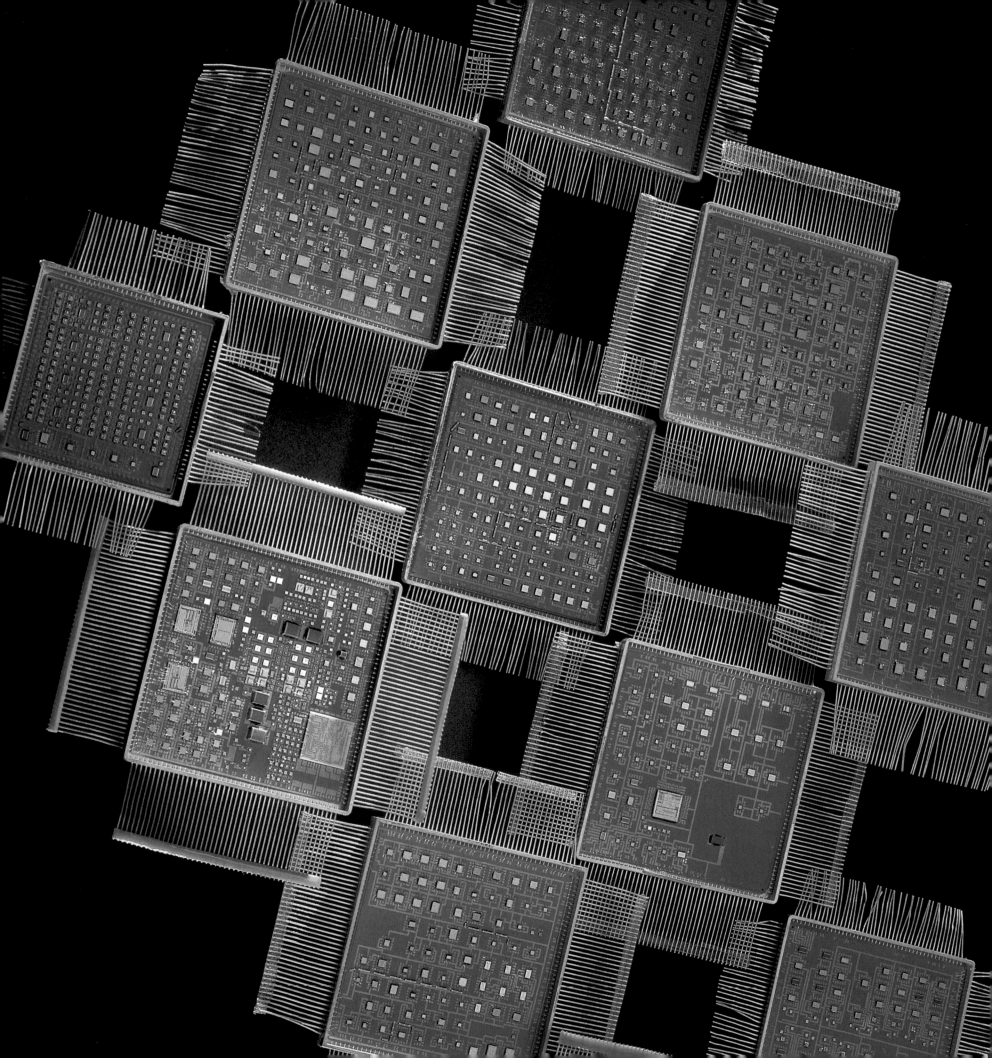

RIU Microelectronic Hybrid FIFO, Milstar Communications Satellite
Late 1980s

"This was a real dog to design, because certain lines could not cross the other lines, and there's ground planes. If you look at the artwork on this thing, it's really brutal. There are ground planes, there are voltage planes, all kinds of neat stuff."

—VINCENT GERACI, MICROELECTRONIC HYBRID DESIGN ENGINEER, 1999

ELECTRONICS and the space age are inseparable. Over the course of the Cold War, spaceflight and military needs pushed innovation in technologies we take for granted today—including the integrated circuit (the "chip") and computers. This device marks one small spot on a vast map of interactions among space, defense, and electronics that have reshaped the way we live.

Despite its dense technical name, this device's function was simple. As the communications satellite received radio signals, the RIU (remote interface unit) coordinated those signals and this unit processed them according to a basic principle: first in, first out (to another electronic device)—hence, the acronym, FIFO. But this "FIFO" (a technology of the mid-1980s to early 1990s), donated to the Museum by Lockheed Martin in 1998, was a "dog" to design—and therein lies its story.

The device is called a hybrid because it combines integrated circuits with other components such as capacitors and resistors. Creating a plan to arrange and connect these elements (the "artwork") was formidable, made more so by a remarkable feature of these hybrids. Beneath the visible surface layer of electrical devices and connections are four additional conducting layers (and seven nonconducting layers)—all contained in its 10-centimeter (4-inch)-square, .63-centimeter (.25-inch)-deep box. The buried conducting layers provide additional pathways for connecting the electrical devices on the hybrid's uppermost layer. The completed hybrid is a clever puzzle in which more than 100 chips and devices are integrated through more than 2,000 connections. The gold-colored wire pins issuing from the four sides of the case connect the hybrid to other devices.

Why and how did such hybrids come to be? Two factors stood out. One was the extreme premium placed on reducing the weight and size of components on spacecraft and satellites—small is better. Rather than put a large circuit board in the satellite for this one specific function, there was this compact, layered hybrid.

The other factor was ruggedness—their ability to perform with high reliability under the extreme conditions of space. This last characteristic had special importance for Milstar. Initiated in the early 1980s and operational today, this military satellite system enables secure communications among the president, the Pentagon, and troops stationed in a war zone. (It is widely used in Iraq and Afghanistan.)

Manufacturing these hybrids combined engineering smarts and artisan craft. General Electric built up a special unit, located at its plant in Valley Forge, Pennsylvania (one of the very first specialized "space" plants created after Sputnik). If FIFO was a "dog" to design, it also was a "dog" to build—and women, for the most part, did the work. A supervising engineer explained why:

> Their dexterity. And they could sit and do a repetitive job … Men would get more frustrated. You're talking about 1-millimeter wire [.039-inch], in some cases .7-millimeter [.027-inch] wire bonding to a pad that's maybe 4-mils [.157 inches]. Most men would go nuts after doing it. Women, for some reason, can sit there and do that day after day. It doesn't seem to bother them … there are differences. I don't care what other people say. Women can do this better than men.

In this "FIFO" we see not just the complex relation between electronics and spaceflight, but a long tradition of needlework and other fine hand-eye work taken on by women.

OPPOSITE: Early Milstar satellites used a variety of microelectronic hybrids to perform specialized functions (the RIU FIFO is bottom, center).

Buran and *Energia* Models
1992

"You cannot get something for nothing."

—Konstantin Feoktistov, designer of the Soviet Vostok, 1988

ABOVE AND TOP: A *Buran* fixed onto an *Energia* rocket moves to the pad (top) and then is placed in an upright, ready-for-launch position (above). The *Buran* had its one and only space launch in November 1988.

IN THE Soviet Union interest in a reusable space plane began in the 1950s. Their designers and managers believed that such a craft ultimately would provide more reliable and efficient access to space than single-use rockets. Their first effort, known as the *Burya*, employed a ramjet. Rocket designers and cosmonauts continued work on a space plane, called *Spiral*, during the 1960s. After several incomplete design projects during the 1970s, the Soviets revived the effort in the 1980s to build the *Buran*. Only two flight-ready spacecraft were manufactured.

At the time of the early U.S. Space Shuttle launches, the Soviets were testing an unpiloted scale model of the *Buran*, called the *Bor*. Amid much international speculation, and after many delays, the Soviet Union launched the *Buran* ("Snowstorm"), its first full-scale reusable space shuttle, on November 15, 1988. Although they tested the *Buran* extensively in the Earth's atmosphere with trained pilots, the maiden, and only, orbital launch was made without a crew. The *Buran* launched strapped onto the *Energia* launch vehicle, the largest among Soviet launch vehicles. It closely resembled the American Shuttle—not by coincidence. Through espionage, the Soviets had obtained the design specifications of the U.S. Shuttle.

Buran's launch occurred during a critical time in Soviet history. Premier Mikhail Gorbachev's policy of perestroika called for a tighter accounting of state expenditures. His policy of glasnost, or openness, allowed for public debate of policies. The response to *Buran* from within the Soviet aerospace community was immediately and resoundingly negative. Engineers, such as Konstantin Feoktistov, who had designed the Vostok spacecraft, wrote Gorbachev a 19-page critique of the program. In his letter, he pointed out that the expense for the reusable shuttle would sap the budget for existing programs—the point of his quote above. Planetary scientist Roald Sagdeev also expressed his disapproval of the program as one that would undercut Soviet expenditures in the space sciences. No flight occurred after *Buran*'s maiden voyage. The government officially cancelled the program after the dissolution of the USSR.

Boris Nikolaevich Yeltsin, the first democratically elected president of Russia, presented these models of the Soviet *Buran* spacecraft and *Energia* launch vehicle to the Smithsonian Institution in June 1992 during a summit in Washington, D.C., with American President George H.W. Bush. These models commemorate the first launches of the *Energia* launch vehicle in May 1987 and the *Buran* shuttle in November 1988.

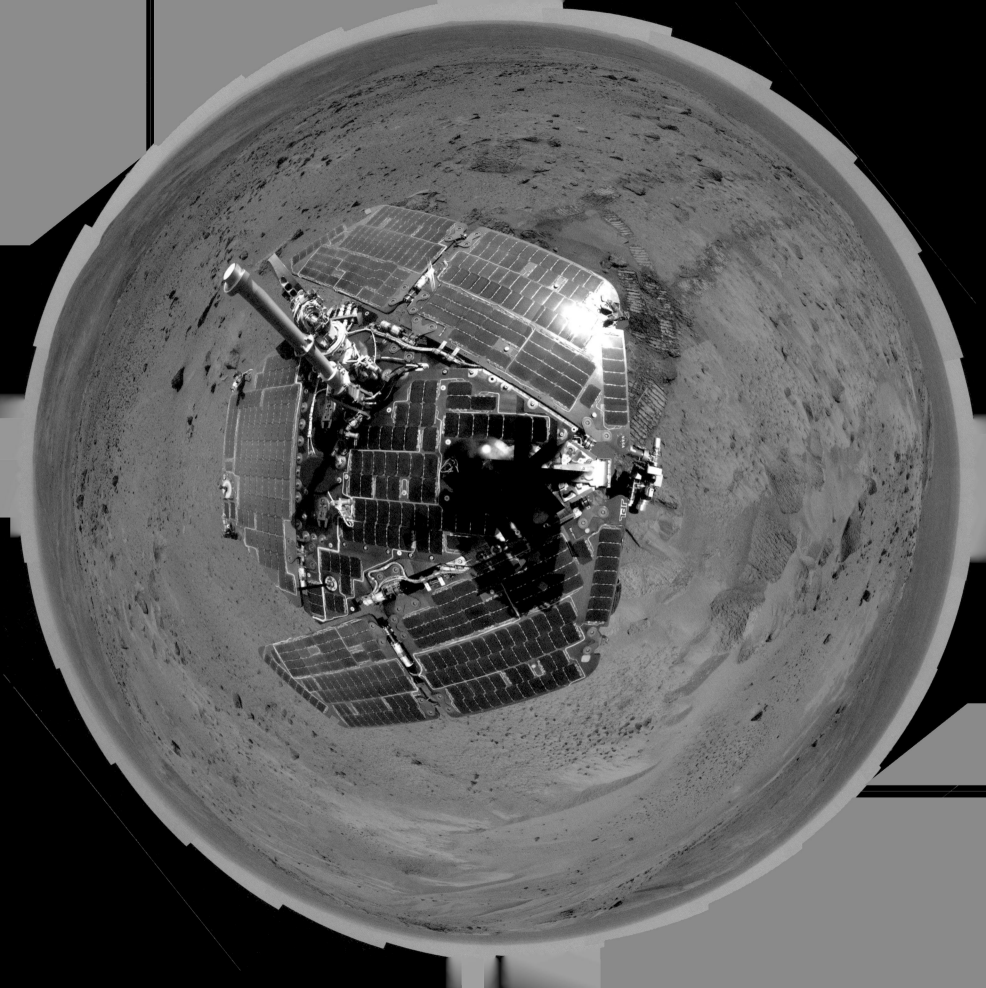

Space in the Global Era, 1989–2007

I N 1960, philosopher and educator Marshall McLuhan introduced the phrase "global village" into the popular lexicon. The evocative metaphor caught on quickly, symbolizing the profound ways in which ground-based radio and television seemed to be changing the condition of human life. The broad reach and omnipresence of these technologies, McLuhan felt, contracted "the world to a village" in which "everything happens to everyone at the same time." For better or worse, "we wear all mankind as our skin."

McLuhan's thoughts, offered just three years into the space age, were neither strictly true, nor fully descriptive of later developments. At that time, radio carried across much of the international landscape, but television was concentrated in the West and in populous regions. And each had a very limited range of programming and viewpoints. But over the ensuing decades, the creation of a rich, pervasive, multifaceted global web of connectivity became a reality—one fundamentally dependent on space-age technologies. The global village, as we experience it in the early twenty-first century, grew out of the placement in Earth orbit of communications, weather, reconnaissance, guidance and navigation, and science satellites numbering, in total over the period, in the thousands. The United States and the Soviet Union first dominated these developments, but then, by the 1990s, tens of other nations joined this space activity, particularly in the area of communications. As this summary suggests, life in the "village" featured more data (and more agendas) than just news and entertainment. Perhaps, more important, such satellites and spaceflight in general encouraged a significant turn in human thought. Through a long record of accomplishments and through especially powerful moments such as the Apollo photographs of

the blue Earth hanging in the black cosmos, spaceflight made the idea of the "global" seem natural, part of the intellectual toolkit for living in the present era.

The development of Earth-orbiting technologies pushed, benefited from, and merged with the information revolution—the 1970s and 1980s confluence of microelectronics, computing, and the Internet. The global became a stage, a place of action, for the powerful and for the average citizen, allowing anyone sitting at home, lolling in their pajamas, to tap into the possibilities of the seemingly endless information superhighway. By the 1990s, spaceflight thus entered, quietly and often unseen, into nearly every corner of our lives—in our homes, our cars, on a walk in the woods, in a credit card transaction at the local store—touching our sense of our selves, our neighbors, and of fellow humans around the world.

As implied by McLuhan's "village," the ubiquity of global-making technologies held the power to create new modes of community. To draw one example from a stream of thousands, consider the death of Pope John Paul II in April 2005. Television via satellite, world-wide in extent, beamed scenes of the faithful and admirers massing for prayers in Rome and other cities around the world. The *Washington Post* noted that "satellites can transmit spiritual values as well as electronic signals … circulating images that brought him [John Paul II] to life … there he was—kneeling in prayer on behalf of those less righteous, laughing at a baby thrust into his arms, listening patiently during a visit to the man who had tried to assassinate him." The coverage linked millions of people, in real time, in a powerful emotional event. And this coverage, this collective, "there at a distance" experience was not extraordinary—it had been happening every day for more than a decade, via satellite television and the Internet, drawing people into common awareness (if not

OPPOSITE: NASA technicians assembled this panoramic view of the Mars Exploration Rover *Spirit* and the surrounding Martian landscape from hundreds of images taken by the rover as it sat at the summit of "Husband Hill."

purpose or agreement), over events trivial and momentous, creating a global "skin" that was not merely metaphorical but real.

This global world had a particular history, one rooted in the Cold War and in the increasing emphasis in the 1970s and 1980s on a pragmatic rather than on a romantic, pioneering view of spaceflight. With the fall of the Berlin Wall in 1989 and the collapse of the Soviet Union in 1991, the Cold War ended and the United States stood as the lone superpower. Years of government-directed effort and a broad base of corporate know-how gave the U.S. a big edge over other nations in the "hardware" and "software" of the new global world. Aerospace firms, Silicon Valley wunderkinds, Microsoft, old-line corporate giants such as General Electric and Motorola, Wall Street financial companies, and America's unparalleled industry of popular entertainment, in large measure, defined and poured content through the ever expanding and ramifying arteries of the globe—at least in the beginning.

THE example of John Paul II's death highlights the unifying possibilities of the new global capabilities. But the real-politik behind them raised tensions and outright opposition. Some, especially in non-Western societies, saw in the new technologies and their content a powerful, unwanted force that corrupted local beliefs and traditions, a tension in extreme form that one writer dubbed a "clash of civilizations." But such reactions reflected genuine concerns. Take Hindi parents in India, as reported by the *Washington Post*, who found it hard to reconcile their beliefs with their teenage children's desire to view via satellite television American-produced MTV, with its free-wheeling, hyper-sexual style. The attractions of and resistance to an American-led global world stimulated tens of countries and indigenous businesses to launch their own communications satellites and thereby reinforce local culture against the U.S. global tide.

America's restless brand of capitalism drew the most attention in the making of the global village, but the U.S. military, too, played an essential part, creating seemingly strange linkages between national security and a rampant transnational consumer culture. Consider the example of the Global Positioning System (GPS), a network of 24 satellites orbiting about 19,000 kilometers (about 12,000 miles) above the Earth designed and operated by the U.S. Air Force. Conceived in the early 1970s and only becoming fully operational in 1995, GPS's history straddled the Cold War and its

market-oriented aftermath. The system provided any soldier or any machine (a ship or airplane, for instance) with information on their exact position anyplace on the planet via signals encoded with highly accurate time data, its profound effects symbolized by its use in guiding "smart" bombs and missiles with deadly, precise accuracy in the post–September 11 conflicts in Afghanistan and Iraq.

In the 1990s, though, GPS became not just a tool for the U.S. military, but for anyone, anywhere in the world—for a hiker in the Rocky Mountains, for Mom and Dad driving the family car, as well as for past and present adversaries: a Russian or Chinese soldier or a terrorist. In its posture toward users, the system was egalitarian in the extreme. Not surprisingly, in the go-go market-driven world of recent years, business and consumer use of GPS vastly outstripped that of the military. Imagine a use for location, tracking, or accurate time information and one found GPS there, often in the intimate contours of daily life—tracking a young child at play, a spouse suspected of having an affair, the wanderings of a teenage driver, each person with a recent-generation cellular telephone. And not just in the U.S.: as *The New York Times* observed, "The world has incorporated our GPS into its daily life as rapidly as Americans took up the ATM banking network." Nothing, perhaps, speaks more to the odd permutations of life in the global era than GPS, its radio signals equally available to friends and foes, to weapons in flight, and to off-the-shelf products offering to meet every consumer need.

GPS was one of many twists that turned Cold War precedents upside down and threw into relief the new global realities. The ironic: the Russians sold at auction at Sotheby's in New York City treasures of the Soviet space program, some of which were purchased by U.S. oil magnate Ross Perot and then loaned to this Museum. The serious: American, Russian, and European firms operating sophisticated reconnaissance satellites provided strikingly detailed images of cities from space (recall newscasts showing before and after space views of Manhattan in September 2001). During the Cold War, such images were among the most closely guarded secrets of state. Now they provided the raw material for Google Earth, allowing anyone with access to a computer to explore nearly every nook and cranny of the planet's land mass. The implications were potentially staggering. In 1999, *The New York Times* noted, "Soon, Hezbollah and Osama bin Laden are going to have access to a one-meter image of … you pick it … the White House, the Congress, the Pentagon."

The new power of the image went hand-in-hand with an enthusiasm for virtual reality, playing into a recalibration of the roles of human and machine space exploration. Computer-generated graphics, video games, and one-world connectivity made remote exploration of the universe and the planets seem natural. In a period of just over 10 years beginning in the 1990s, NASA orbited a string of "great" observatories that investigated fundamental questions of the universe, achieving far-reaching scientific success. Launched in 1990, the Hubble Space Telescope has been the most well-known, in part for the flawed mirror that initially blurred its vision and its subsequent dramatic repair by Shuttle astronauts. The only one of the great observatories that viewed the universe in the visible light spectrum, Hubble connected with the public because of its images—dramatic shots of cosmic objects and events never before seen. Awe-inspiring, majestic, and, for some, sacred, the Hubble images dramatized basic questions of existence—of the universe and of life.

BUT public enthusiasm for space exploration via machine reached a crescendo with NASA's two missions to Mars that landed rovers on the surface of the red planet, *Sojourner* in 1997 and the Mars Exploration Rovers, *Spirit* and *Opportunity* in 2004. *Sojourner*, the United States' first automated rover, looked toylike as it inched along the Martian surface but was scientifically and technically clever, a triumph for the engineers at NASA's Jet Propulsion Laboratory. The first successful U.S. journey to Mars in more than 20 years, the mission generated intense public interest—a result of the novelty of a machine explorer and a long-running fascination with the possibility of discovering evidence of life on the planet. *Sojourner's* exploratory pluck (a combination of automatic "intelligence" and video game–like control from Earth), stunning images, and the ability to follow the action promptly via the Internet gave the experience a completeness and excitement unmatched since Apollo.

When the space age began, commentators and advocates expected the transforming effects of flight beyond the Earth to come primarily from the deeds of frontier-exploring astronauts and cosmonauts. The Moon landings gave this view credence and in the years after a series of national commissions and presidential announcements tried to sustain this vision, but with an increasing mismatch between expectations and political commitment. After the end of the Cold War, human space exploration bore the brunt of this "go but stop" dynamic. In the U.S., it achieved remarkable accomplishments, including tens of Shuttle flights; successful collaborations with the Europe, Canada, Japan, and other nations; a veritable merger with the Russian space program; and the partial completion of the International Space Station. Despite such achievements, as measured against long-standing visionary rhetoric, the whole effort seemed anemic, drained of the stimulating effects of true exploration. The Shuttle *Columbia* accident, in 2003, resulting in the deaths of seven astronauts, only heightened the tension between the romantic view and extant realities. The public still cared deeply for their astronauts but, in the global era, human space exploration was, as one newspaper noted, "increasingly fueled by international diplomacy or cash [and] is threatening to turn from a calling into a currency."

These disenchantments with the state of human space exploration highlighted the dramatic reorientation of spaceflight since Sputnik and the Apollo years. By the 1990s, the fault line in American policy seemed clear. The nation valued idealism, but the culture gave greater weight to the pragmatic—to the workings of profit-producing markets and to the immediate needs of national security. The passing of time, too, brought to the fore NASA's odd status in the American political tradition. Since the earliest days of the republic the government had encouraged exploration and pioneering, through policy and funding, but it had never set up a permanent bureaucracy dedicated to those goals. In the American experience, citizens drove pioneering, part of the nation's core ethic of individualism and belief in the power of private markets and dollars to create new worlds. Human space exploration, with its extraordinary expenses, seemed an exception to this impulse—until the early twenty-first century when inventor Bert Rutan designed SpaceShipOne and sent the first "private" astronauts into space. His effort to make astronauts out of consumers heralded, in his own words, "an era of personal spaceflight first described by the visionary science fiction writers of the 1940s and 1950s … and will provide the foundation for the human colonization of space."

After 50 years of the space age, we have circled "back to the future." The driving forces of spaceflight now track long-standing tensions in American life—between romantic ideals and pragmatism, between the private and the governmental. But we go forward from a new vantage. The space age has remade us: our outlook, our everyday lives, and our expectations of the future are forever changed.

Pegasus XL Launch Vehicle
2004

*"Pegasus will be launched into space [for its first test] from the same
NASA B-52 used to drop the X-15 research rocket aircraft."*

— *Aviation Week*, 1988

NAMED after the flying winged horse of Greek mythology, Pegasus XL is an improved model of the world's first aircraft-deployed launch vehicle (LV). A modified Lockheed L-1011 or other aircraft carries the three-stage solid-fuel rocket to its launch altitude of 12,000 meters (39,000 feet). Then, the Pegasus releases and free-falls for five seconds in a horizontal position to achieve a safe distance from the aircraft. Its first-stage rocket motor ignites, followed by the other stages. The vehicle's delta wing fins give additional lift as it travels through the upper atmosphere. In a little over 10 minutes, it reaches orbital velocity and ejects its satellite into orbit.

Pegasus was designed as a low-cost vehicle for launching lightweight satellites—450 kilograms (1,000 pounds) or less. As an air-launched vehicle, it does not require expensive ground facilities, elaborate launch preparations, or numerous personnel. It can be launched over the ocean, away from population centers, with spent stages falling safely into the sea.

In 1987, Dr. Antonio L. Elias, the son of a Spanish diplomat, joined the Orbital Sciences Corporation and remembers conceiving the idea of an air-launched LV while doodling in a "boring meeting" in a hotel not far from the Museum's present Udvar-Hazy Center in Virginia. One of the company's main lines of work was developing upper-stage rockets for LVs, and planners had been looking for new business opportunities. On his yellow notepad, Elias—trained as an aeronautical engineer, with a PhD from the Massachusetts Institute of Technology—sketched a picture of "a little F15 aircraft with a rocket and satellite on the rocket." He then passed the drawing to his two colleagues at the table, who both raised their eyebrows. They knew this was "it," and the air-launched concept, not another upper-stage booster, became their focus.

Later, in March 1988, Elias patented the idea. The patent features drawings of the vehicle's configuration, including one of it on the carrier plane, and a typical launch and flight profile calculated by Elias.

The idea of an air-launched LV was not new; in 1956, the Italian Aurelio Robotti had proposed such a scheme, and early Space Shuttle concepts also used the idea. As Elias developed his concept, he was unaware of this prior work, but did know about an Air Force ASAT (anti-satellite) aircraft-deployed missile that intercepted and destroyed an in-orbit satellite in 1985. For Elias, the crucial idea in the ASAT project was that a rocket launched from a plane could reach space readily. His company received Elias's concept enthusiastically, and from 1987 to 1992 he led the technical team that developed it.

The idea worked. On April 5, 1990, at NASA's Dryden Research Center in California, Orbital Sciences launched the first Pegasus, which was carried aloft beneath a NASA B-52 aircraft. While Pegasus benefited from some government help, it was the first privately developed launch vehicle—previous U.S. rockets had been developed through federal funding.

Since that first flight, Pegasus has orbited 70 satellites in more than 30 missions. The Museum's Pegasus is the XL (extra-length) version, donated by Orbital Sciences in 2004. Its first-stage motor was used in XL ground tests, and its wings were flown into space and recovered.

ABOVE: A NASA B-52 aircraft carries a Pegasus on the first test of the new launch vehicle in November 1989.

Salyut Flight Suit
Late 1980s

"Space to the Children!"

—SLOGAN FOR EXPANDING THE SPACE PROGRAM DURING THE LATE SOVIET PERIOD, 1989

THIS FLIGHT suit, although not used in space, is an example of the type of clothing worn on board the Salyut space station during the Soviet Union's long-duration missions from the early 1970s through early 1980s. A lightweight, synthetic coverall, the suit was designed for comfort and convenience on board the space station; it did not provide any life support. The design tapers at the ankles and wrists to eliminate loose material that might catch in machinery inside the space station. Handy zippering allowed cosmonauts to don and remove these in-flight suits very quickly, in case of an emergency.

As a sequel to Salyut, the Soviet Union launched the base component of the *Mir* space station in 1986 and planned for a rapid expansion of the station through the addition of modules. During the late 1980s, the Soviet Union aspired to enlarge *Mir* and diversify the corps of cosmonauts serving on the station to include people from a variety of occupations. They even considered developing a second space station, *Mir 2*, which would rotate about its axis while in orbit, creating its own artificial gravity. Space advocates pushed these plans despite increasingly tight national budgets within the Soviet Union. They hoped that by expanding the appeal of the program to the next generation, they could gain funding for human spaceflight.

Although this training suit was originally designed for the Salyut program that ended in the early 1980s, a female Russian journalist wore it in preparation for a mission to *Mir* and donated it to the Museum in 1998. She had been selected, along with several other journalists, to inaugurate a diversified cosmonaut corps. These prospective journalist-cosmonauts went through physical and medical screening, followed by months of physical and classroom training. However, in 1990, the Soviet government dashed their hopes when it decided to send Japanese TV journalist Toyohiro Akiyama to *Mir*. The reason was financial: Akiyama's employer made a reported $10 million contribution to the Soviet space program. Soon after, the Soviets selected the finalists for the journalist program from among their trainees, but then announced its termination. This signaled an end to plans for significantly expanding *Mir*, for an artificial gravity space station, and for Soviet space "villages" populated with people performing a range of occupations.

EGRET
Late 1980s

"There is no other phenomenon like this in any other area of astronomy."

—GERALD FISHMAN, NASA SCIENTIST, 1991

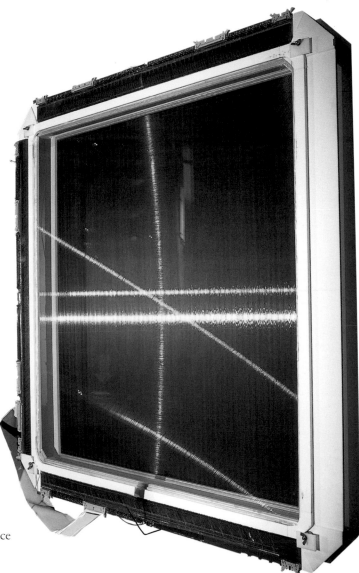

DURING Super Bowl XXVII in January 1993, as the Cowboys buffaloed the Bills, something happened in deep space which, as reported by the *Washington Post*, made astronomers that day's "biggest winners." A "monstrously powerful" burst of gamma radiation lit up the universe for a brief moment. A satellite, waiting in orbit for just such an event, recorded the burst, allowing scientists to examine its physical characteristics and dissect its meaning. By April, scientists realized that the gamma radiation represented a whole new class of celestial fireworks, revealing a universe far more violent than ever imagined. That satellite was the Compton Gamma Ray Observatory (CGRO), named in honor of the famous physicist Arthur Holly Compton. Launched by the Shuttle *Atlantis* in 1991, it weighed 17,000 kilograms (17 tons), the largest scientific satellite ever flown, and operated into the year 2000. The Museum preserves but one small element of that huge endeavor, a flight-spare piece of one of the detectors, transferred from NASA in 1994.

High-energy physicists think big when it comes to their tools. They use magnetized race tracks that dwarf the Indianapolis speedway in size to accelerate particles to very high speeds and then crash the particles into detectors the size of apartment buildings. In concept, CGRO was no different, except that the cosmos was its accelerator. Most of its bulk came from a grouping of four large chambers (its "apartment buildings"). In space, they collected naturally occurring radiation and high-speed particles, sifting and sorting them by energy and point of origin.

One chamber, called EGRET, for "Energetic Gamma-Ray Experiment Telescope," captured gamma rays in the 20 million to 30 billion electron volt energy range—typically over 10,000 to 150,000 times more energetic than diagnostic medical x-ray photons and at least 15 billion times more energetic than visible sunlight photons.

Known as a large spark chamber "telescope," EGRET detected gamma rays and then determined their originating direction—it was a tool for mapping the distribution of gamma rays in the universe.

Gamma-ray detectors on early satellites had reported time and again bursts of gamma ray energy in deep space. But the observations were too sporadic and the detectors inadequate to determine the source of the energy. Scientists developed CGRO mainly to produce a systematic survey of gamma radiation, as well as to look for and record transient bursts, localize their positions, and determine spatial distribution. After only four months of observations, data from CGRO's four detectors hinted that these gamma-ray bursts occurred evenly about the entire sky. They were not local phenomena, but deep space events of enormous energies, far beyond even cataclysmic supernovae.

Then, on January 31, 1993, came the "Super Bowl Burst" that EGRET reported. Once the data were analyzed it turned out that at the peak of the burst, it was at least 1,000 times brighter than any other gamma ray source in the sky. Strikingly, gamma-ray bursts seem to produce more energy in 10 seconds than the Sun emits in its lifetime! But gamma bursts of this magnitude are different from those produced by our Sun or other astronomical objects. How such energy is produced is still unknown. One theory is that these energies are released from extremely hot matter as it is annihilated and disappears into super-massive black holes at the centers of unusually active galaxies. After the Super Bowl one of the CGRO scientists said, "It's probably safe to assume that we'll have to rewrite the textbook on gamma-ray bursts." More than a decade later, gamma-ray bursts remain an active field of study.

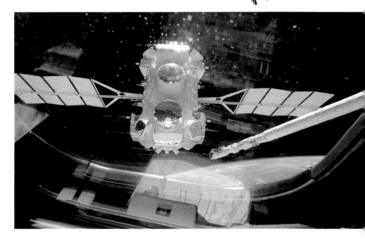

ABOVE: The Gamma Ray Observatory is visible through a flight deck window of the Shuttle *Atlantis* during its release from the payload bay in December 1991.

Hubble Space Telescope Mirror Flight Backup
1980s

"The initial shock is wearing off, and people are beginning to think about clever work-arounds ..."

—RAY VILLARD, SPACE TELESCOPE SCIENCE INSTITUTE, 1990

"WE'RE stubborn and clever," said NASA's chief scientist, Lennard Fisk, in June 1990 at a packed news conference called to address the discovery that the Hubble Space Telescope, newly in orbit, was flawed. A much anticipated scientific milestone, the telescope promised to provide the clearest views of the heavens ever seen. But after NASA expended billions of dollars and 25 years of effort, the telescope arrived in orbit with blurred vision. The news of the flaw came as a "punch in the stomach" to many astronomers and drew instant recriminations from Congress, the media, and the world of science. As Fisk spoke to reporters, he was adamant, "We're going to make it work."

Fisk was right—but it took many weeks for NASA and astronomers to identify the problem conclusively: the wonderful 240-centimeter (94-inch) mirror at Hubble's heart, manufactured by the Perkin-Elmer Corporation of Danbury, Connecticut, a reliable and prestigious optical firm, was ground and polished to the wrong formula. The press mocked the mirror as a "fun-house mirror," but its flaw was miniscule by everyday standards. A casual observer could never discern a difference between the flawed mirror and its correctly polished twin—a backup copy built by Kodak on display at the Museum. The difference in shape between the two amounted to less than two microns, or 1/50th the width of a human hair. To make this point, the Museum displays a human hair next to the mirror.

As small as two microns may be in everyday life, in optical physics it can make a monumental difference. To make the telescope work properly, this error had to be corrected. As Fisk predicted, NASA found a way, designing a set of corrective optics that space-walking astronauts installed on the telescope in 1993. Since that time, Hubble has made good on its promise to provide the sharpest wide-field optical images of celestial scenery ever seen by humanity. If NASA had not corrected the flaw, Hubble still would have worked but fulfilled only about 60 percent of its planned scientific mission.

The Corning Glass Works in Canton, New York, fabricated the mirror blanks using a special silicon/titanium oxide formula designed to make the mirrors almost completely insensitive to temperature changes. To limit weight and ensure sufficient rigidity to withstand the Shuttle launch, each blank combined two thick disks of glass fused to thin honeycomb support structures.

The Museum's blank then moved to the Kodak Apparatus Division, located in Rochester, New York, and the flight blank went to Perkin Elmer. Engineers ground each mirror to successively finer tolerances using computer-assisted testing routines. For the flight mirror, the final step was to aluminize the mirror's surface, to make it reflective and thereby allow the telescope to capture and analyze incoming light. Kodak did not go through this final step with the Museum's backup mirror—explaining (as seen here) the absence of a shiny surface. If NASA had chosen initially to fly the Kodak backup, in all likelihood the "flaw" would never have been part of Hubble's famous legacy.

ABOVE: In 1984, technicians at Perkin-Elmer's Optical inspect the Hubble primary mirror—the flaw in the mirror was not detected until 1990, after launch.

Impacts and Influences

Looking for Life
in a Vast Universe

Search for Extraterrestrial Intelligence, SETI, 1960–present

SPACE enthusiasts have always counted among their members a dedicated minority, for whom the main purpose of space exploration is to make contact with alien civilizations residing on other worlds. From what we understand of the laws of physics, sending spaceships across the cosmos is impractical. But *receiving* radio waves, which travel at the speed of light, from the vast reaches of space is practical. The field of radio astronomy is dedicated to capturing and analyzing such signals using radio telescopes positioned around the world.

Beginning around 1960, Frank Drake and other radio astronomers proposed that radio telescopes also could be used to detect signs of intelligent life in the universe. These signs might be radio emissions inadvertently released (as with our television shows for the past 60 years) from a distant planet or intentionally beamed our way from an extraterrestrial civilization interested in making contact with us. This inaugurated the Search for Extraterrestrial Intelligence, or SETI. The undertaking attracted much support in the science community as an empirical quest to pursue one of the great questions of the human condition, "Are we alone in the universe?" And, not surprisingly, this effort drew strong public interest.

Still, the practical implementation of this search has not been easy. To what part of the radio spectrum do we tune the receiver? How do we scan the entire sky? Above all, how do we know that a signal we receive is truly from an intelligent source and not the result of a natural phenomenon? That last criterion has been the most difficult to satisfy—and is essential to SETI's public credibility. For example, in the mid-1960s, astronomers at Cal Tech catalogued a radio source, C.T.A.-102, that varied at a rapid and regular interval. A Soviet physicist suggested it might be an artificial signal, and the pop group the Byrds even recorded a song about it. But C.T.A -102 was found to be a natural object, a "quasar."

With the advent of cheap personal computers, SETI supporters created a new approach, called SETI@Home. A privately funded institute arranges for observing time on the giant 350-meter (1,150-foot) dish at Arecibo, Puerto Rico. The astronomers concentrate on a frequency around 1,420 MHz, which is relatively uncontaminated by man-made signals (such as cellular telephones) or by interstellar noise. Here is the key part: owners of personal computers download sections of the captured data over the Internet and process them in the "background"; i.e. when their computer is not doing much work. The power of today's inexpensive PCs and the availability of high-speed Internet connections make SETI@Home the equivalent of a single giant supercomputer. PC owners by the thousands and from around the world have signed on—who would not want to be the first to receive a message from another civilization?

The SETI@Home project is ongoing and continues to find willing participants and sources of private funding, although NASA and other government agencies have steered clear of the project. SETI@Home's development of software to distribute complex problems to home computers proved a genuine innovation. This technique is now being applied to other computationally intensive problems, such as decoding genetic information or constructing models of complex proteins.

After processing signals since 1965, no other civilizations have been found—but plenty of people are still looking.

ABOVE: Two views of the Allen Telescope Array (supported by philanthropist Paul Allen) located at Hat Creek Radio Observatory in northern California.

"C.T.A. 102, year over year receiving you / Signals tell us that you're there, we can hear you loud and clear. / On a radio telescope, science tells us that there's hope / Life on other planets might exist."

—THE BYRDS, "C.T.A. 102"
   (© 1996 SONY MUSIC ENTERTAINMENT, INC.)

Hubble Telescope Charge-Coupled Device
1970s

"It is possible to conceive of press photography in the future using CCD cameras."

—JAMES CHUNG, FUJI FILM, 1980

ABOVE: Hubble's CCD technology helped take this image of the Whirlpool M51 galaxy. To the right, galaxy NGC 5195 is moving on a path that will carry it behind M51.

IN THE late 1960s, two scientists at Bell Laboratories in New Jersey developed a new type of integrated circuit—one that converted light into electricity. They soon realized their design might be used to sense and record images and act like an electronic form of photographic film. Thus was born the charge-coupled device (CCD). Now ubiquitous in digital cameras, the technology took years to develop. But from the moment of its discovery, the CCD excited astronomers, who quickly looked for ways to use it in telescopes. It soon became a vital part of one of astronomy's most important projects—the Hubble Space Telescope.

An astronomer's job is to collect radiation from celestial sources, record it, amplify it, and study it. In the late 1950s and 1960s, as space exploration got underway, this basic set of tasks grew more complicated for astronomers who conducted research with orbiting telescopes. Developing devices for capturing and returning data presented a special problem. Standard photography was out of the question. Other technologies, such as television, proved either unreliable or insufficiently sensitive for the special needs of space-based astronomy. An ideal detector or "eye" had to meet a key challenge: record very faint objects, typically the most valuable for research. The CCD seemed the answer as it was sensitive, reliable, and suited for sending data back as electronic signals.

Unfortunately, the first commercial imaging CCDs produced in the mid-1970s fell short of these expectations. A decade later, though, stimulated by commercial and military applications, CCDs possessed a quantum efficiency (a sensitivity scale used by astronomers) 50 times greater than photography. Their capacity, too, measured in pixels, continued to grow larger. As early as 1973,

astronomers began to think about using the CCD as an astronomical imager. In the spring of that year, teams of astronomers working with Jet Propulsion Laboratory (JPL) and Goddard Space Flight Center secured funding to support a development study by Texas Instruments to build larger pixel arrays for astronomical use. Larger arrays allowed researchers to image large areas in high resolution.

The prospects for CCDs seemed promising enough that astronomers sought to include them in the much-anticipated Hubble. Three teams competed to build the wide field and planetary camera (WFPC), Hubble's prime imaging camera. Two teams proposed using CCDs, and one a refined image orthicon TV tube. At that point, the 70 mm orthicon could produce a larger image than any CCD. To overcome this disadvantage, the winning team from JPL proposed using two arrays, each containing four CCDs. Each set of data captured by the elements of the array could then be recombined electronically to produce a wide-field image equivalent to the 70 mm TV tube. CCDs used in this way produced some of Hubble's most striking and scientifically important images of the universe.

The Museum's artifact is a CCD array from the Hubble program, a flight spare used for testing and evaluation. Built by Texas Instruments, it consists of an 800x800 pixel array, mounted in a quartz-windowed pressurized framework by the JPL. It is identical to the CCD arrays in the original WFPC flown on the telescope and was transferred to the Museum by Goddard Space Flight Center in 1979. In 1993, NASA flew a repair mission to Hubble, replaced the WFPC with a new version, and brought the original camera back to Earth. That historic instrument, also in the Museum collection, includes two flown CCD detectors, but they remain inside their electronics and optical housings.

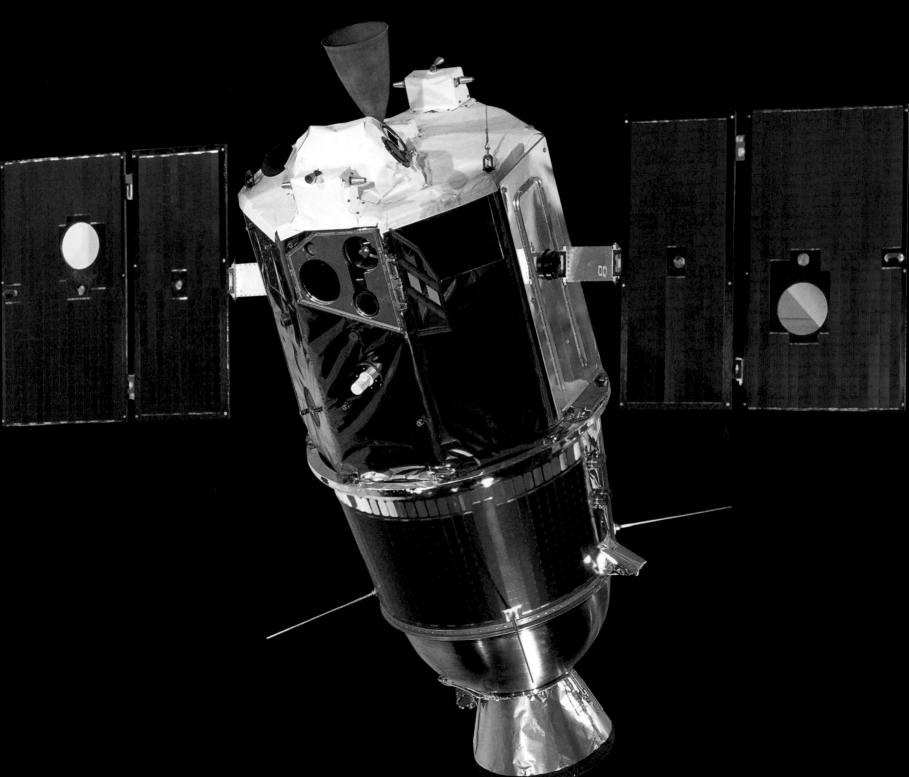

Clementine
Early 1990s

"An abundant supply of water on the Moon would make establishment of a self-sustaining lunar colony much more feasible and less expensive than presently thought ... scientists have suspected that ... water [meteorites] could migrate to permanently dark areas at the lunar poles.... These suspicions appear to be correct."

—Paul D. Spudis, Clementine deputy chief scientist, 1996

I n 1994, the United States returned to the Moon for the first time since Apollo 17, more than 20 years earlier. The Clementine spacecraft, sponsored by the Department of Defense's (DoD) Strategic Defense Initiative Organization with support from the Naval Research Laboratory and NASA, reflected the new cost-conscious 1990s. With a "cheaper, better, faster" philosophy, the mission's primary purpose was to evaluate the performance of commercial "off the shelf" sensors and components under space conditions. Its secondary purpose was science, primarily related to mapping. The mission, though, produced a surprise: the detection of ice embedded deep in a crater of the Moon's South Pole astounded the public and sparked intense debate among scientists of the validity of the discovery.

The Clementine spacecraft arrived at the Moon in February 1994. Lunar mapping took approximately two months, but as the spacecraft prepared to leave and conduct a flyby of the near-Earth asteroid 1620 Geographos, a thruster malfunctioned, ending the mission. Before the mishap, Clementine's suite of cameras, snapping images over a range of wavelengths (known as multispectral imagery), recorded approximately 1.5 million images of over 99.9 percent of the Moon's surface.

The acquisition of multispectral imagery reenergized lunar science, especially when data returned by Clementine indicated that ice may exist at the Moon's South Pole. Scientists, led by chief project scientist and lunar geologist Gene

Shoemaker, asserted that a patch of ice about 7.6 meters (25 feet) thick and the diameter of a small pond might well exist on the South Pole. They constructed a mosaic of some 1,500 images of the lunar South Pole, revealing a depression named the South Pole-Atkin Basin estimated at 4 billion years old and some 1,500 miles wide. By far the deepest impact crater in the solar system, this area never saw the Sun and gave off data readings suggestive of large concentrations of ice. Temperatures never exceeded -230°C (-383°F), making it a perfect storehouse for water brought by comets to the Moon. Shoemaker quipped, "This is the place on the Moon where you would go to get ice for your cocktail." Excitement over this discovery spurred the team developing the Lunar Prospector, a small, spin-stabilized craft sent to the Moon in 1998 to "prospect" the lunar crust and atmosphere for minerals, water ice, and certain gases.

Scientists remain divided over the potential of ice on the Moon. Imagery from Lunar Prospector seemed to confirm Clementine's data indicating lunar ice, but when the spacecraft crashed into the presumed icebed on July 31, 1998, at the end of its mission, it yielded no confirming evidence that any water, frozen or otherwise, might be on the Moon. "It certainly would have been nice to find some sort of lunar skating rink, or thick layers of ice, but it looks like it's just not there," said Museum scientist Bruce Campbell. The jury is still out.

The object in the Museum's collection is an engineering model of Clementine transferred from the Naval Research Laboratory in 2002.

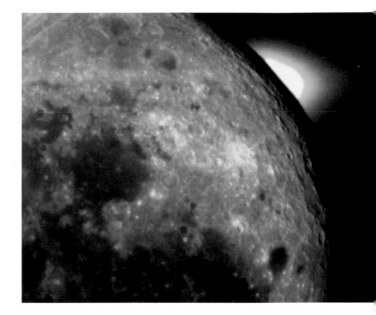

ABOVE: Clementine's Attitude Determination Camera captured this image of the Moon illuminated by earthshine. The edge of the Sun is visible on the right.

Babylon 5 Jumpgate Pin
1993–1999

"If you saw a <> in an email from someone, it was like a secret handshake among fans."*

—ELANA, *Babylon 5* JUMPGATE PIN CREATOR

SCIENCE fiction fans have reputations for being uncommonly devoted. This frontplate from a fan-created pin is memorabilia from *Babylon 5*, a space-themed television show that aired from 1993 to 1999. The show offered adult science fiction that merged with the Internet age, spawning a virtual fan community that thrived online. Chatting with the show's creator and other fans fostered a new kind of participatory fandom that ultimately kept *Babylon 5* on the air.

J. Michael Straczynski's *Babylon 5* is set in 2257 on the last Babylon space station, a place offering "the last hope for a galaxy without war." Unlike most producers, who usually write episode by episode, or season by season, Straczynski developed the show's five-year story arc (with a beginning, a middle, and an end) long before production began. Although individual episodes could stand independently, seasons functioned as chapters in a "five-year-long novel for television." Plots were not always resolved at the end of the hour. Characters died. Characters changed based on their experiences. The arc's rich foreshadowing (and literary and historical allusions) inspired a loyal fan base.

Even more significant, *Babylon 5* spawned a fan culture that used online communication (in addition to newsletters, conventions, and collectible items) to share their affection for the show. Straczynski (known online as JMS) maintained close contact with fans. Beginning as early as November 1991, Straczynski participated in Web conversations with fans. As he explained, "By talking to the fans, and appearing at cons [conventions]…I've been able to find out what the fans want, and to combine that with my own personal vision." He even kept a "Best of the Nets"

area on his wall, posting printouts of particularly useful chats. Online resources became the backbone of the show's fan culture, marking "Babylon 5" as new and different—and emblematic of its times.

Fans repaid Straczynski's attention by creating an active, loyal community. During B5's first season, "internetted" fans adopted the ASCII symbol <*>, which resembled a "jumpgate" (the "portal into hyperspace" used in the B5 universe for interstellar travel), as the emblem of international B5 fandom. Like the colon and a right parenthesis that became a sideways "smiley" (the first e-mail "emoticon"), the three-keystroke symbol <*> became an easy way to identify other B5 fans. The symbol is still used in signature files and on various B5 newsgroups, majordomo lists, online newsletters, and Web pages.

As befitted B5's virtual community, when a fan named Elana developed the first jumpgate pins, she sent Straczynski a prototype for his personal inspection. When he pronounced it "quite nice," Elana began producing the pins for sale, knowing "that this Pin had passed THE most difficult quality test possible within the B5 fan world." The lack of a "pinback" on the jumpgate pin held in the Museum's collection identifies it as one of Elana's first batch. The B5 pin frontplate came as part of a memorabilia collection that also contained computer printouts of Straczynski's online postings.

This pin exemplifies how a fan culture that was created using new modes of communication allowed immediacy despite distances, strengthening relationships among people who had never met in person. Indeed, in the end, it was *Babylon 5*'s determined fan base that allowed Straczynski's vision to be realized. When Warner Brothers cancelled the show after four seasons, a vigorous fan campaign found it a home at the cable network TNT for its fifth season, allowing Straczynski to complete his vision.

Mars 96 Lander
Early 1990s

"Is there life on Mars?"

"No, not there either."

—RUSSIAN SAYING POPULAR IN THE SOVIET PERIOD

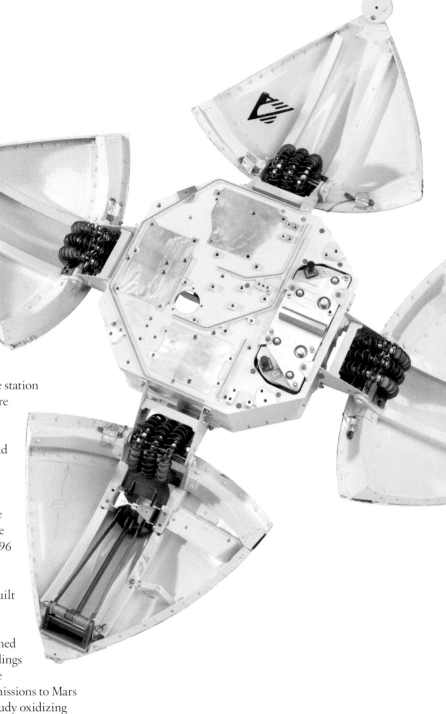

PLANETARY scientists agree on one thing: exploring Mars is extremely difficult. Only robotic missions to Mars have flown, recording as many failures as successes. Success for the Soviet Union—then for Russia— proved particularly elusive. In nine tries, no mission performed as planned—*Mars 96*, launched in 1996, was the 10th and last in a series of frustrating efforts.

In the 1960s and 1970s, the Soviets launched seven of these ill-fated missions to Mars. During the 1970s the USSR's frustrated planetary scientists shifted their full attention to Venus—ultimately with much greater success. The Soviet Union began launching the Venera series of probes in 1961, and by the early 1970s, the series began to yield a trove of scientific data. This ambitious program culminated with Vega, a dual mission to the planet and comet Halley. Basking in their success on Venus, Soviet scientists and engineers resolved to renew their efforts to explore Mars. In 1988, their first attempts aimed to study Mars's moons, with the Phobus 1 and 2 missions. Neither spacecraft, though, reached its targets. Undaunted, the Russians (continuing the Soviet program) planned a series of additional missions to Mars, with the first probe designated *Mars 92*. Financial problems slowed the effort and *Mars 92* became *Mars 96*.

The Russians intended the *Mars 96* mission to be a multinational undertaking. Planning began in the 1980s but stalled. The end of the Cold War and improved relations with former enemies helped sustain the planning for the mission as its projected launch date slipped from 1992 to 1996. France, the newly reunified Germany, and the United States all participated.

Mars 96 used a complex array of instruments to study the planet. The device pictured here was a working engineering model for a *Mars 96* landing probe. The design called for *Mars 96*, as it orbited around Mars, to release a probe capable of landing on the surface of Mars. *Mars 96* failed to reach Mars, troubled by navigation or communication problems (the cause was not clear)—the same set of technical difficulties that savaged previous Soviet probes to the planet.

Nevertheless, this small surface station is an excellent example of a hardware design that Lavochkin Company of Russia refined during 35 years of building planetary probes. Small and simple, with a spring-release petal design, the actual station carried experiments from France, Finland, Germany, and the U.S. Each package fit into one of four small bays on the lander. In preparation for the *Mars 96* mission, the company sold this engineering model to NASA's Jet Propulsion Laboratory (JPL). JPL built the Mars Oxidation Experiment (MOx), which occupied one of the lander's bays. Scientists at JPL designed the MOx experiment to pursue findings from experiments performed by the Viking landers, NASA's successful missions to Mars in the mid-1970s. They hoped to study oxidizing agents in the Martian soil and atmosphere, the presence of which scientists had inferred from the results of Viking biology experiments.

JPL transferred this engineering version of the Russian lander to the Museum in 2001, after *Mars 96* failed. Notice that JPL inserted its backup MOx instrument into the structure (the shiny "box" just inside the right petal).

Magellan T. Bear, STS-63
1995

"He will be the only astronaut, so far at least, to orbit the earth in a Ziploc bag. But then, no one wants fur floating all over the cabin."

—*People*, FEBRUARY 1995

EVEN the most well-traveled rarely experience trips on a nuclear submarine, a night in the Lincoln Bedroom of the White House, or a flight into space. A bright blue stuffed children's toy named Magellan T. Bear is the proud owner of a special passport recording his travels to those locations and many more around the globe.

With the encouragement of a librarian passionate about students learning math, science, and geography, students at Elk Creek Elementary School in Pine, Colorado, sent Magellan T. Bear on a journey over land, under the sea, and beyond the atmosphere. What began as an around-the-world flight with United Airlines evolved into other fantastic adventures as the students imagined new travels for Magellan, and for what they might learn along the way. In his exclusive bear-sized flight suit, Magellan orbited the Earth as an "education specialist" in the SPACEHAB module aboard the Space Shuttle *Discovery* during eight days in February 1995.

Through his unique travels, Magellan became an informal ambassador for education—especially for NASA. He helped the librarian, Penny Wiedeke, her students, and NASA to enhance educational ties among children across the globe and encouraged the space agency's commitment to educating and inspiring young people. The bear became symbolic of a way to ready a new generation for space exploration by drawing students into careers in science, technology, engineering, mathematics—like the "Teacher in Space" program of the mid-1980s. This blue bear flew on a Space Shuttle, not just for fun and adventure, but to serve as a symbol for education programs aimed at improving the country's poor performance in science and math. NASA placed itself at the forefront of that effort by using the uniqueness of spaceflight to encourage students in their studies.

The 1995 flight of Magellan T. Bear coincided with a pivotal moment in NASA's educational programs and in national policy discussions. At the encouragement of the presidential administrations of George H.W. Bush and Bill Clinton, teachers and administrators worked to establish standards for students to achieve in most academic subjects, especially science, math, and history. While aimed at raising student achievement, the development of specific guidelines for classroom education provoked intense debate. Though NASA stood outside the fray over standards, it worked at adjusting its own educational outreach programming to better meet the needs of teachers and students.

The space agency took full advantage of the Internet's rapid growth by developing high-tech and innovative teaching materials and programs accessible in an electronic format, beginning with the launch of the NASA education web site in 1996. These electronic resources complemented a range of educational activities, including programming for NASA TV, school visits by astronauts and scientists, and other projects bringing NASA's work directly to school children. These initiatives evolved into a large-scale effort that reached every state in the nation and sought "to inspire the next generation of explorers … as only NASA can."

After his high-flying travels, teacher Penny Wiedeke and colleague Jerry Williams retired Magellan T. Bear from active status as an education specialist and donated him to the Museum to take a special place in the *How Things Fly* gallery. The education initiatives he came to symbolize continue, reflecting NASA and the country's belief that nothing sparks student interest as much as astronauts and space exploration.

Magellan GPS "Trailblazer" Receiver
Mid-1990s

"When the [Gulf] War broke out [in 1990] there were a limited number of military receivers in the DoD inventory. This led the DoD to purchase thousands of GPS civilian receivers ... [mostly] from Trimble Navigation and Magellan Systems.... Close to 90 percent of the GPS receivers used in the war were of the commercial sort."

—THE GLOBAL POSITIONING SYSTEM, 1995

IN U.S. history, one rarely finds an instance in which a military technology, actively in use, also becomes a mass market global consumer product—one available to anyone, anywhere on the Earth. This "Trailblazer" receiver, donated to the Museum by the Magellan Corporation, was available for sale in the mid-1990s. It gave consumers access to the signals of the Global Positioning System of satellites developed and operated by the U.S. military. The GPS allows a person—military or consumer—anywhere in the world to determine their location within tens of feet. For hikers and other outdoor enthusiasts, this handheld GPS receiver became as essential as a canteen of water or a sturdy pair of boots and was part of the first wave of a consumer trend. Now practically anything that moves might use GPS—from cars, trucks, ships, and airplanes, to cell phones, to children tagged by their parents.

For the military, the uses of precise location information are similar but different: to guide cruise and ballistic missiles to enemy targets, to give soldiers on the ground their position, and to allow ships and submarines to navigate across the oceans. These military needs inspired GPS's global coverage. The military began deploying the GPS in 1979 but did not fully complete the system until 1995. Even before it was fully operational, GPS proved its military worth in the Gulf War, and it has since been integrated into nearly all current military systems and operations—just as in the consumer world.

At the heart of GPS is a constellation of 24 satellites, orbiting at an altitude of about 19,000 kilometers (12,000 miles) and carrying accurate atomic clocks that broadcast precise time information. Each satellite also sends information about its position, determined by Air Force ground tracking stations. By measuring the delay in receiving time signals from four or more satellites, a receiver can locate its position in three dimensions. A receiver does not need a sensitive dish antenna or an atomic clock of its own; the satellites provide enough redundancy to give accurate time, and they broadcast using a technique that a simple antenna can receive.

But how did GPS become both a military and consumer phenomenon? At its inception, GPS, with its accuracy and worldwide coverage, was viewed primarily as a military tool, with important geopolitical implications. A specific incident sparked discussions of civilian use, and the policy evolved over nearly a decade.

On September 1, 1983, a Korean airliner strayed off course and began flying over restricted territory of the Soviet Union, where it was shot down by a Soviet jet fighter. All on board the airliner perished. It seemed an avoidable tragedy. Many in the world community felt that if the crew of that airliner had a working GPS unit on board, they might have avoided straying in Soviet airspace. At the urging of Congress, President Reagan ordered that the GPS system be made available to civilian as well as military users. At first, the military allowed civilians to use only a degraded signal; this too was relaxed by President Clinton a decade later. Those presidential directives, combined with the advances in microelectronics and software that have occurred at a furious pace, gave us this "Trailblazer," from a company with the appropriate name "Magellan."

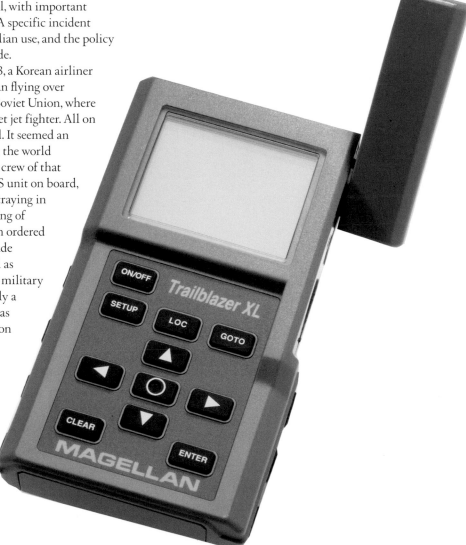

Soyuz Toilet; *Mir* Toilet
1970s to present; 1986–2001

"What kind of toilets are on board spacecraft, and how do they work?"

—A COMMON QUESTION

ABOVE: Soyuz toilet

OPPOSITE: *Mir* toilet, configured for use by females

MANAGING bodily waste has been an issue since humans began flying in space. Cosmonaut Yuri Gagarin relieved himself just prior to boarding the Vostok capsule before his first human spaceflight. After waiting hours in position for his launch, Alan Shepherd had no alternative but to urinate in his spacesuit just before the first American suborbital flight. In each case, spacesuit designers did not provide for a urine receptacle in the suit. But these still Earth-bound men benefited from gravity and their enclosed spacesuits—once the men relieved themselves, the waste products stayed in place.

In orbit, the problem is different. Astronauts and cosmonauts often traveled with their spacesuits partially open or off. In microgravity, going to the bathroom created the likelihood of human wastes floating freely in the spacecraft. To overcome this safety and health issue, engineers developed two solutions to waste management: diapers and suction. Soviet and Russian designers preferred suction as a solution.

The receptacles pictured here are Soviet-designed and -built human waste disposal units—or, plain and simple, toilets. One was designed for use on board the Soyuz spacecraft. Still employing the same basic design since its inaugural flight in

1967, Soyuz has had the longest operational life of any human-rated spacecraft. The other was designed for use on board the *Mir* space station, which was the longest-occupied space station (over 12 years of continuous occupation after its launch in 1986). The Russian company Zvezda manufactured both at the Museum's request.

The Soyuz toilet was for use by male cosmonauts. Built into the spacecraft, cosmonauts used it during the two-day flight to the *Mir* space station. The funnel and cup are at the end of a hose that the cosmonaut inserts into his spacesuit for appropriate use. The suction from the compressor deposits all waste materials into the gray tank. Once the cosmonauts have arrived at their orbiting destination, they can remove the tank and dispose of it.

In contrast, engineers adapted this *Mir* toilet to the female anatomy, so that it could be used by both sexes. Most of the differences between the Soyuz and *Mir* toilets reflect the distinct character of each spacecraft, not differences between sexes. While on board the station, cosmonauts routinely did not wear spacesuits, and thus could position themselves over the toilet, which features a lid, much like our Earth-bound toilets. Once the cosmonaut positioned herself correctly, she then activated the suction and proceeded.

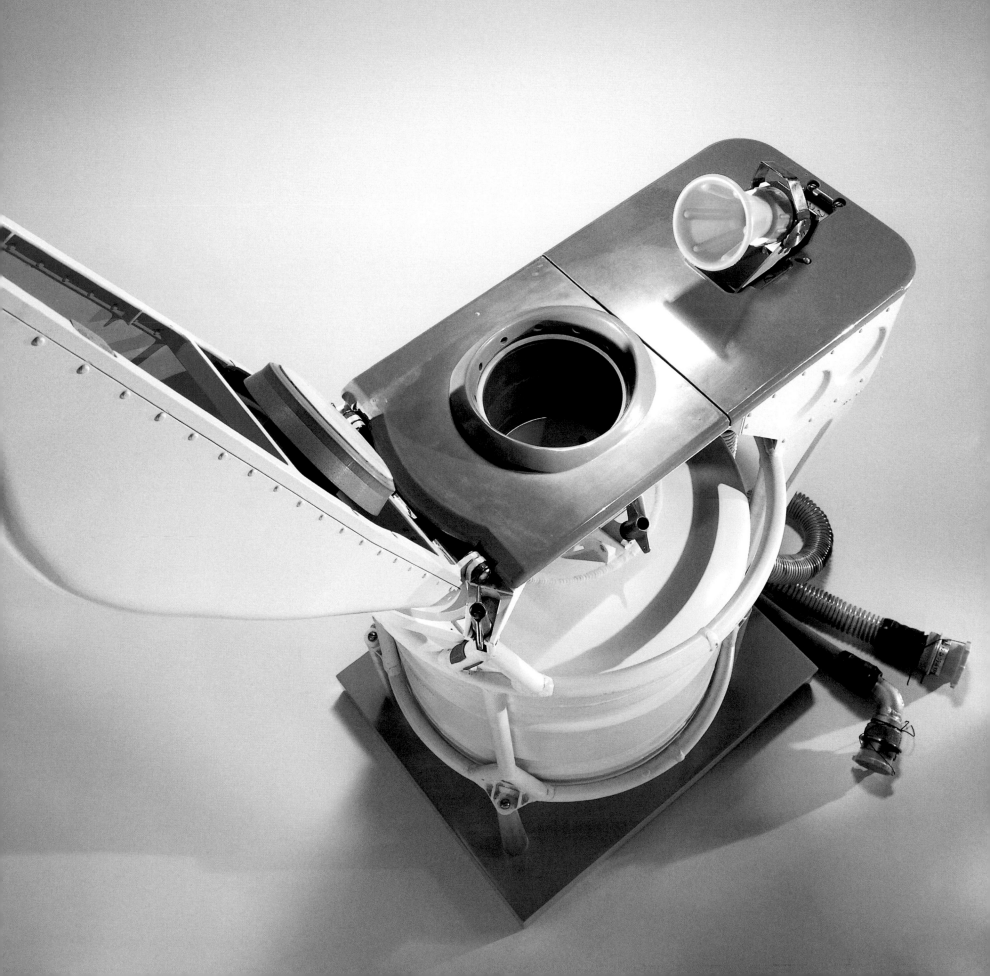

Shannon Lucid's Penguin-3 Spacesuit
1996

"Down here, you know where your arm is, but in space that's not always the case."
—VICTOR SCHNEIDER, NASA SCIENTIST FOR SPACE PHYSIOLOGY, 1997

AMERICAN astronaut Shannon Lucid wore this flight suit during her American-record-breaking six-month stay on board the Russian *Mir* Space Station (March–September 1996). During her long period in microgravity, this suit helped to moderate the debilitating effects of prolonged weightlessness on the skeletal muscular system. Dr. Lucid's mission as a member of the *Mir 21* crew was one of a series of joint Russian-American missions in preparation for the construction of the International Space Station.

The Penguin-3 suit is a snug-fitting, full-length, long-sleeved jumpsuit that zips vertically to the crotch, with horizontal zip pockets on either side of the chest and just below the waist. The suit is made of a synthetic fabric with elastic inserts at the collar, waist, wrists, ankles, and along the vertical sides of the suit. The inside of the suit contains a system of elastic, straps, and buckles that can adjust the fit and tension of the suit. When astronauts and cosmonauts tighten the tension on these bands, the suit exerts pressure on the body and simulates Earth-like gravity. This feature aids in the retention of muscle mass and calcium in the bone—the two physical areas greatly affected by prolonged weightlessness. As the suit adds constant pressure to the body, an

astronaut suffers less from the effects of microgravity. The design and construction of the suit reflect more than two decades of research on combating the effects of weightlessness on the human body. The Russian company JSC (Joint Stock Company) Zvezda (Star) manufactured this suit and all Soviet and Russian spacesuits, as well as much of the life-support equipment of the Soviet and Russian manned spaceflight programs.

When Dr. Lucid trained to use the spacesuit, she also underwent a complementary, rigorous exercise program, a system of preparation the Soviets and Russians refined since the beginning of space station activity in 1971. She demonstrated the effectiveness of the Penguin-3 suit and exercise program when after six months on *Mir* she returned to Earth, and "normal" gravity, and emerged from the Space Shuttle *Atlantis* able to walk.

This suit bears a patch with Dr. Lucid's name above the left upper pocket. It also displays the mission patch for *Mir 21*, the NASA logo patch, and flags of the United States and the Russian Federation. After the mission, the Russians gave the suit to NASA, who in turn transferred it to the Museum in 1996.

ABOVE: Astronaut Shannon Lucid communicates with ground control from the core module of the *Mir* Space Station during her stay in 1996.

Space Acceleration Measurement System (SAMS) Crew Logbook
1997

"For every action, there is an equal and opposite reaction."

—NEWTON'S THIRD LAW OF MOTION, 1687

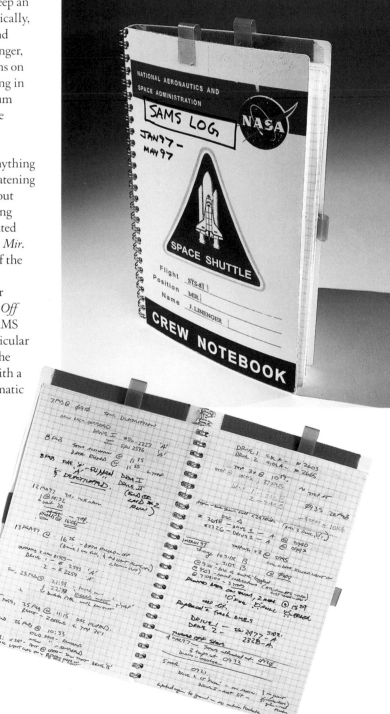

MICROGRAVITY, or weightlessness, is one of the joys of spaceflight. No matter how busy they are, astronauts always take a few minutes to float or fly or tumble playfully, unconstrained by gravity. It is easy for them to demonstrate Newton's third law of motion by spinning themselves around a screwdriver or shooting away from a wall with a gentle fingertip push.

Microgravity is not all about fun and games, however. Scientists are quite serious about using it as a unique laboratory for fundamental research. Without the normal influence of terrestrial gravity, fluids, mixtures, crystals, materials, fire, and living things behave differently. Studying those differences can enhance the understanding of basic biology, chemistry, and physics and also help solve a variety of problems in the materials and life sciences.

Since the early 1980s, investigators have loaded the Space Shuttle and space stations with hundreds of microgravity experiments. For maximum success, these experiments require that the environment be as still as possible. But Newton's third law of motion makes that difficult to guarantee. Small vibrations and occasional jolts occur as the air conditioning system hums, the attitude control thrusters fire, and the crew members work out on their treadmill.

To monitor the vibration environment aboard the Shuttle, NASA developed a sensor system called the Space Acceleration Measurement System (SAMS). It is a set of three triaxial accelerometer sensor heads attached by long cables to a recording system that keeps a timed electronic log of miniscule vibrations and larger disturbances inside the spacecraft. This record can then be used to characterize the actual microgravity environment in which the experiments were performed and thus help to interpret their results or to redesign the experiments. SAMS also flew on the Russian space station *Mir* and the International Space Station.

One of the crew members' duties is to keep an eye on the monitor—checking SAMS periodically, changing data discs, resetting the recorder, and noting anything worth noting. Jerry M. Linenger, an American astronaut who spent five months on *Mir* in 1997, recorded such routine monitoring in this SAMS logbook, transferred to the Museum from NASA in 2000. His entries are what one might expect of such a housekeeping task, dutiful but not terribly interesting.

In fact, Linenger's stay aboard *Mir* was anything but routine. He had the unexpected life-threatening adrenaline rush of fighting a fire that broke out early in his mission. If that were not harrowing enough, a few weeks later an arriving automated Progress supply vehicle nearly slammed into *Mir*. Surely that impact would have sent SAMS off the chart, and might well have doomed the crew.

Jerry Linenger described these and other experiences on *Mir* quite vividly in his book *Off the Planet*, published in 2000. The official SAMS crew logbook serves well as a record of a particular technical task. His postflight reflections, on the other hand, are a personal account written with a keen eye for detail; he tells a fascinating, dramatic story of his time in space.

Microgravity is the salient fact of the spaceflight experience. It affects everything from the way objects move in space to the movement of fluids within the body, the quality of crystals and alloys formed in space experiments, the behavior of liquids and fire, and certainly the astronauts' sensations and perceptions. Jerry Linenger's records of his time in weightlessness—both this logbook and his memoir—suggest a new spin on Newton's third law as it might apply to writing about spaceflight: for every routine observation, there is an equally extraordinary one.

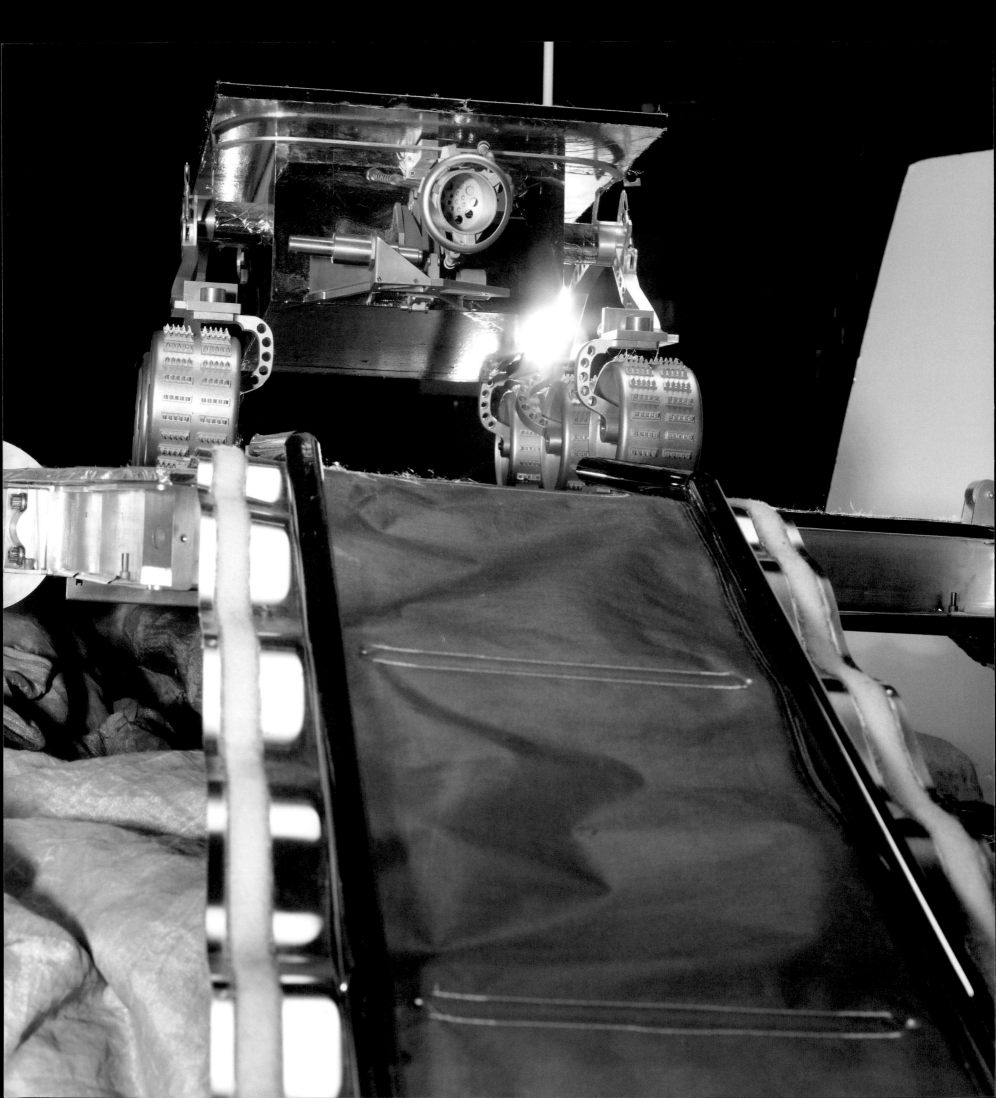

Mars Pathfinder
1996

"[It was a] rock festival on Mars."

—L. Jaroff, *Time*, July 21, 1997

I N August 1996, a team of NASA and Stanford University scientists announced that a Mars meteorite found in Antarctica contained possible evidence of ancient Martian life. Scientists hypothesized that the 1.9-kilogram (4.2-pound), potato sized rock (identified as ALH84001) formed as an igneous rock about 4.5 billion years ago, when Mars was much warmer and probably possessed oceans hospitable to life. Then, about 15 million years ago, a large asteroid hit the red planet and jettisoned the rock into space, where it remained until crashing into Antarctica around 11,000 B.C. According to some experts, ALH84001 appeared to contain fossil-like remains of Martian microorganisms 3.6 billion years old. The findings electrified the scientific world and excited the public, boosting support for an aggressive set of missions to Mars to investigate these tantalizing possibilities.

By chance, Mars Pathfinder stood ready—the first mission to the red planet since Viking 20 years before. Underway at the time of the ALH84001 discovery, Pathfinder became an urgent priority for NASA. Launched in December 1996, the probe successfully landed on Mars on July 4, 1997.

Nearly everything about Pathfinder generated excitement. It exemplified a new NASA spacecraft philosophy of "cheaper, better, faster" and featured two noteworthy innovations. As the craft descended through the Martian atmosphere, it used inflated airbags to cushion landing, allowing the probe to bounce along the surface until coming to rest. As the airbags deflated, three solar panels unfolded, revealing the probe's most innovative component—a small, 10.5-kilogram (23-pound) solar-powered robotic rover named *Sojourner*.

The solar panels, positioned close to the surface, provided a ramp for *Sojourner* to venture onto Mars. It proceeded to take close-up images of the surface using two color cameras on the front and a black-and-white camera on the rear. The rover also contained a rear-mounted Alpha Proto X-ray Spectrometer that provided bulk elemental composition data on surface soils and rocks. In addition, the stubby wheels of the rover provided information about the physical characteristics of the surface soil and rocks. It discovered important new data about rocks washed down into the Ares Vallis flood plain, an ancient outflow channel in Mars's northern hemisphere. Projected to operate for 30 days, the rover worked for nearly three months, capturing far more data on the atmosphere, weather, and geology of Mars than scientists expected. NASA lost communications with Pathfinder on September 27, 1997. In all, the Pathfinder mission returned more than 1.2 gigabits (1.2 billion bits) of data and more than 10,000 pictures of the Martian landscape. While this scientific harvest did not answer the questions posed by ALH84001, it did deeply enrich our understanding of the planet.

The mission caught the imagination of the public—especially the activities of the plucky *Sojourner* rover. Its semi-autonomous "behaviors" made it a favorite of the public, who followed the mission with great interest via the World Wide Web. Twenty Pathfinder "mirror" sites recorded 565 million hits worldwide during the period of July 1 to August 4, 1997. The highest volume of hits in one day occurred on July 8, when a record 47 million hits were logged.

The lander in the NASM collection is a backup acquired from the Jet Propulsion Laboratory in 2000. The rover is a full-scale model built by the Jet Propulsion Laboratory for the Museum. The rocker-bogie, wheels, and frame are constructed of aluminum and similar to the actual flight vehicle; however, the scientific navigation and communication instruments are made of resin.

Spacelab
1983–1998

"President Reagan and Chancellor Helmut Kohl of West Germany today hailed the flight of the space shuttle Columbia *carrying the European-built Spacelab as an impressive example of international cooperation and perhaps a major step toward further American-European joint space ventures."*

—Joint telecast to the Spacelab 1 crew, reported in *The New York Times*, December 6, 1983

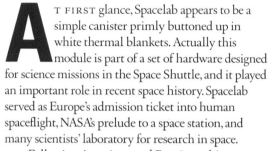

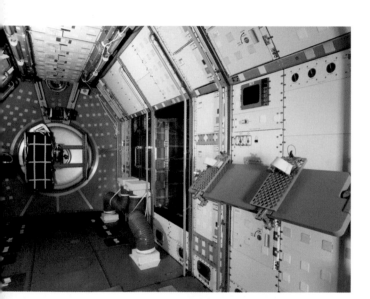

A T FIRST glance, Spacelab appears to be a simple canister primly buttoned up in white thermal blankets. Actually this module is part of a set of hardware designed for science missions in the Space Shuttle, and it played an important role in recent space history. Spacelab served as Europe's admission ticket into human spaceflight, NASA's prelude to a space station, and many scientists' laboratory for research in space.

Following American and Russian achievements in space, European leaders in science, industry, and government moved to develop their own capabilities to join the spacefaring community. A multinational European enterprise began to produce rockets and scientific payloads, but human spaceflight beckoned, and with it the possibility of carrying out laboratory research in space.

In 1973, NASA reached a partnership agreement with the European Space Research Organization (a precursor to the European Space Agency, ESA) to develop a research laboratory suite to operate in the Shuttle's payload bay. In this collaboration, the Europeans built the hardware, selected astronauts for training and flights with American astronauts, and were entitled to their own missions. NASA mentored the hardware development process, trained the astronauts, and managed the flights. The primary benefits of this international cooperation were that the Europeans gained experience and stature in human spaceflight, NASA added a large research capability to the Shuttle, and all partners became prepared for future projects together.

From 1983 until 1998, the Shuttle flew more than 30 Spacelab missions, 16 of them with a laboratory module. A pressurized cylindrical room connected by a tunnel to the crew cabin, the laboratory was outfitted with computers, workstations, stowage lockers, supplies, equipment, and scores of experiments that varied from mission

to mission. The Spacelab module in the Museum flew nine times, first on the Spacelab 1 mission in 1983 and last on the dual Microgravity Science Laboratory missions in 1997.

Scientists were the heart of the Spacelab missions. Hundreds of researchers worked together on selected flight experiments. Guest scientists, called payload specialists, joined members of the astronaut corps to carry out the onboard research. Laboratory missions typically had seven-member crews for intense round-the-clock operations. Crew members talked directly to scientists on the ground while conducting their experiments, a valuable new protocol.

Some Spacelab missions had a diverse research program, from astronomy to zoology. Others were dedicated to life sciences, physical sciences, atmospheric or celestial observations, and even neurology research. Over the years, hundreds of investigators from institutions around the world participated in research conducted on Spacelab missions. When the Spacelab program ended in 1999, NASA and ESA had met the managerial, technical, and political challenges of a complex technology development program across national lines, currencies, languages, and customs. The Europeans gained the necessary expertise for human spaceflight by developing hardware, flying 10 of their astronauts, and being involved in mission operations. With those credits, they became full partners in the space station development that began in the 1980s and culminated in the International Space Station.

As often happens, the ambitious goals of the Spacelab program met with realities. Not as many missions flew as expected, international tempers sometimes flared, and some research results were disappointing. Yet Spacelab was a hallmark accomplishment—for ESA, for NASA, and for doing scientific research in space.

ABOVE: French Space Agency (CNES) payload specialist Jean-Jacques Favier performs two experiments at once. While he prepares a sample for the Advanced Gradient Heating Facility, he also wears instruments that measure upper body movement.

TOP: Interior view of Spacelab.

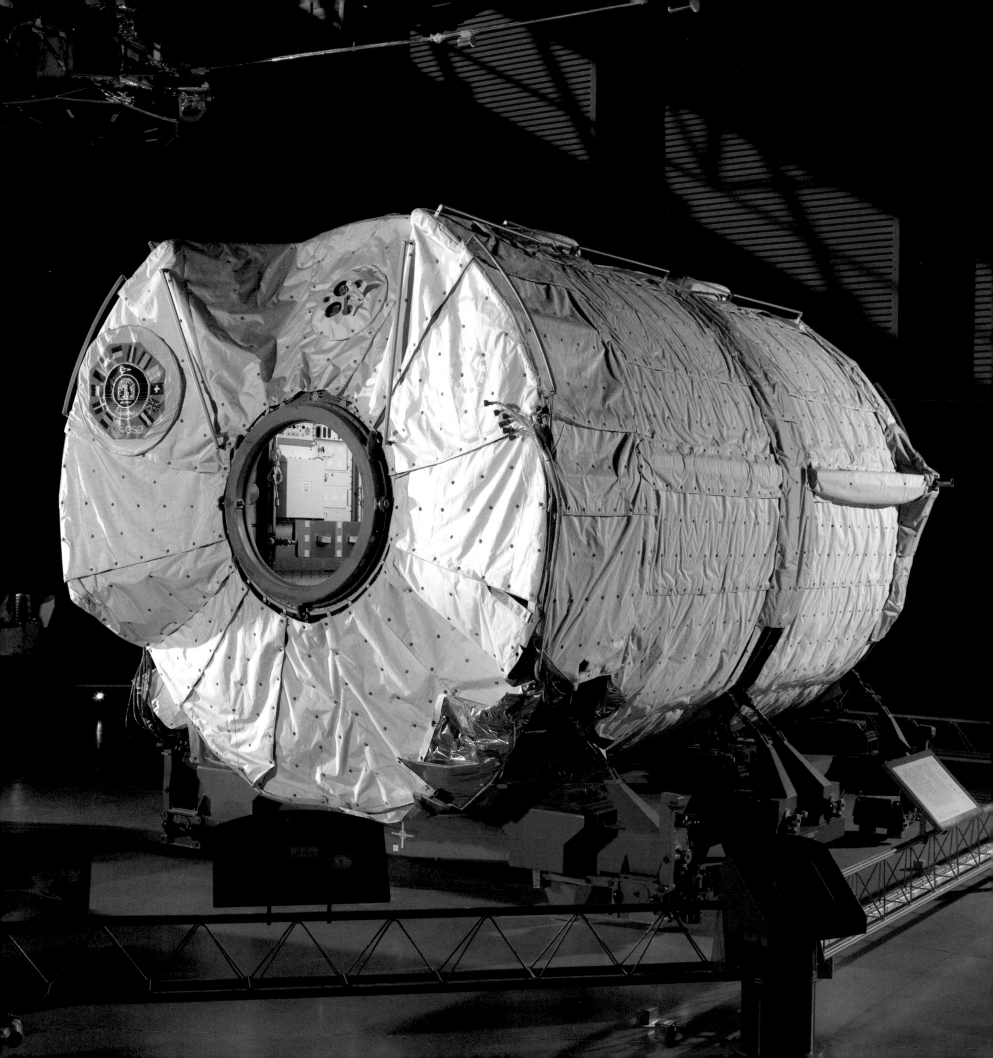

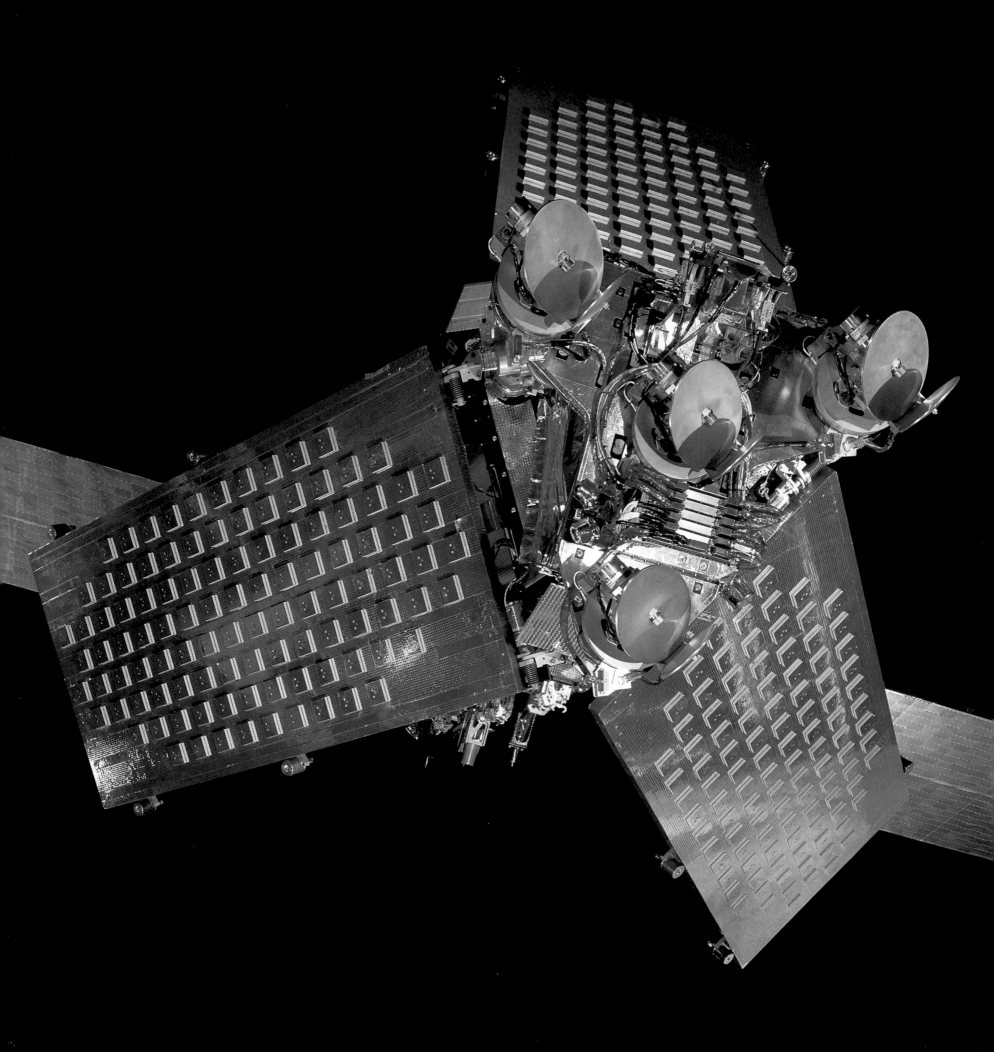

Motorola Satellite for Iridium
1996

"One World … One Telephone"

—Marketing tagline for the Iridium Satellite Telephone, mid-1990s

This satellite is the heart of a space-based communications system called Iridium, conceived, designed, and built by Motorola. Inaugurated in November 1998, under the auspices of Iridium LLC (a "startup" company formed by Motorola), this complex space system allows callers using handheld mobile phones and pagers to communicate anywhere in the world—a first in the history of telephony.

This simple rundown, though, hides a "Hollywood" story. Iridium was the boldest and most expensive private space initiative yet undertaken—a test of the post–Cold War notion that markets might replace government as the drivers of space exploration and development. Its development unfolded like a larger-than-life script—a tale of big ideas, big money, boundless optimism, clever achievements, and then reversals of fortune, heartbreak, collapse, and, rebirth.

In 1988, Motorola engineers Ken Peterson, Raymond Leopold, and Bary Bertiger conceived the Iridium idea: a digital, wireless telephone and paging service that covered the entire Earth—a feat made possible by a constellation of 66 satellites in low-Earth orbit. No one had ever used low-Earth orbits for a communications system, nor attempted to master the numerous technical challenges of building, launching, and operating 66 satellites as a telephone network in the sky. As a supranational enterprise, Iridium had its own country code—like the U.S. or China—and had to overcome the technical and bureaucratic hurdles of meshing its system with phone networks around the world.

In the 1990s, the end of the Cold War created strange bedfellows. Communist China and Russia, the United States' Cold War archenemies, joined the venture as investors, as did other countries and corporations from around the world. *Wired*, the go-go magazine for technological enthusiasts, dubbed the undertaking the "United Nations of Iridium" and wryly noted (with reference to Superman), "It's a bird, it's a phone, it's the world's first pan-national corporation able to leap geopolitical barriers in a single bound." After the collapse of the Berlin Wall in 1989, Iridium emerged as a potent symbol of the role of communications in transcending national borders and connecting people around the world—part of a technological vanguard that heralded liberal democratic values and a global shift toward market economies. Through satellites and the Internet, new streams of information ran through and over political boundaries, whether individual nations welcomed it or not.

In November 1998, after much hard work, many innovations, and the expenditure of billions of dollars, Iridium started service. The system worked (with a few glitches), allowing users with a phone a bit bigger than a cell phone (detractors described it as "brick-sized") to call Antarctica, Mt. Everest, or Washington, D.C., with equal ease. Expectations were high that Iridium would catch on in a world in which people could not seem to get enough "anytime, anywhere" communication.

But a mere nine months after its debut, Iridium filed for bankruptcy. There were too few customers to pay off the enormous investment costs. Motorola prepared to de-orbit the satellites and incinerate them in the Earth's atmosphere. But then a government rather than a private market voice pushed to save the system—the U.S. Department of Defense (DoD). DoD participated in planning for Iridium and was an early customer—and for good reason. The system's global coverage served the far-flung geographical needs of the military services. DoD encouraged a new group of investors to buy the bankrupt enterprise. Under the reborn company, called Iridium Satellite, DoD has used satellite telephony extensively in operations in the post–September 11 wars in Afghanistan and Iraq.

This satellite was the first of the Iridium series built by Motorola, probably manufactured in late 1996. The company donated it to the Museum in 1998.

ABOVE: In Kosovo, in 1999, refugees line up to make phone calls via Iridium, which provided service when ground-based telephony was destroyed in the war.

Eileen Collins's Clothing, Space Shuttle Mission STS-93
1999

"Eventually, having women in these roles won't be news anymore. It will be accepted and expected. I'm setting a precedent for women to follow. With that in mind, I want to do the best job that I can."

—EILEEN COLLINS, PILOT ASTRONAUT AND SHUTTLE MISSION COMMANDER, 1999

EILEEN COLLINS has worn many uniforms in her career: the blues of a collegiate member of the United States Air Force Reserve Officers Training Corps (ROTC), blue day and dress uniforms as an active-duty officer in the Air Force, plus her sage green flight suits in 30 different types of aircraft.

Since becoming an astronaut in 1991, Collins has performed a variety of roles at NASA, each with its own uniform—the blue coveralls for T-38 flights and public appearances, the bright orange Shuttle launch-entry suit, and the comfortable polo shirts and trousers preferred for crew attire in space. The only one Eileen Collins has not donned is the bulky white extravehicular activity pressure suit, because she has not ventured outside on spacewalks. She was busy in the cockpit, piloting the Shuttle and commanding the mission.

Eileen Collins entered the pilot astronaut "pipeline" almost as soon as it opened to women. She graduated from college in 1978, the same year NASA selected the first group of astronauts to include women, and she completed Air Force pilot training a year later. During the 1980s she advanced through Air Force assignments and in 1990 achieved two milestones: graduation from Test Pilot School and selection as an astronaut. Eileen Collins was the first woman chosen to fly the Space Shuttle—perhaps the only woman pilot at that time to have all the necessary qualifications.

Collins's first two flights as a pilot occurred in 1995 and 1997, the first a rendezvous mission with the Russian space station *Mir* and the second a docking with *Mir*. In 1999, Collins commanded the mission to deploy the Chandra X-Ray Observatory, and in 2005 she commanded the return-to-flight mission more than two years after the *Columbia* tragedy. As the first woman pilot and first woman commander of Space Shuttle missions, Col. Eileen Collins has flown *Discovery* (twice), *Atlantis*, and *Columbia*.

Collins has reflected openly on her pioneering role and her approach to spaceflight command—and clothing plays a part. Of all her uniforms, she especially likes the Shuttle crew wear. The trousers came as standard issue from the NASA inventory, but the shirts were the crew's choice. As a group, they decide beforehand what styles and colors to order for their in-flight wardrobe, and on a few days in orbit everyone wears the same shirts. The clothing is unisex, differing only in size. There are no badges of rank, no nametags, and no way to distinguish a commander's shirt from a rookie's. All members of the crew are a team, all essential, all responsible for mission success. Eileen Collins believes it fervently and leads from that premise.

Eileen Collins wore this shirt on the STS-93 mission aboard *Columbia* in July 1999, her first mission as commander. Collins donated the shirt and NASA her trousers to the Museum in 2000. Except for the Chandra X-Ray Observatory emblem, it is an anonymous uniform for a woman whose name will be long remembered in the annals of space history.

ABOVE: Astronaut Eileen Collins reviews a checklist on the *Columbia* mid-deck during her 1999 mission as the first woman to command a Shuttle.

Shuttle Radar Topography Mission (SRTM) Payload
2000

"'SRTM made possible use of Global Positioning Satellite–guided missiles in U.S. military effort against Afghanistan,' said a spokesman for Globalsecurity.org, defense research organization. 'This is a wonderful technology … Without the elevation data, the bomb would be useless.'"

—*Satellite Week*, OCTOBER 15, 2001

D URING the 1990s and early 2000s, video games and computer generated graphics made the experience of moving through virtual 3-D spaces commonplace. With such techniques, artists immersed a welcoming public in mythical places (as in the *Star Wars* films), or in real, but inaccessible environments, like NASA's "fly-bys" of the surfaces of Venus or Mars. Such images are so omnipresent in contemporary life that we tend to look at modeling of artificial and real worlds in the same way. But, of course, there is an important difference: real world modeling often is intended for real world applications and requires real data.

Enter the Shuttle Radar Topography Mission (SRTM) payload, carried into orbit aboard the Shuttle *Endeavor* in February 2000. SRTM enabled Shuttle astronauts to collect data essential in creating a highly detailed, three-dimensional map of more than 70 percent of the Earth's surface. The crucial data pertained to measuring the elevation of Earth's surface features—from valleys to mountains. When combined with existing cartographic information (latitude and longitude), modelers now could create authentic 3-D depictions of much of the Earth's surface—as single images (as seen in this rendering of Mt. McKinley) or as movies. Before SRTM, no consistent set of data on Earth's elevation features existed (surprisingly, the U.S. had better elevation data for the planet Venus, acquired by the Magellan spacecraft in 1990).

But why collect this data? A confluence of factors brought about the SRTM mission—distinctive of a post–Cold War world. The idea for the mission originated with Department of Defense's National Imagery and Mapping Agency (now called the National Geospatial-Intelligence Agency), who wanted the data to program into computers used for terrain navigation for planes, cruise missiles, tanks, and training simulators. The data (in a somewhat lower resolution form) also is useful to Earth scientists, land use planners, air traffic controllers, and businesses—for example, a cell phone company analyzing where to place a cellular tower. Prior to the end of the Cold War, a defense program that combined military, civilian, and commercial uses would have been unusual. Only a few in the post–Cold War world batted an eye at a mission designed to support enhanced cruise missile performance and better cell phone coverage.

To acquire its data, the SRTM used a novel hardware system that featured a main antenna located in the Shuttle payload bay, a folding mast that extended 60 meters (197 feet) from the Shuttle, and then another antenna system positioned at the end of the mast (the Museum artifacts include everything but the payload bay antenna). This dual antenna system—the largest rigid structure flown in space—produced, through interferometry (a technique for combining the information obtained from the two antennas), a three-dimensional mapping of the Earth. After the Shuttle astronauts completed their 10-day mission, they stowed all the hardware and returned it to Earth.

NASA's Jet Propulsion Laboratory transferred the artifacts to the Museum in 2003. When JPL, AEC Able (the contractor who built the ingenious mast system), and Museum staff inspected the artifact, the mast canister had remained unopened since its return from space. With the canister lid removed and the mast partially extended, they noticed evidence of the mast's time in low-Earth orbit: atomic oxygen at the outer reaches of the atmosphere had collided with the structure, leaving numerous small pits, visible to the naked eye.

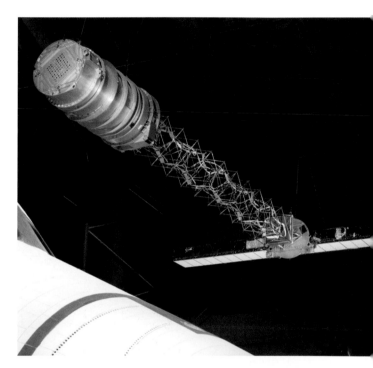

ABOVE: Data from SRTM helped produce this topographic image of Kamchatka Peninsula in Russia. The green is "false" color used to indicate elevation; the snow and sky were added for aesthetics.

U-2 Microwave Radiometer
1975–1979

"After I could trust the data, the question was, what does it all mean?"

—George Smoot, quoted by John Noble Wilford, 1992

ABOVE: The WMAP satellite, a scientific successor to the U-2 radiometer, produced this map of the universe as it appeared 300,000 years after the Big Bang. The red and yellow blotches indicate tiny temperature fluctuations in the early universe; the white lines (drawn in by scientists) suggest the emergence of physical structure.

In 1948, the British astronomer Fred Hoyle coined the name "Big Bang" as a joke—but it seemed aptly descriptive for a theory that said the universe came into existence through an unimaginably hot and violent primordial explosion. Scientists now regard it as the most successful account for understanding the origin and development of the universe. Its primary tenet is that the universe started as a dimensionless point that began to expand outward through the "Big Bang," and continues to do so today. In its earliest period, the universe was extremely hot, but as it continued to expand it cooled. Despite this cooling, theory predicts that the original heat still exists as very low-energy radiation (microwave radiation), uniformly permeating the cosmos.

Though it had been predicted to exist in the 1940s, in 1965, Arno Penzias and Robert Wilson inadvertently detected this radiation—called the cosmic microwave background radiation (CMBR)—thereby giving credence to the Big Bang theory. But important questions still remained—particularly how today's universe developed—and in response, physicists began a number of research initiatives.

In 1992, physicist George Smoot offered some insight into this question. Using instruments he and a team of scientists built for the Cosmic Background Explorer (COBE) spacecraft, launched in 1989, Smoot demonstrated that the CMBR was not perfectly smooth, but ever so slightly rippled with tiny temperature fluctuations. This implied that the early universe had "structure"—areas dense with radiation, but others less so. These differences, in theory, created the distributions of vast "empty" spaces and concentrations of matter and energy seen today in planets, stars, nebulae, and galaxies.

The COBE instruments had limitations, however—they could discern structure in the CMBR but not measure it with precision. Subsequent instruments, notably the *Boomerang* balloon telescope and the Wilkinson Microwave Anisotropy Probe (a spacecraft launched in 2001), achieved seven times greater resolution. These studies provided for the first time details of early cosmic structure and helped pinpoint the age of the universe at 13.7 billion years.

Smoot's interest in the problem of how structure developed in the universe began well before COBE. After Penzias and Wilson's discovery of the CMBR, Smoot started to develop an instrument to provide more refined data. Pictured here is a set of electronic ears called a microwave radiometer designed and built by Smoot and his colleagues at the University of California, Berkeley—the university's Lawrence Berkeley Laboratory donated it to the Museum in 1996.

From 1975 to 1979, it flew 16 times on a NASA U-2 aircraft and detected the Earth's motion with respect to this background radiation. Smoot and his team then used the U-2 device to develop a prototype for a more sophisticated instrument that flew on the COBE mission. Called the differential microwave radiometer (DMR), it featured three sets of dual radiometers that enabled the spacecraft to measure radiation from every direction.

This allowed Smoot to produce "all-sky" maps that revealed the first-ever evidence that the background radiation field was not perfectly smooth, but in fact was lumpy, with hot spots and cool spots fluctuating by extremely tiny amounts of temperature around the mean of 2.73 degrees above absolute zero. Smoot and his team sifted the data time and again to be sure the incredibly subtle effects were real. They made no announcements until they were sure of the answer to his "What does it all mean" query: "When I was sure," Smoot says, "I felt it in my bones—that what we were seeing wasn't just me wishing." These computer-generated false color images were widely publicized, giving Smoot instant celebrity.

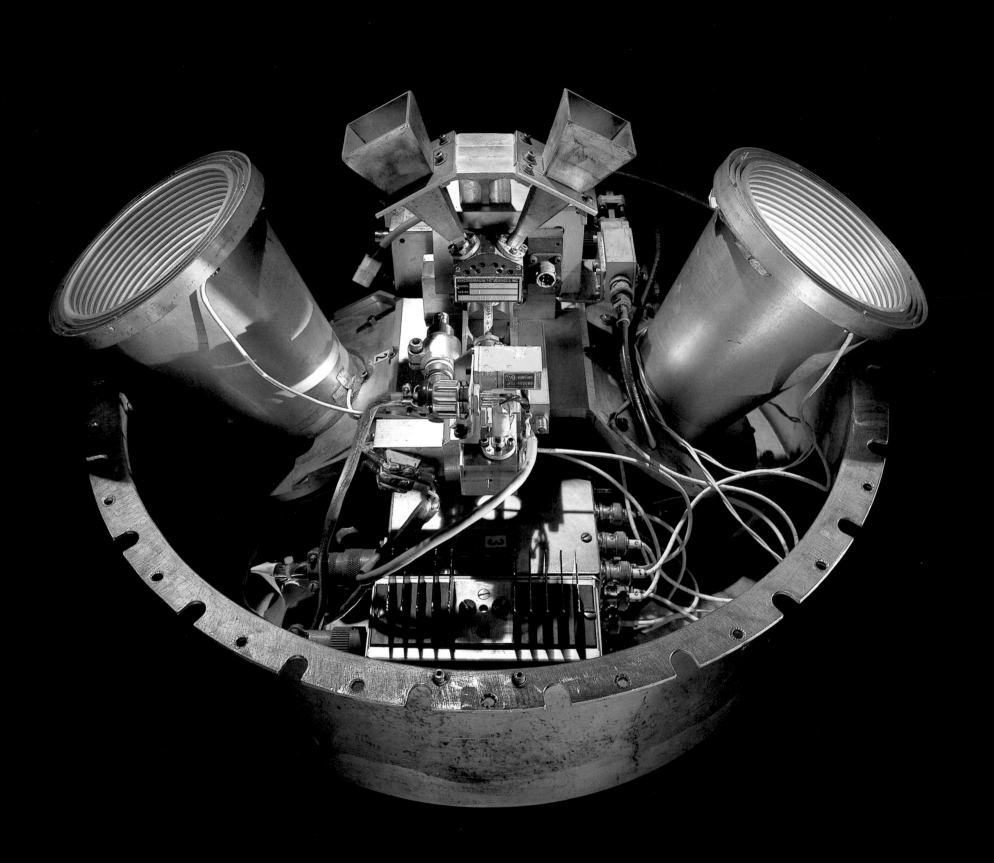

Astronaut Pamela Melroy's Scrunchie
2000

"Gimme a head with hair, long beautiful hair / Shining, gleaming, streaming, flaxen, waxen /
Give me down to there hair, shoulder length or longer . . ."

—LYRICS BY JAMES RADO AND GEROME RAGNI, FROM THE MUSICAL *Hair*, 1968

SOMETIMES an ordinary object has a good story—like this scrunchie. Scrunchies are fabric-covered elastic loops that long-haired girls and women use to keep their hair tidy and out of the way in a ponytail or bun. Scrunchies are readily found in the hair products aisle of stores. They are also, on occasion, found in space.

Hair was not an issue when the only astronauts were male, typically with short military haircuts. However, when women astronauts flew on the Space Shuttle, their longer hair drifted about their heads like halos—a vivid demonstration of weightlessness and of the subtle differences that women's presence brought to spaceflight. NASA, though, worried that such glorious, zero-gravity hair might affect operations in space.

Some of the women astronauts already wore short hairstyles, others cut theirs before a mission to simplify hair care, and others left theirs long. NASA did not want to impose conformity to a strict hair-length standard, but the longer hair raised some new questions about hygiene and safety. Was billowing hair a nuisance to other crew members in the crowded cabin? Might stray long hairs float around the cabin and clog the air filters? Might free-flowing hair tangle on switches or snag in equipment?

In 1990, NASA issued a policy guideline that astronauts pull back long hair in a ponytail, braids, or other manner to keep it from floating free. Loose hair was permitted briefly when no work was in progress—for example, to take photos of the "halo" effect. Otherwise, crew had to confine long hair during a mission. To date, this guideline has affected only the women astronauts; no male astronauts have sported long hair.

The mundane act of using scrunchies, clips, barrettes, and other accessories to manage hair is an ordinary fact of life in space as it is on Earth. It is also an emblem of small but noticeable changes that occur when women enter a new workplace. The same kinds of questions arose when women became factory workers, where long hair posed safety concerns. Long hair is often confined for a more professional look in offices and other environments, especially those where men traditionally predominate.

Col. Pamela A. Melroy gave this scrunchie to the Museum after she had flown twice as a Space Shuttle pilot astronaut on two missions to the International Space Station in 2000 and 2002. Pictures from the mission show her before, during, and after using it to secure her hair in space. She says that hair management in space is a good analogy for managing any aspect of one's uniqueness on a mission: "You balance hygiene, practicality, and a few moments of fun with professionalism and consideration for the rest of the crew."

Skylab Dart Game; Space Shuttle CD Player
1973; 1998

"Then it came time to play 'Zero-g 3-D tennis,' … A wadded-up ball of omnipresent gray tape was the ball, and two of our Flight Data File books were the rackets … we perfected a gymnastic racket sport that had us both laughing hysterically and sweating profusely."

—SCOTT PARAZYNSKI, STS-86 (1997)

MOST OF US can only dream of spaceflight. We may imagine the astronaut experience as magical, but once in space astronauts confront a very demanding reality. They face pressures of completing mission objectives with care and skill, as they adhere to rigid eating and sleeping schedules in a confined space. They contend with microgravity. In the midst of these challenges, the astronauts, like everyone else, need "down time." With no chance to step outside for fresh air or a walk, NASA and the astronauts find other ways to relieve stress and relax.

In the early years of spaceflight, when missions lasted only hours to a few days, astronauts focused solely on their technical and scientific activities and time to eat and sleep. As missions lengthened to a week or more, spare time became not only inevitable, but also desirable for the mental benefit of the astronauts. As they traveled to the Moon, Apollo astronauts included a few amusements such as playing cards and books in their personal preference kits, but time for reading or other relaxation came as schedules and mission tasks permitted.

With missions to the first U.S. space station, Skylab, the philosophy of crew activity changed to incorporate recreational time into astronauts' schedules. That practice continued for Space Shuttle flights in Earth orbit and aboard the International Space Station. Diversions such as this "low-tech" Velcro dart game (identical to one stowed on Skylab) and "high-tech" personal electronic music systems like this CD player flown on the Shuttle permitted astronauts to enjoy traditional games or listen to their favorite music in the unique environment of space. Both artifacts were transferred from NASA to the Museum.

Crew selection teams have always been mindful of the psychological impact of confinement in a spacecraft. Mercury astronaut candidates underwent a variety of mental tests, and astronaut candidates today still go through similar scrutiny. As astronauts undertook extended missions, psychological researchers quickly discovered that the demands of living and working in space could seriously affect mental and emotional health. As astronaut crews became more diverse—including different professions, women, and other nationalities and races—planners placed more emphasis on proper mental health assessments and including recreational time into mission plans to minimize social isolation or group tension.

Crews of Skylab, the Space Shuttle, and later *Mir* and the International Space Station had time for distractions within their scheduled responsibilities. Astronauts often commented on the benefits of tending to animals and plants aboard the spacecraft. Even casually observing Earth from their lofty orbital position has been shown to relax the astronaut's mind. NASA encouraged astronauts to strike a balance between needs for privacy and group activity. Bunks on *Mir* and the International Space Station became havens for personal reminders of home, books, photographs, music devices, and even DVD movies. On long missions, crews found time to play together by improvising games such as catching floating peanuts on Skylab or zero-g soccer and tennis on the Shuttle and space station. Such built-in opportunities for relaxation allowed astronauts to maintain a sense of emotional well-being under the strains of spaceflight.

Dennis Tito's Sokol KV-2 Spacesuit
2001

"Personally, I have had the time of my life. I have achieved my dream."

—Dennis Tito, May 6, 2001

CALIFORNIA businessman Dennis Tito became the first space tourist on April 28, 2001. After months of training and preparations in Moscow and at the Cosmonaut Flight Training Center in Star City, Russia, and the Johnson Space Center in Houston, Tito rode the Soyuz TM-32 from Baikonur, Kazakhstan, to spend six days on board the International Space Station (ISS). He paid the Russians a reported $20 million for this adventure.

During the 1970s and 1980s, the Soviet government offered foreign cosmonauts a ride to their Salyut and *Mir* space stations. Initially, the flights were symbolic, conferring political prestige. But toward the end of the program, governments and private organizations financed the flights on behalf of specific specialists for scientific research. After the collapse of the Soviet Union, Russia transformed the guest cosmonaut program into a moneymaking venture.

Dennis Tito's flight marked the first time that a private individual paid his or her own way to fly in space. As an aerospace engineer-turned-businessman, Tito dreamed of spaceflight for many years. As partner in the ISS, the National Aeronautics and Space Administration initially questioned the safety

of the arrangements and resisted his flight. But Tito eventually flew, thoroughly enjoying his experience and paving the way for future space tourism. South African Mark Shuttleworth followed Tito into space the following year and used his flight as a means to encourage the development of high-tech industry in his country. In 2005, American engineer and entrepreneur Greg Olsen used his flight to promote his own research in optics.

The Soviet spacesuit and life-support manufacturer Zvezda designed the Sokol ("Falcon") spacesuit in the early 1970s to protect cosmonauts' lives during emergencies. Cosmonauts wear the suit during the most dangerous periods of spaceflight— during launch and landing. This spacesuit was custom-made for Mr. Tito from existing components and previously tested parts. During the mission, the suit bore patches and flags that represented his mission. After the flight, technicians removed the patches and presented Tito with the complete set of his flown patches as souvenirs. The plugs and tubes extending from the suit connect to life-support systems built into the Soyuz spacecraft. Tito purchased this suit from Zvezda after his flight and then donated it to the Museum along with his gloves and patches.

LEFT: Dennis Tito practices in a Soyuz TM-32 capsule at the Cosmonaut Flight Training Center at Star City, Russia, in preparation for his trip to the International Space Station.

Impacts and Influences

Simulating the Space Experience
for Everyday Citizens

Walt Disney World Resort's "Mission SPACE," 2003

O N JANUARY 3, 2004, when *Spirit*, one of two large rovers, landed on Mars, NASA administrator Sean O'Keefe declared, "We're back … and we're on Mars." Three months later, his words joined inspirational quotes by philosophers, astronauts, and President John F. Kennedy on the walls of "Mission SPACE," Walt Disney World Resort's newest Epcot attraction. Connecting the $100 million thrill ride to NASA's real-life Mars missions followed a long history of Disney collaborations with actual space explorers. Just as Walt Disney himself worked with Wernher von Braun in the 1950s to promote early human spaceflight (and Disneyland) through television broadcasts and futuristic rides, so "Mission SPACE" endorses Mars exploration through a high-tech spaceflight simulation.

Beginning in 1955 with a trio of space-themed episodes for the "Disneyland" television program as well as the "Rocket to the Moon" ride in "Tomorrowland," Disney theme parks regularly included space attractions. In 1967, NASA helped to convert Disneyland's "Rocket to the Moon" into "Flight to the Moon." At Walt Disney World Resort in Florida, however, by the time "Flight to the Moon" opened in December 1971, real lunar landings had undercut its futuristic appeal. The attraction's replacement, "Mission to Mars," took visitors on an imagined trip to the red planet. To keep the attraction ahead of current-day developments, travel occurred via an imaginary "space jump" or "space warp."

In 2003, "Mission SPACE" revived the realistic space adventure attraction—this time employing twenty-first-century technologies to simulate NASA-like missions with high fidelity. Inside the attraction, four massive human centrifuges each carry 10 capsules containing four "astronaut trainees" assigned to specific roles. After some "training," the new crew executes a simulated launch, mission, and reentry, complete with increased g forces and brief weightlessness. Disney's Imagineers consulted NASA to ensure accuracy. The extreme experience has made some riders ill (each capsule carries motion sickness bags), and two people have died from previously undiagnosed health conditions after exiting the attraction.

"Mission SPACE" reinforces its credibility by placing visitors in the timeline of actual space exploration. Visitors to the "International Space Training Center" learn that their space travel in 2036 builds on actual space history achievements. In the attraction's courtyard, a Moon sphere features colored markers showing the sites of 29 human and robotic lunar landings between 1959 and 1976.

As in the 1950s, a national television broadcast once again linked Disney's commercial entertainment to real space travel. On August 16, 2004, Apollo astronaut Buzz Aldrin co-hosted *ABC Saturday Night at the Movies* with actor Jerry O'Connell. Broadcasting from inside "Mission SPACE," the real-life Moonwalker and the Hollywood Mars astronaut offered behind-the-scenes glimpses of the attraction between segments of the film *Mission to Mars* (2000). Aldrin's presence lent authority to Disney's depictions of dramatized human space exploration.

Just as Disney influenced public opinion regarding human space travel in the 1950s, "Mission SPACE" seeks to foster enthusiasm for current space exploration goals. Visiting Mars represents a recurring element of American space policy: from President George H.W. Bush's Space Exploration Initiative in 1989, to the popular Mars Pathfinder rover *Sojourner* in 1997 or the larger rovers *Spirit* and *Opportunity* in 2004 and 2005, and President George W. Bush's "Vision for Space Exploration."

ABOVE: Exterior view of Disney's "Mission: SPACE" attraction.

"Being in space is one of the most exciting things I've ever done and I'd like to think that other people can share in that."

—SHUTTLE ASTRONAUT RHEA SEDDON
ON "MISSION SPACE," AUGUST 4, 2003

Origami Crane
2003

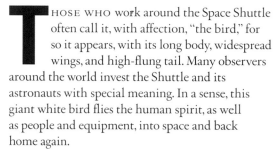

"That cranes may journey at such [high] altitudes, disappearing from the sight of earthbound mortals, may account for their near-sacred place in the earliest legends of the world as messengers and harbingers of highest heaven."

—Peter Matthiessen, *The Birds of Heaven: Travels with Cranes*, 2001

THOSE WHO work around the Space Shuttle often call it, with affection, "the bird," for so it appears, with its long body, widespread wings, and high-flung tail. Many observers around the world invest the Shuttle and its astronauts with special meaning. In a sense, this giant white bird flies the human spirit, as well as people and equipment, into space and back home again.

But twice—once during launch and once on the way to landing—the Space Shuttle fractured in flight. Each time the world mourned the loss of seven astronauts.

In the early February days after the 2003 *Columbia* tragedy, many visitors to the Museum brought mementos in remembrance of the fallen astronauts. The tributes began with flowers wrapped in cellophane and tied with ribbons: red carnations, mixed bouquets, and yellow roses of Texas. Soon there were flags, toy Space Shuttles, mission patches, candles, poems, scripture, angels, teddy bears, and children's notes and drawings.

This origami crane is one of the most poignant items from the spontaneous shrine that arose in Space Hall. It is delicate and small enough to cup in one's hand, crafted by the ancient art of paper folding. As an item left in remembrance, it is rich in symbolism.

In ancient and modern cultures, the crane has served as an emblem of hope and aspiration. For some, it signifies loyalty, longevity, health, good fortune, wisdom, or a happy marriage. Like the dove, it has widespread contemporary meaning as a symbol of peace—in this world and hereafter.

We do not know the exact reason a visitor left this crane, but a handwritten note placed with it is suggestive: "To the seven brave members of *Columbia* who gave the ultimate sacrifice in pursuit of knowledge; and to the many following them who will find the courage to take up the cause once again." These words imply both sorrow and affirmation.

The impromptu shrine to the *Columbia* astronauts communicated a deep public connection to the Space Shuttle and astronauts—for many, a bond intensified in time of tragedy. The crane and other objects testified to space exploration's power as a cultural symbol of patriotism, curiosity, dedication, and courage. As the select few to travel to the space frontier, astronauts often bear the hopes and dreams of people around the world and assumed the role of humanity's delegates, even heroes.

The small paper crane left to mourn the passing of astronauts carries a wish of peace for their souls and a hope for courage to come. Perhaps this crane also pays tribute to the bird that carried them. It is possible to see in the two birds both the power and fragility of flight. Likewise, the tokens of remembrance left in sorrow quietly express a widespread hope, inspired by flight into space.

Shenzhou 5 Model
2003

"Space technology does not belong to the rich countries alone."

—Zhang Houying, scientific director, Shenzhou program, March 2003

For 40 years, the United States and the Soviet Union (later Russia) belonged to an exclusive club of two who had the capability to launch humans into space. Even though a score of other countries achieved the ability to launch their own satellites, none took the next step: to mount their own, independent program to send humans into space, a feat accorded the highest prestige in the international community. For nations that considered undertaking human spaceflight, in each case governments decided that costs outweighed the possible benefits of enhanced international prestige.

On October 15, 2003, China showed that it placed a high value on human space exploration. With the launch of Shenzhou 5, China became the third nation to launch a human into Earth orbit. Chinese pilot Yang Liwei became the first Chinese national to orbit the Earth. In popular parlance, Yang is the first taikonaut, a term derived from the Chinese root word for "space."

The Chinese motivations for launching Shenzhou 5 paralleled those of the USSR and U.S. in making their first human spaceflights. Taikonaut Yang's orbital flight did not occur as the result of

direct geopolitical competition (as did the flights of Yuri Gagarin and John Glenn). But in sending a human into space, China sought to demonstrate that it, too, has capabilities and resources commensurate with those of the superpowers of the Cold War.

Benefiting from a rich history of spacecraft development, the Chinese utilized, in particular, Soviet and Russian knowledge as they designed and built their spacecraft. The Chinese openly acknowledged that they purchased *Soyuz* spacecraft from the Russians and that their scientists, engineers, and taikonaut candidates trained on and adapted the hardware to their own purposes. In one such adaptation, the Chinese modified a section of the *Soyuz* that cosmonauts sat in during a spaceflight mission. In the Chinese version, taikonauts can detach this section and place it into orbit as an independent satellite, with its own power source, capable of performing scientific missions.

This model currently is the best representation of China's spaceflight achievement in the Museum. The government of China gave the model as a diplomatic gift for the American people.

RIGHT: In 1999, Shenzhou 3, a test vehicle, returned safely from space, landing in the Nei Mongol region of China. It prepared the way for the successful Shenzhou 5 human spaceflight mission in 2003.

中 国 航 天
CHINA AEROSPACE

SpaceShipOne
2004

"I want commercial flights to the moon within my lifetime."

—Burt Rutan, SpaceShipOne designer, 2004

ABOVE: *White Knight*, a specially designed carrier aircraft, lifts SpaceShipOne up to its launch altitude, for the first of its Ansari X Prize flights in September 2004.

Like many of our air- and spacecraft, SpaceShipOne is a winner, a record-setter, and, to its admirers, a harbinger of the future. It arrived at the Museum barely a year after carrying a person into space and winning the coveted X Prize as the first privately owned piloted craft to reach space. Unlike the Mercury, Gemini, Apollo, and Space Shuttle vehicles developed by the United States government, this one was a private enterprise.

SpaceShipOne was a joint venture by legendary aircraft designer Burt Rutan and investor-philanthropist Paul G. Allen, cofounder of Microsoft. For years, Rutan turned ideas into uniquely innovative winged vehicles, and then he applied the same creativity to a spacecraft. Allen, captivated by new ideas that solve important problems and improve people's lives, sought to anticipate the future and hasten its arrival. In SpaceShipOne, Rutan and Allen committed themselves to a new vision of spaceflight— one shaped by private enterprise, not by government programs.

SpaceShipOne's creators had a straight-forward but ambitious goal: to make spaceflight safe and accessible for the public, opening space for commercial tourism. As Paul Allen said in 2005, "I saw SpaceShipOne as a great opportunity to … demonstrate convincingly that private space exploration could someday be within the reach of individual citizens." It represented their first step.

In appearance and operation, SpaceShipOne is unlike any other spacecraft. With a fuselage faintly resembling the bullet-shaped Bell X-1 rocket plane, SpaceShipOne has distinctive swept wings with tail fins. These design features were the heart of a unique, three-part flight profile. For the first part of a flight, the spacecraft was tucked under a long-winged carrier aircraft called *White Knight*. At 15,240 meters (50,000 feet, or 10 miles), the plane released SpaceShipOne and the spacecraft pilot ignited a hybrid rocket motor—solid rubber fuel burned with liquid nitrous oxide—to reach Mach 3 and 54,864 meters (180,000 feet, or 34 miles). The vehicle then coasted to an altitude of more than 100,000 meters (328,000 feet, or 62 miles) for a suborbital arc through space.

On the way to apogee, the pilot reconfigured the craft. The twin tails and about a third of the wing tilted up. This "shuttlecock" mode helped SpaceShipOne maintain its stability and brake as it reentered the atmosphere. While the pilot enjoyed the view, the vehicle began its descent. After deceleration, the pilot lowered the wings and tail back into position for atmospheric flight and glided to a runway landing.

Pilot Mike Melvill took SpaceShipOne to 10,0000 meters (328,000 feet, or 62 miles) on June 21, 2004, and to 103,000 meters (338,000 feet, or 64 miles) on September 29. Brian Binnie flew it to 113,000 meters (370,000 feet, or 70 miles) on October 4, 2004. For the last two flights, Burt Rutan, Paul Allen, and the SpaceShipOne team won the $10 million Ansari X Prize for significant advancement toward commercial spaceflight.

SpaceShipOne is displayed in the Museum's central *Milestones of Flight* gallery, near the 1903 Wright Flyer and Charles Lindbergh's Ryan NYP *Spirit of St. Louis*, the Bell X-1 *Glamorous Glennis*, the Mercury *Friendship 7*, and the Apollo 11 command module *Columbia*. Its presence there affirms that milestones of flight continue in this new century. If Rutan, Allen, and others succeed, it may soon be possible for many more people to experience the adventure of spaceflight.

Index

Page numbers in *italics* refer to photographs.

Credits

Essay and Sidebar Text

PAUL CERUZZI: 26, 66, 82, 97, 106, 117, 118, 148, 168, 174, 197, 221, 229

MARTIN COLLINS: 53, 60, 74, 103, 167, 209, 239, 241

TOM CROUCH: 17, 37, 100, 110, 111

JIM DAVID: 46, 52, 57, 59, 61, 80, 81, 89, 108, 169, 186, 187

DAVID DEVORKIN: 24, 25, 49, 50, 151, 152, 162, 182, 219, 220, 222, 242

ROGER LAUNIUS: 86, 113, 137, 141, 179, 180, 200, 225, 235

CATHY LEWIS: 43, 45, 65, 78, 79, 170, 172, 173, 202, 206, 210, 218, 227, 230, 232, 246, 249

VALERIE NEAL: 158, 161, 188, 191, 194, 195, 196, 198, 201, 204, 205, 233, 236, 240, 244, 248, 250

ALLAN NEEDELL: 123, 124, 127, 133, 135, 142, 143, 145, 146, 149

MIKE NEUFELD: 16, 22, 36, 67, 77, 90, 94, 105

JENNIFER SKOMER: 138, 163, 164, 185, 203, 228, 245

MARGARET WEITEKAMP: 20, 44, 64, 85, 98, 99, 102, 107, 136, 175, 176, 181, 226, 247

FRANK WINTER: 14, 19, 21, 28, 30, 33, 55, 56, 62, 114, 192, 217

AMANDA YOUNG: 34, 76, 93, 121, 128, 130, 132

Artifact Photography - Smithsonian National Air & Space Museum

(T=Top, C=Center, B=Bottom)

MARK AVINO: 35, 92, 98, 111, 120, 129, 132T, 146T, 232, 246

DON HURLBERT: 37

ERIC LONG: 1, 2-3, 14, 17, 19T, 23, 24T, 27, 29, 30T, 31, 42, 45, 47, 48, 51, 52, 53T, 56, 57T, 58, 60T, 61T, 65T, 66, 67, 69, 75, 76, 77, 80, 81, 84, 85, 87, 88, 91, 95, 96, 100, 101, 102T, 103T, 104, 107, 109, 112, 115, 119, 122, 125, 126, 131, 133T, 136, 137, 140, 142C, 143, 144, 147, 148, 149, 150, 153, 159, 160, 163, 165, 166, 169, 171, 172, 173, 174, 177, 178, 180T, 183, 184, 190, 194T, 195T, 197, 201T, 203T, 204T, 207, 208, 211, 218, 219T, 220T, 223, 224, 226, 228, 229, 230, 231, 233T,B, 238, 240T, 243, 244, 245T,B, 248, 249

CHUCK MOORE: 176

DANE PENLAND: 6, 15, 16T, 18, 21C, 54, 55C, 63, 78, 79, 83, 106, 116, 134, 135T, 139, 162T, 168, 186, 187B, 189, 193, 196, 199, 202, 205T, 216, 227, 234, 235, 236T, 237, 241T, 245, 251

CAROLYN RUSSO: 32T, 33

COURTESY OF DANIELE MELGIOVANNI/SCIENCE MUSEUM: 25

Historical Images

NASA (NATIONAL AERONAUTICS AND SPACE ADMINISTRATION): 10, 38, 49, 50, 55, 59T, 62, 68, 70, 76C, 79T, 82, 103B, 105, 113, 114T,B, 117, 121, 123, 124B, 127, 130, 132B, 133B, 135B, 138, 141, 142T, 145T,B, 146B, 151, 154, 161T,B, 162B, 164, 170T, 179, 180B, 182, 188, 191, 192T,C,B, 194B, 195B, 198, 201B, 203B, 204B, 205B, 212, 217, 220B, 222, 225, 232C, 236B, 240B, 241B, 242

NASM (NATIONAL AIR AND SPACE MUSEUM): 9, 16B, 18B, 19B, 20, 21T, 22, 24B, 26, 30B, 33B, 43, 57B, 60B, 61B, 89T,B, 118, 152, 157, 170B, 187T, 206, 210T,B, 219B

ADDITIONAL SOURCES: Collection, Kevin Osborn 28, Collection, Randy and Yulia Liebermann 36T,B, 99T,C,B, 200T,C,B, ©2006 Genlyte Group Inc. 44, Time Life Pictures/ Getty Images 46, 53B, 59B, 64, Collection, Charles O. Hyman 65, 181T,B, Courtesy of Lance Ginner/ Project Oscar 74, Printed by permission of the Norman Rockwell Family Agency, ©1965 The Norman Rockwell Family Entities 100, 101, National Archives and Records Administration 108B, Courtesy of Hamilton Sunstrand 128, Quantum Press/Doubleday ©1984 Point 175, Rick Forster/ SETI Institute 220T,B, Courtesy of Iridium Satellite 239, © Disney Enterprises, Inc. 246, Sinodefense.com 249, Jim Canpbell/ Aero News Network Inc. 250